SYSTEMS BIOLOGY - THEORY, TECHNIQUES AND APPLICATIONS

SYSTEMS BIOLOGY

AN INTEGRATED PLATFORM FOR BIOINFORMATICS, SYSTEMS SYNTHETIC BIOLOGY AND SYSTEMS METABOLIC ENGINEERING

Systems Biology - Theory, Techniques and Applications

Additional books in this series can be found on Nova's website under the Series tab.

Additional e-books in this series can be found on Nova's website under the e-book tab.

SYSTEMS BIOLOGY - THEORY, TECHNIQUES AND APPLICATIONS

SYSTEMS BIOLOGY

AN INTEGRATED PLATFORM FOR BIOINFORMATICS, SYSTEMS SYNTHETIC BIOLOGY AND SYSTEMS METABOLIC ENGINEERING

BOR-SEN CHEN
AND
CHIA-CHOU WU

nova publishers
New York

LIBRARY OF CONGRESS CATALOGING-IN-PUBLICATION DATA

Systems biology : an integrated platform for bioinformatics, systems synthetic biology and systems metabolic engineering / editors, Bor-Sen Chen and Chia-Chou Wu (National Tsing Hua University, Taiwan).
pages cm. -- (Systems biology - theory, techniques and applications)
Includes index.
ISBN 978-1-63321-588-7 (hardcover)
1. Systems biology. I. Chen, Bor-Sen.
QH324.2.S9718 2014
570.1'13--dc23

2014028259

Published by Nova Science Publishers, Inc. † New York

CONTENTS

Preface		**ix**
Chapter 1	Introduction to Systems Biology	**1**
	Abstract	1
	Introduction	1
	1.1. The Role of Systems Biology in Modern Biology Research	2
	1.2. The Role of Bioinformatics in Systems Biology	2
	1.3. The Role of Systems Biology in Synthetic Biology and Metabolic Engineering	4
	1.4. The Role of Systematic Analysis and Mathematic Modeling in Systems Biology	5
	1.5. Future Challenges of Systems Biology	6
	Conclusion	7
	References	7
Chapter 2	Preliminary Mathematics	**13**
	Abstract	13
	Introduction	13
	2.1. Linear System Model	13
	2.2. Matrix Norms	17
	2.3. Statistical Inference	21
	2.4. Statistical Hypothesis Testing	26
	2.5. Stochastic Systems	28
	2.6. Numerical Optimization	32
Chapter 3	Reverse Engineering in Systems Biology	**37**
	Abstract	37
	Introduction	37
	3.1. The Least Square Parameter Estimation Method	38
	3.2. The Maximum Likelihood Parameter Estimation Method	40
	3.3. The Recursive Least Square Parameter Identification Algorithm	42
	3.4. System Model Order Determination	43
	3.5. Nonlinear Parameter Estimation Methods	44

Chapter 4	Systems Biology Approach to GRNs and PPI Networks	**47**
	Abstract	47
	Introduction	47
	4.1. Construction of GRNs via Microarray Data	48
	4.2. Construction of PPI Networks	51
	4.3. Construction of Integrated GRN and PPI Cellular Networks	52
	4.4. Interspecies PPI Network between Host and Pathogen	57
	4.5. Stochastic Spatial-Temporal Dynamic Model for Gene/Protein Interaction Network in Early Drosophila Development	63
	4.6. Conclusion	83
	4.7. Appendix	83
	References	85
Chapter 5	Measurement of System Characters of Biological Networks via Microarray Data	**93**
	Abstract	93
	Introduction	93
	5.1. Measurements of Network Robustness and Sensitivity Estimation via Microarray Data Using a Systems Biology Approach	94
	5.2. Tradeoff between Network Robustness and Network Response Ability during Aging Process	98
	5.3. Gene Response Ability of Individual Genes in the Aging-Related Regulatory Network in the Young, Aged, and CR Groups	100
	5.4. Network-Based Biomarker Determination via Sample Microarray Data Using a Systems Biology Approach	104
	5.5. Conclusion	113
	5.6. Appendix	114
	References	116
Chapter 6	Network Robustness, Filtering, and Transduction Ability Analysis of Stochastic Biological Networks	**121**
	Abstract	121
	Introduction	121
	6.1. Network Robustness Analysis of Stochastic GRNs	122
	6.2. Filtering Analysis of Stochastic GRNs	125
	6.3. Transduction Ability Analysis of Signal Transduction Pathways	128
	6.4. Conclusion	141
	6.5. Appendix	141
	References	149

Chapter 7	Systems Synthetic Biology	**153**
	Abstract	153
	Introduction	153
	7.1. Design of Specifications-Based Systems Synthetic Biology	154
	7.2. Robust Synthetic Gene Network Design via Library-Based Search Methods	165
	7.3. Conclusion	169
	7.4. Appendix	169
	References	172
Chapter 8	Systems Metabolic Engineering	**175**
	Abstract	175
	Introduction	175
	8.1. Robust Biochemical Circuit Design in Metabolic Networks	176
	8.2. Multipurpose Circuit Control Design of Metabolic Networks	183
	8.3. Robust Circuit Design of Stochastic Dynamic Metabolic Networks	191
	8.4. Isobutanol Biosynthetic Pathway in E. coli for Biofuel via Systems Synthetic Biology	201
	8.5. Conclusion	215
	8.6. Appendix	215
	References	220
Chapter 9	Systems Biology Approach by Stochastic Dynamic Nash Games	**225**
	Abstract	225
	Introduction	225
	9.1. Robust Model Matching Control of Immune Systems under Environmental Disturbances: Dynamic Game Approach	226
	9.2. Robust Synthetic Biology Design: Stochastic Game Theory Approach	248
	9.3. The Stochastic Evolutionary Game on A Population of Biological Networks under Natural Selection	264
	9.4. Conclusion	295
	9.5. Appendix	295
	References	313
Chapter 10	Concluding Remarks and Future Challenges in Systems Biology	**321**
	Abstract	321
	Introduction	321
	10.1. Future Challenges in Systems Biology	322
	10.2. The Future Prospect of Systems Biology	324
	10.3. Conclusion	324
	References	324

Authors' Contact Information **325**

Index **327**

PREFACE

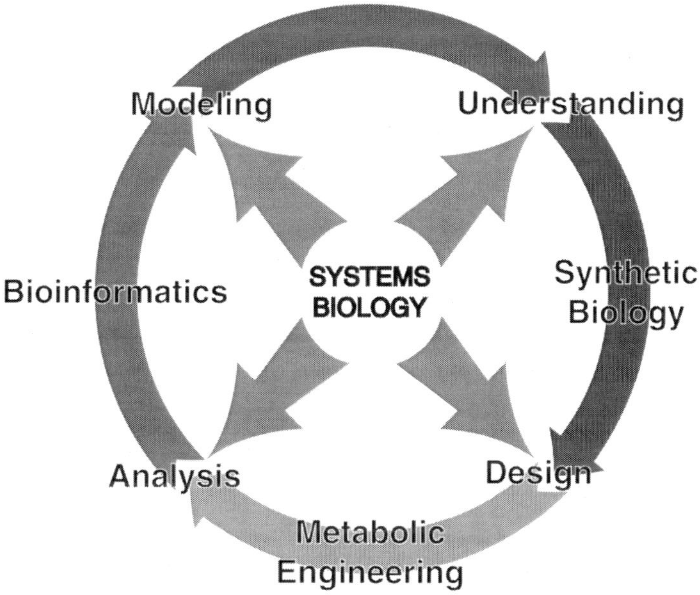

Systems biology is a new area of study in biology that seeks to incorporate physics, chemistry, computer science, engineering, and mathematics to improve the current understanding of life. Systems biology aims to achieve a system-level understanding of living organisms and apply this knowledge in fields such as synthetic biology, metabolic engineering, and medicine. Certain network properties help provide insight for a system-level understanding of living organisms. These include: (i) system structure and mechanisms such as gene regulation, protein interactions, signaling, and metabolic pathways; (ii) system dynamics derived from system identification and analysis, providing an understanding of stability, robustness, and transduction ability; (iii) system control methods at different network levels, which provide an understanding of systematic mechanisms to robustly control system states, minimize malfunctions, and provide potential therapeutic targets in disease treatment; and (iv) systematic design methods for the modification and construction of biological networks with desired behaviors, which provide system design principles for system simulations in synthetic biology and systems metabolic engineering.

The amount of data produced by the study of biological systems at different scales, from molecules to ecosystems, is growing exponentially. As this increase in information presents challenges to some areas of biology, systems biology researchers employ bioinformatics technologies to handle large amounts of data from local production to storage in publicly accessible and integrative depositories. Systems biology combined with bioinformatics, synthetic biology, and metabolic engineering promises to advance and transform our ability to increase biological understanding, program biological functions, and render the engineering of biological circuits or pathways faster and more predictable. A large-scale, integrative, multidisciplinary approach is needed for systems biology to flourish.

This book describes current developments in systems biology, synthetic biology, and metabolic engineering for engineers and researchers. Biological research is currently undergoing revolutionary changes due to the integration of powerful technologies. Systems biology sits at the heart of a new integrative 21st century paradigm. The flowchart at the beginning of the preface highlights the research cycle for bioinformatics, systems biology, synthetic biology, and metabolic engineering for analyzing biological data, modeling dynamic equations to understand biological systems, and designing synthetic biological systems with applications for metabolic engineering. We also discuss challenges and future prospects for systems biology as a platform integrating bioinformatics, systems synthetic biology, and systems metabolic engineering.

Chapter 1

INTRODUCTION TO SYSTEMS BIOLOGY

ABSTRACT

Systems biology is a large-scale, integrative, and multi-disciplinary approach in modern biological research. It unites bioinformatics, synthetic biology, and metabolic engineering through modeling, understanding, analysis, and design as highlighted in the cyclic flowchart at the beginning of the preface. In the first chapter, we will briefly introduce present developments and future challenges of systems biology.

INTRODUCTION

The nature of traditional biology is descriptive. Still, biology research retains a unique philosophy which tends towards reductionism, hypothesis formation, discrimination, etc. Reductionism is the approach of "divide and conquer"; it is based on making simplifications. Under reductionist assumptions, no interactions between subsystems exist and thus, biological phenomena are additive. Impacts of reductionism on modern biology are reflected in the discovery of the structure of DNA in the 1950s, genetic engineering in the 1970s, and genome projects in the 1990s. The main reason reductionism is effective is that simplification helps reveal common features of apparently different phenomena. Reductionist thinking facilitates hypothesis formation and discrimination. Reductionist approaches always break up a box to analyze the parts inside, in hopes of understanding how the box operates. Systems-synthetic approaches provide alternatives to reductionism. A systems approach analyzes the signal radiating from the box to predict what is inside the box and how it operates. A synthetic approach is a reductionist approach in reverse; parts are used to synthesize a desired biological system that fulfills a particular purpose. This approach offers scientists a useful way to test their understanding of the complex functional genetic and molecular networks that mediate life processes.

1.1. THE ROLE OF SYSTEMS BIOLOGY IN MODERN BIOLOGY RESEARCH

The Human Genome Project and high-throughput experimental methodologies such as microarray chromatin-immunoprecipitation DNA chips (ChIP-chips) have led to the development of biology as an increasingly information-rich science encompassing transcriptomes, proteomes, metabolomes, interactomes, and so forth [1, 2]. Some have suggested that systems biology is nothing more than a new name for integrative physiology, which has been practiced for 50 years or more. However, owing to new technologies, biologists have been able to collect data at a rate that was unimaginable a decade ago. The context of biology has profoundly changed over the past two decades, and this change provides a powerful new framework for systems biology that moves it far beyond classical integrative physiology. A systems biology approach implies that any level of a biological system can be analyzed with respect to the system's structure in terms of its dynamics, method of control, and method of design. Systems biology involves genomic, transcriptomic, proteomic, and metabolic investigations from a systematic perspective, and therefore has become the frontier of modern biological research. The large amount of new -omics data cannot be understood without a network or systems viewpoint and highly sophisticated computational analyses [3-11].

Systems biology in modern biological research (Figure 1.1) requires powerful computational tools to mine large-scale data sets of information on genetics, proteins, DNA–protein binding, metabolism, and so forth. These tools are used to construct dynamic system models to interpret specific mechanisms of certain cellular phenotypes (or behaviors) from a system (or network) perspective [12-17]. In constructing a dynamic system model, system identification technologies (i.e., reverse-engineering schemes) are needed to estimate parameter values of dynamic models and biological network order [18-44]. Approaches using synthetic biology and metabolic engineering have recently been developed to design synthetic genetic networks with the aim of inducing specific cellular functions in host cells [45-51]. These synthetic genetic circuits and engineered metabolic pathways can be designed based on system models and mechanisms to investigate cellular behaviors [52-64]. These technologies can then be employed to investigate the models and mechanisms outlined by systems biology. Discrepancies between the behavior of synthetic genetic networks and the behavior predicted by the models and mechanisms of systems biology can then be used as feedback to modify the models using the methodologies of systems synthetic biology and systems metabolic engineering. Based on the role of systems biology as an integrative platform (Figure 1.1), this book describes current developments in bioinformatics, systems synthetic biology, and systems metabolic engineering. It discusses how systems biology can serve to unify bioinformatics, systems synthetic biology, and systems metabolic engineering in the future.

1.2. THE ROLE OF BIOINFORMATICS IN SYSTEMS BIOLOGY

Bioinformatics is crucial for genome-wide analyses to understand cellular physiology at different levels (i.e., genome, transcriptome, proteome, and metabolome) The various disciplines within bioinformatics provide invaluable information concerning global cellular status for systems biology, systems synthetic biology, and systems metabolic engineering, allowing for thorough analysis of the cell. Genomic information in bioinformatics represents the entire genetic makeup of the organism [12-17], and comparative genomic analysis can contribute to creating desirable cellular phenotypes in systems synthetic biology or systems metabolic biology through targeting and engineering genetic circuits. Transcriptome profiling involves the use of DNA microarrays to decipher the expression levels of thousands of genes under various biological conditions [14, 15]. The results can be used to select candidate genes for modification based on a systematic analysis of regulatory genes responding to genetic variations and environmental changes, or to identify novel factors for enhancing heterologous product secretion in metabolic pathways [65-69]. Proteome profiling is also useful in obtaining transcriptome profile data at the protein level. The metabolome comprises information on metabolites present within or outside the cell under specified conditions [61-64]. In synthetic circuit engineering, it is expected to contribute significantly to an understanding of the cell and of metabolic pathways. In this book, specific examples of recent advances in the application of bioinformatics to systems synthetic biology and systems metabolic engineering are reviewed.

Although bioinformatics information (e.g., data on genetics, protein binding, and metabolism) is available, systems dynamics modeling in systems biology requires several stages to help understand underlying molecular mechanisms of genetic regulatory (GR) networks [70], protein–protein interaction (PPI) networks [25-29], and metabolic networks [61-64] under various biological conditions. At the first stage, a putative (candidate) GR or PPI network is created through large-scale integration of knowledge from publications, databases, and high-throughput data (from data mining or deep curation). The actual physiological GR or PPI network of a cell can be identified based on the candidate network and dynamic modeling with system identification methods (reverse-engineering schemes [34]) using specific gene or protein expression microarray data [18-28]. For example, GR networks (GRNs) with dynamic modeling using microarray data have been constructed for cell cycles [18, 23, 24], environmental stressors [28, 44], photosynthesis [71], aging [34], and cancer [39]. PPI networks have been constructed for cancer [3, 9, 33, 39, 72], inflammation [41], biofilm formation [43], and infection by *Candida albicans*. Comparing PPI networks of healthy and cancerous cells can provide network markers to investigate the systematic mechanisms of cancer [72]. The integration of GR and PPI cellular networks provides deeper insight into actual biological networks and is more predictive than a non-integrative approach [73]. A systematic and efficient method is based on coupling dynamic models and statistical assessments for assembling different kinds of -omics data from microarrays to construct integrated cellular networks [44]. This method is powerful and flexible, and can be used at different cellular levels to investigate cellular machinery under different biological conditions and for different species. Coupling dynamic models of an entire integrated cellular network are very useful for theoretical analyses and for further experiments in the field of network biology and synthetic biology.

1.3. The Role of Systems Biology in Synthetic Biology and Metabolic Engineering

Synthetic biology is the engineering of biological systems to fulfill a particular purpose. The main goal of this nascent field is to design and construct systems with desired behaviors [51]. It is possible to build living machines using off-the-shelf genetic devices by employing many of the same strategies that electrical engineers use to manufacture computer chips [47-49]. Synthetic biology envisions a redesigning of natural biological systems and construction of functional "genetic circuits" using a set of powerful biotechniques for automated synthesis and assembly of DNA molecules into genes and microbial genomes [47]. It is predicted that synthetic biology will have important applications in biotechnology, metabolic engineering, and medicine. It may revolutionize how we conceptualize and approach the engineering of biological systems [49]. Furthermore, as illustrated in Figure 1, synthetic genetic circuits can be employed to confirm, revise, and improve network mechanisms derived using a systems biology approach. Synthetic biology is therefore expected to contribute significantly to a better functional understanding of metabolic pathways and other complex biological systems.

However, development of synthetic gene networks is still difficult. Most newly created gene networks are nonfunctional because of intrinsic parameter fluctuations, environmental disturbances, and functional variations in intra- and extracellular contexts. For this reason, a method of design based on dynamic models for robust synthetic gene networks has become an important topic in synthetic biology [52-60, 69]. These system dynamics-based design methods lead to systems synthetic biology.

Metabolic engineering has merged heterologous genes into pathways, generating a myriad of non-native biochemical products including isoprenoids, hydroxylacids, biofuels, polypeptides, and biopolymers [65-68]. Synthetic biologists developed synthetic tools to engineer genetic devices capable of performing complex biological functions, such as sensing cell states, counting cellular events, and implementing computational logic. These tools have been applied to modify and control metabolic pathways in several organisms. They consist of one or more parts that have been combined to perform a complex function and provide metabolic engineers with novel ways of exerting cellular control over heterologous production pathways. Some synthetic biological devices with potential relevance in metabolic engineering include orthogonal inducible promoters, light-sensitive promoters, state sensors, spatiotemporal controllers, and logic gates, as well as promoter and ribosome binding site (RBS) libraries [36, 63]. Since metabolic engineering attempts to control and manipulate cellular metabolism through heterologous pathways to maximize production of a desired molecule, metabolic engineers are in need of elegant methods for gathering bio-information about cells, their environment, and modulating gene expression in response [37, 61, 62, 64]. Hence, synthetic biology devices promise to be a useful addition to the metabolic engineer's toolbox. Some synthetic devices have already been used to increase product titers, however many remain largely untested in an industrial setting, and biological complexity makes their application a feat of engineering [36, 37, 74]. Continuous work with these devices can help elucidate design rules and aid the development of system dynamics models that facilitate their integration into metabolic industrial processes [36, 37, 74], thereby leading to the development of systems metabolic engineering.

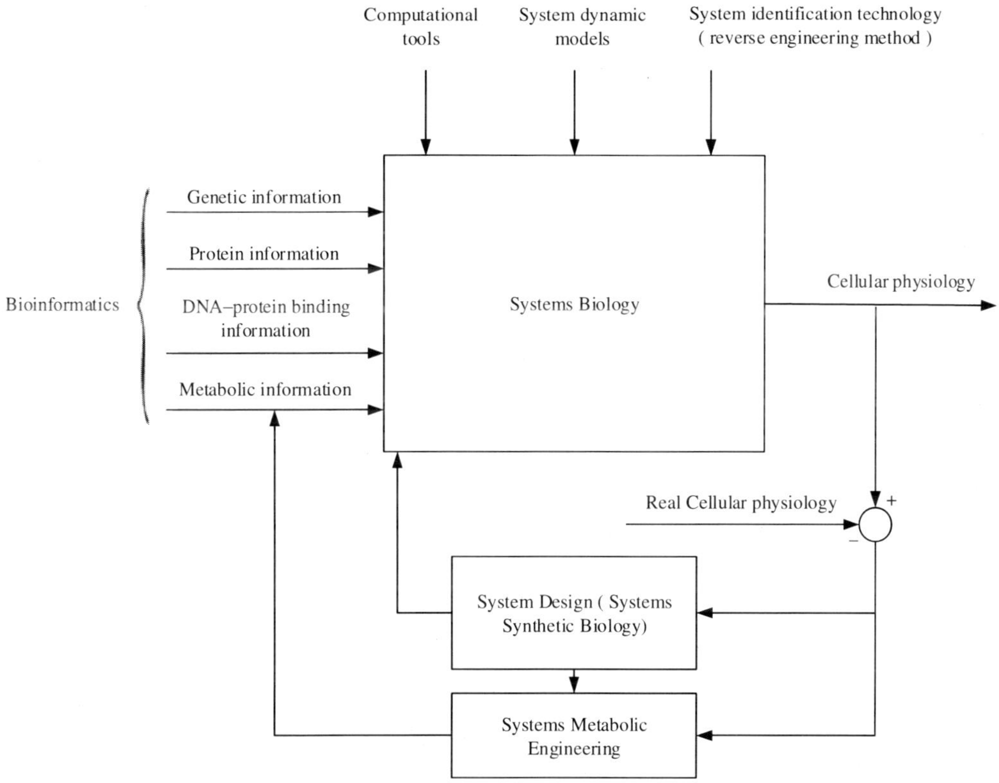

Figure 1.1. The role of systems biology as an integrated platform in modern biological research. Systems biology integrates information on genetics, proteins, DNA-protein binding, and metabolism with system dynamics modeling and system identification technology to develop models and mechanisms for the interpretation of cellular phenotypes and behaviors. Powerful computational tools are necessary for mining large-scale data sets. Synthetic gene circuits are designed based on system models and mechanisms to investigate specific desired physiological behaviors. Discrepancies between real and desired cellular behaviors are used as feedback to adjust system models and mechanisms. Systems biology is thus in a position to integrate bioinformatics, systems synthetic biology, and systems metabolic engineering.

1.4. THE ROLE OF SYSTEMATIC ANALYSIS AND MATHEMATIC MODELING IN SYSTEMS BIOLOGY

Several mathematical techniques have been developed to analyze the systematic properties of complex biological networks. For example, system sensitivity in response to various parameter variations has been analyzed to determine properties affecting network robustness or fragility. System sensitivity analysis not only reveals the robust stability of a biological network against various perturbations, but may also provide information about network controllability [7-9]. A biological network's system response is a measure of its response to environmental signals or disturbances [28, 34]. From the perspective of system theory, robustness with intrinsic system variation and the ability to respond to external stimuli are two important and complementary system characteristics in evaluating system performance [33, 34]. A biological system that is robust to considerable intrinsic parameter fluctuations is less responsive to environmental disturbances, and vice versa. A systems investigation of the aging-related gene network using microarray data found that network robustness increases and network response ability decreases during the aging process [34]. A genetic system's sensitivity to environmental molecular noise is considered indicative of its filtering ability [42]. Systems biology allows for measurement of system characteristics as well as capabilities of signal transduction pathways [8]. Similarly, flux amplification of a metabolic pathway can be estimated using both system dynamics models and signal gain methods.

Nonlinear biological networks operate under different conditions of cellular homeostasis and homeodynamics. This makes systems biology studies on complex biological models based on phenotypic landscapes highly informative. These studies help discover possible equilibrium points (phenotypes) and dynamic behaviors such as bifurcation, oscillation, robust stability, and phase drifts to other equilibrium points (phenotype transition). Bifurcation analysis and phase-plane analysis of nonlinear dynamic networks can be useful in predicting system behavior under intrinsic parameter changes. Through a systems biology approach and dynamic modeling [38-42], network robustness and noise filtering ability can be improved with feedback, redundancy, and modular schemes. This explains the presence of many feedback loops, redundant genes, and modular structures at different network levels. A unifying mathematical framework based on nonlinear stochastic dynamic models [75-77] was recently proposed to describe the different levels of stochastic biological networks under parameter fluctuations, genetic variations, and environmental disturbances [29, 30, 59]. The phenotypic robustness criteria of biological networks in systems, evolutionary, ecological, and synthetic biology were investigated from a systematic perspective based on robust stabilization and filtering ability. Network robustness can confer intrinsic robustness for tolerating intrinsic parameter fluctuations, genetic robustness for buffering genetic variations, and environmental robustness for resisting environmental disturbances. It was found that if the sum of these is less than or equal to the network robustness, then the phenotype is robust at different network levels in systems, evolutionary, ecological, synthetic, and metabolic biology. These phenotypic criteria at different levels are useful for the design of synthetic and metabolic systems. A systems biology approach based on dynamic models can clearly provide systematic insight into behaviors at these different network levels and provides a design platform to improve synthetic and metabolic system robustness, filtering ability, and transduction ability, which are discussed in detail in the following chapters.

1.5. Future Challenges of Systems Biology

The cycle of systematic research highlighted on the cover page and the depiction of the role of systems biology in Figure 1.1 indicate that systems biology faces three major challenges. First, new tools are necessary for systematical perturbation experiments and monitoring signal processes and functions in both cells and organisms. Second, large data sets are required to construct interaction network maps linking genes to each other and to their biological functions. Third, dynamic network models that link genes, proteins, and other biomolecules to cellular and organismal function are required. The big challenge for systems biology is to start with drafted dynamic models. These are typically described by differential equations and are used to propose the most informative perturbation tools and biosensors, which can then generate new bioinformatic data sets. These data sets can be used to construct the real dynamic model to improve the understanding of the underlying theory, and this knowledge can be used for synthetic biological design with applications in metabolic engineering. The systems biology approach is a quest to find such converging cyclic loops between theory, experiment, and application. This requires a new type of research using detailed knowledge of organismal, cellular, and molecular mechanisms and mathematical expertise, as well as creative ideas to confront these issues.

CONCLUSION

Systems biology provides an integrated platform for bioinformatics, synthetic biology, and metabolic engineering. It allows for modeling, analysis and design at different levels of biological networks.

REFERENCES

[1] Joyce, AR; Palsso, BO. "The model organism as a system: integrating 'omics' data sets," *Nature Reviews Molecular Cell Biology*, vol. 7, 198-210, Mar 2006.

[2] Ideker, T; Bafna, V; Lemberger, T. "Integrating scientific cultures," *Molecular Systems Biology*, vol. 3, Apr 2007.

[3] Chuang, HY; Lee, E; Liu, YT; Lee, D; Ideker, T. "Network-based classification of breast cancer metastasis," *Molecular Systems Biology*, vol. 3, Oct 2007.

[4] Oti, M; Snel, B; Huynen, MA; Brunner, HG. "Predicting disease genes using protein-protein interactions," *Journal of Medical Genetics*, vol. 43, Aug 2006.

[5] Peri, S; Navarro, JD; Amanchy, R; Kristiansen, TZ; Jonnalagadda, CK; Surendranath, V; et al. "Development of human protein reference database as an initial platform for approaching systems biology in humans," *Genome Research*, vol. 13, 2363-2371, Oct 2003.

[6] Zhang, SH; Jin, GX; Zhang, XS; Chen, LN. "Discovering functions and revealing mechanisms at molecular level from biological networks," *Proteomics*, vol. 7, 2856-2869, Aug 2007.

[7] Voit, EO. *Computational analysis of biochemical systems : a practical guide for biochemists and molecular biologists*. New York: Cambridge University Press, 2000.

[8] Chen, BS; Wu, CC. "On the calculation of signal transduction ability of signaling transduction pathways in intracellular communication: systematic approach," *Bioinformatics*, vol. 28, 1604-1611, Jun 15 2012.

[9] Kitano, H. "Opinon - Cancer as a robust system: implications for anticancer therapy," *Nature Reviews Cancer*, vol. 4, 227-235, Mar 2004.

[10] Stelling, J; Sauer, U; Szallasi, Z; Doyle, FJ; Doyle, J. "Robustness of cellular functions," *Cell*, vol. 118, 675-685, Sep 17 2004.

[11] Storey, JD; Xiao, WZ; Leek, JT; Tompkins, RG; Davis, RW. "Significance analysis of time course microarray experiments," *Proceedings of the National Academy of Sciences of the United States of America*, vol. 102, 12837-12842, Sep 6 2005.

[12] Cherry, JM; Adler, C; Ball, C; Chervitz, SA; Dwight, SS; Hester, ET; et al. "SGD: Saccharomyces Genome Database," *Nucleic Acids Research*, vol. 26, 73-79, Jan 1 1998.

[13] Gasch, AP; Spellman, PT; Kao, CM; Carmel-Harel, O; Eisen, MB; Storz, G; et al. "Genomic expression programs in the response of yeast cells to environmental changes," *Molecular Biology of the Cell*, vol. 11, 4241-4257, Dec 2000.

[14] Harbison, CT; Gordon, DB; Lee, TI; Rinaldi, NJ; Macisaac, KD; Danford, TW; et al. "Transcriptional regulatory code of a eukaryotic genome," *Nature*, vol. 431, 99-104, Sep 2 2004.

[15] Teixeira, MC; Monteiro, P; Jain, P; Tenreiro, S; Fernandes, AR; Mira, NP; et al. "The YEASTRACT database: a tool for the analysis of transcription regulatory associations in Saccharomyces cerevisiae," *Nucleic Acids Research*, vol. 34, D446-51, Jan 1 2006.

[16] Stark, C; Breitkreutz, BJ; Reguly, T; Boucher, L; Breitkreutz, A; Tyers, M. "BioGRID: a general repository for interaction datasets," *Nucleic Acids Research*, vol. 34, D535-9, Jan 1 2006.

[17] Ashburner, M; Ball, CA; Blake, JA; Botstein, D; Butler, H; Cherry, JM; et al. "Gene ontology: tool for the unification of biology. The Gene Ontology Consortium," *Nat Genet*, vol. 25, 25-9, May 2000.

[18] Chen, HC; Lee, HC; Lin, TY; Li, WH; Chen, BS. "Quantitative characterization of the transcriptional regulatory network in the yeast cell cycle," *Bioinformatics*, vol. 20, 1914-27, Aug 12 2004.

[19] Chen, BS; Wang, YC; Wu, WS; Li, WH. "A new measure of the robustness of biochemical networks," *Bioinformatics*, vol. 21, 2698-2705, Jun 1 2005.

[20] Chang, WC; Li, CW; Chen, BS. "Quantitative inference of dynamic regulatory pathways via microarray data," *BMC Bioinformatics*, vol. 6, Mar 7 2005.

[21] Lin, LH; Lee, HC; Li, WH; Chen, BS. "Dynamic modeling of cis-regulatory circuits and gene expression prediction via cross-gene identification," *BMC Bioinformatics*, vol. 6, Oct 18 2005.

[22] Chang, YH; Wang, YC; Chen, BS. "Identification of transcription factor cooperativity via stochastic system model," *Bioinformatics*, vol. 22, 2276-2282, Sep 15 2006.

[23] Wu, WS; Li, WH; Chen, BS. "Computational reconstruction of transcriptional regulatory modules of the yeast cell cycle," *BMC Bioinformatics*, vol. 7, Sep 29 2006.

[24] Wu, WS; Li, WH; Chen, BS. "Identifying regulatory targets of cell cycle transcription factors using gene expression and ChIP-chip data," *BMC Bioinformatics*, vol. 8, Jun 8 2007.

[25] Chen, BS; Wang, YC. "On the attenuation and amplification of molecular noise in genetic regulatory networks," *BMC Bioinformatics*, vol. 7, Feb 2 2006.

[26] Chen, BS; Chen, WH; Wu, HL. "Robust H-2/H-∞ Global Linearization Filter Design for Nonlinear Stochastic Systems," *IEEE Transactions on Circuits and Systems I-Regular Papers*, vol. 56, 1441-1454, Jul 2009.
[27] Chen, BS; Wu, WS; Wang, YC; Li, WH. "On the Robust Circuit Design Schemes of Biochemical Networks: Steady-State Approach," *IEEE Transactions on Biomedical Circuits and Systems*, vol. 1, 91-104, Jun 2007.
[28] Chen, BS; Chen, PW. "Robust Engineered Circuit Design Principles for Stochastic Biochemical Networks With Parameter Uncertainties and Disturbances," *IEEE Transactions on Biomedical Circuits and Systems*, vol. 2, 114-132, Jun 2008.
[29] Chen, BS; Lin, YP. "A Unifying Mathematical Framework for Genetic Robustness, Environmental Robustness, Network Robustness and their Trade-off on Phenotype Robustness in Biological Networks Part I: Gene Regulatory Networks in Systems and Evolutionary Biology," *Evolutionary Bioinformatics*, vol. 9, 43-68, 2013.
[30] Chen, BS; Lin, YP. "A Unifying Mathematical Framework for Genetic Robustness, Environmental Robustness, Network Robustness and their Tradeoff on Phenotype Robustness in Biological Networks Part II: Ecological Networks," *Evolutionary Bioinformatics*, vol. 9, 69-85, 2013.
[31] Wang, YC; Huang, SH; Lan, CY; Chen, BS. "Prediction of Phenotype-Associated Genes via a Cellular Network Approach: A Candida albicans Infection Case Study," *Plos One*, vol. 7, Apr 11 2012.
[32] Kuo, ZY; Chuang, YJ; Chao, CC; Liu, FC; Lan, CY; Chen, BS. "Identification of Infection- and Defense-Related Genes via a Dynamic Host-Pathogen Interaction Network Using a Candida Albicans -Zebrafish Infection Model," *J Innate Immun*, vol. 5, 137-52, 2013.
[33] Tu, CT; Chen, BS. "New measurement methods of network robustness and response ability via microarray data," *Plos One*, vol. 8, e55230, 2013.
[34] Tu, CT; Chen, BS. "On the Increase in Network Robustness and Decrease in Network Response Ability During the Aging Process: A Systems Biology Approach via Microarray Data," *IEEE/ACM Trans Comput Biol Bioinform*, Mar 19 2013.
[35] Csete, ME; Doyle, JC. "Reverse engineering of biological complexity," *Science*, vol. 295, 1664-9, Mar 1 2002.
[36] Lee, KH; Park, JH; Kim, TY; Kim, HU; Lee, SY "Systems metabolic engineering of Escherichia coli for L-threonine production," *Molecular Systems Biology*, vol. 3, 149, 2007.
[37] Park, JH; Lee, SY; Kim, TY; Kim, HU. "Application of systems biology for bioprocess development," *Trends Biotechnol*, vol. 26, 404-12, Aug 2008.
[38] Chen, BS; Chang, YT; Wang, YC. "Robust H infinity-stabilization design in gene networks under stochastic molecular noises: fuzzy-interpolation approach," *IEEE Trans Syst Man Cybern B Cybern*, vol. 38, 25-42, Feb 2008.
[39] Chu, LH; & Chen, BS. "Construction of a cancer-perturbed protein-protein interaction network for discovery of apoptosis drug targets," *BMC Syst Biol*, vol. 2, 56, 2008.
[40] Chen, BS; Chang, CH; Chuang, YJ. "Robust model matching control of immune systems under environmental disturbances: Dynamic game approach," *Journal of Theoretical Biology*, vol. 253, 824-837, Aug 21 2008.

[41] Chen, BS; Yang, SK; Lan, CY; & Chuang, YJ. "A systems biology approach to construct the gene regulatory network of systemic inflammation via microarray and databases mining," *BMC Medical Genomics*, vol. 1, Sep 30 2008.

[42] Chen, BS; Chen, PW. "On the estimation of robustness and filtering ability of dynamic biochemical networks under process delays, internal parametric perturbations and external disturbances," *Mathematical Biosciences*, vol. 222, 92-108, Dec 2009.

[43] Wang, YC; Lan, CY; Hsieh, WP; Murillo, LA; Agabian, N; Chen, BS. "Global screening of potential Candida albicans biofilm-related transcription factors via network comparison," *BMC Bioinformatics*, vol. 11, Jan 26 2010.

[44] Li, CW; Chen, BS. "Identifying Functional Mechanisms of Gene and Protein Regulatory Networks in Response to a Broader Range of Environmental Stresses," *Comparative and Functional Genomics*, 2010.

[45] Gardner, TS; Cantor, CR; Collins, JJ. "Construction of a genetic toggle switch in Escherichia coli," *Nature*, vol. 403, 339-42, Jan 20 2000.

[46] Elowitz, MB; Leibler, S. "A synthetic oscillatory network of transcriptional regulators," *Nature*, vol. 403, 335-8, Jan 20 2000.

[47] Hasty, J; McMillen, D; Collins, JJ. "Engineered gene circuits," *Nature*, vol. 420, 224-230, Nov 14 2002.

[48] McDaniel, R; Weiss, R. "Advances in synthetic biology: on the path from prototypes to applications," *Curr Opin Biotechnol*, vol. 16, 476-83, Aug 2005.

[49] Endy, D. "Foundations for engineering biology," *Nature*, vol. 438, 449-53, Nov 24 2005.

[50] McAdams, HH; Arkin, A. "It's a noisy business! Genetic regulation at the nanomolar scale," *Trends in Genetics*, vol. 15, 65-69, Feb 1999.

[51] Batt, G; Yordanov, B; Weiss, R; Belta, C. "Robustness analysis and tuning of synthetic gene networks," *Bioinformatics*, vol. 23, 2415-2422, Sep 15 2007.

[52] Chen, BS; Chang, CH; Lee, HC. "Robust synthetic biology design: stochastic game theory approach," *Bioinformatics*, vol. 25, 1822-1830, Jul 15 2009.

[53] Chen, BS; Wu, CH. "A systematic design method for robust synthetic biology to satisfy design specifications," *BMC Systems Biology*, vol. 3, Jun 30 2009.

[54] Chen, BS; Chen, PW. "GA-based Design Algorithms for the Robust Synthetic Genetic Oscillators with Prescribed Amplitude, Period and Phase," *Gene Regul Syst Bio*, vol. 4, 35-52, 2010.

[55] Chen, BS; Wu, CH. "Robust Optimal Reference-Tracking Design Method for Stochastic Synthetic Biology Systems: T-S Fuzzy Approach," *IEEE Transactions on Fuzzy Systems*, vol. 18, 1144-1159, Dec 2010.

[56] Wu, CH; Zhang, WH; Chen, BS. "Multiobjective H-2/H-infinity synthetic gene network design based on promoter libraries," *Mathematical Biosciences*, vol. 233, 111-125, Oct 2011.

[57] Wu, CH; Lee, HC; Chen, BS. "Robust synthetic gene network design via library-based search method," *Bioinformatics*, vol. 27, 2700-2706, Oct 1 2011.

[58] Chen, BS; Hsu, CY; Liou, JJ. "Robust Design of Biological Circuits: Evolutionary Systems Biology Approach," *Journal of Biomedicine and Biotechnology*, 2011.

[59] Chen, BS; Lin, YP. "A Unifying Mathematical Framework for Genetic Robustness, Environmental Robustness, Network Robustness and their Trade-offs on Phenotype Robustness in Biological Networks. Part III: Synthetic Gene Networks in Synthetic Biology," *Evolutionary Bioinformatics*, vol. 9, 87-109, 2013.

[60] Chen, BS; Hsu, CY. "Robust synchronization control scheme of a population of nonlinear stochastic synthetic genetic oscillators under intrinsic and extrinsic molecular noise via quorum sensing," *BMCSystems Biology*, vol. 6, Oct 26 2012.

[61] Kim, PJ; Lee, DY; Kim, TY; Lee, KH; Jeong, H; Lee, SY; et al. "Metabolite essentiality elucidates robustness of Escherichia coli metabolism," *Proc Natl Acad Sci U S A*, vol. 104, 13638-42, Aug 21 2007.

[62] Segre, D; Vitkup, D; Church, GM. "Analysis of optimality in natural and perturbed metabolic networks," *Proc Natl Acad Sci U S A*, vol. 99, 15112-7, Nov 12 2002.

[63] Park, JH; Lee, SY; Kim, TY; Kim, HU. "Application of systems biology for bioprocess development," *Trends in Biotechnology*, vol. 26, 404-412, Aug 2008.

[64] Chandran, D; Copeland, WB; Sleight, SC; Sauro, HM. "Mathematical modeling and synthetic biology," *Drug Discovery Today: Disease Models*, vol. 5, 299-309, 2008.

[65] Zhou, S; Iverson, AG; Grayburn, WS. "Engineering a native homoethanol pathway in Escherichia coli B for ethanol production," *Biotechnology Letters*, vol. 30, 335-342, Feb 2008.

[66] Shlomi, T; Eisenberg, Y; Sharan, R; Ruppin, E. "A genome-scale computational study of the interplay between transcriptional regulation and metabolism," *Molecular Systems Biology*, vol. 3, Apr 2007.

[67] Jamshidi, N; Palsson, BO. "Formulating genome-scale kinetic models in the post-genome era," *Molecular Systems Biology*, vol. 4, Mar 2008.

[68] Bonneau, R; Facciotti, MT; Reiss, DJ; Schmid, AK; Pan, M; Kaur, A; et al. "A predictive model for transcriptional control of physiology in a free living cell," *Cell*, vol. 131, 1354-1365, Dec 28 2007.

[69] Kuepfer, L; Sauer, U; Parrilo, PA. "Efficient classification of complete parameter regions based on semidefinite programming," *BMC Bioinformatics*, vol. 8, Jan 15 2007.

[70] !!! INVALID CITATION !!!

[71] Yao, CW; Hsu, BD; Chen, BS. "Constructing gene regulatory networks for long term photosynthetic light acclimation in Arabidopsis thaliana," *BMC Bioinformatics*, vol. 12, p. 335, 2011.

[72] Wang, YC; Chen, BS. "A network-based biomarker approach for molecular investigation and diagnosis of lung cancer," *BMC Medical Genomics*, vol. 4, Jan 6 2011.

[73] Wang, YC; Chen, BS. "Integrated cellular network of transcription regulations and protein-protein interactions," *BMC Systems Biology*, vol. 4, Mar 8 2010.

[74] Shiue, E; Prather, KLJ. "Synthetic biology devices as tools for metabolic engineering," *Biochemical Engineering Journal*, vol. 65, 82-89, Jun 15 2012.

[75] Chen, BS; Zhang, WH. "Stochastic H(2)/H(infinity) control with state-dependent noise," *IEEE Transactions on Automatic Control*, vol. 49, 45-57, Jan 2004.

[76] Zhang, WH; Chen, BS; Tseng, CS. "Robust H-infinity filtering for nonlinear stochastic systems," *IEEE Transactions on Signal Processing*, vol. 53, 589-598, Feb 2005.

[77] Zhang, WH; Chen, BS. "State feedback H(infinity) control for a class of nonlinear stochastic systems," *SIAM Journal on Control and Optimization*, vol. 44, 1973-1991, 2006.

Chapter 2

PRELIMINARY MATHEMATICS

ABSTRACT

Systems biology is based on mathematical modeling, data analysis and system design. This creates a need for differential equations and linear algebra for use in dynamic models of GRNs, PPI networks, and metabolic pathways, statistics and stochastic processes for data analysis and parameter estimation, and optimization theory for system design in synthetic biology and metabolic engineering. Some useful mathematical methods for systems biology are introduced in this chapter.

INTRODUCTION

In this chapter, linear mathematical models and related systematic and mathematic analysis are introduced, followed by some statistical testing, detection, and estimation methods for statistical inference and hypothesis testing in bioinformatics and systems biology. Stochastic systems in systems and synthetic biology and their stability and H_∞ robust disturbance attenuation for GR and PPI networks are discussed. Finally, some numerical optimization technologies by recursive searching schemes and genetic (evolutionary) searching algorithms for optimal design in systems, synthetic, and metabolic engineering are introduced.

2.1. LINEAR SYSTEM MODEL

In the modeling of biological systems with n species in Figure 2.1, for example, a metabolic pathway with n metabolites, the dynamic model of the biological system can be represented by the following dynamic equation

$$x_i(k+1) = a_{i1}x_1(k) + \cdots + a_{in}x_n(k) + v_i(k), i = 1,\ldots,n \tag{2.1}$$

where $x_i(k)$ denotes the state variable of the i-th species (for example, the concentration of the i-th metabolite in metabolic pathway) at the present time k. a_{ij} are real numbers to denote the dynamic interaction from the j-th species to the i-th species at the next time $k+1$. $v_i(k)$ denotes the random noise due to modeling error or environmental disturbance.

Then the dynamic model of the linear biological system with n species in (2.1) can be augmented as

$$\begin{bmatrix} x_1(k+1) \\ \vdots \\ x_n(k+1) \end{bmatrix} = \begin{bmatrix} a_{11} & \cdots & a_{1n} \\ \vdots & \ddots & \vdots \\ a_{n1} & \cdots & a_{nn} \end{bmatrix} \begin{bmatrix} x_1(k) \\ \vdots \\ x_n(k) \end{bmatrix} + \begin{bmatrix} v_1(k) \\ \vdots \\ v_n(k) \end{bmatrix} \tag{2.2}$$

which can be represented by linear state space dynamic model

$$x(k+1) = Ax(k) + v(k) \tag{2.3}$$

where $x(k)=[x_1(k) \cdots x_n(k)]^T$ denotes the state vector of linear biological system in Figure 2.1. A denotes the system matrix of biological system, and $v(k)$ denotes the noise vector.

If the linear biological system is described by the continuous equation, then the linear state space dynamic model in (2.3) must be modified

$$\tfrac{d}{dt}x(t) = Ax(t) + v(t) \tag{2.4}$$

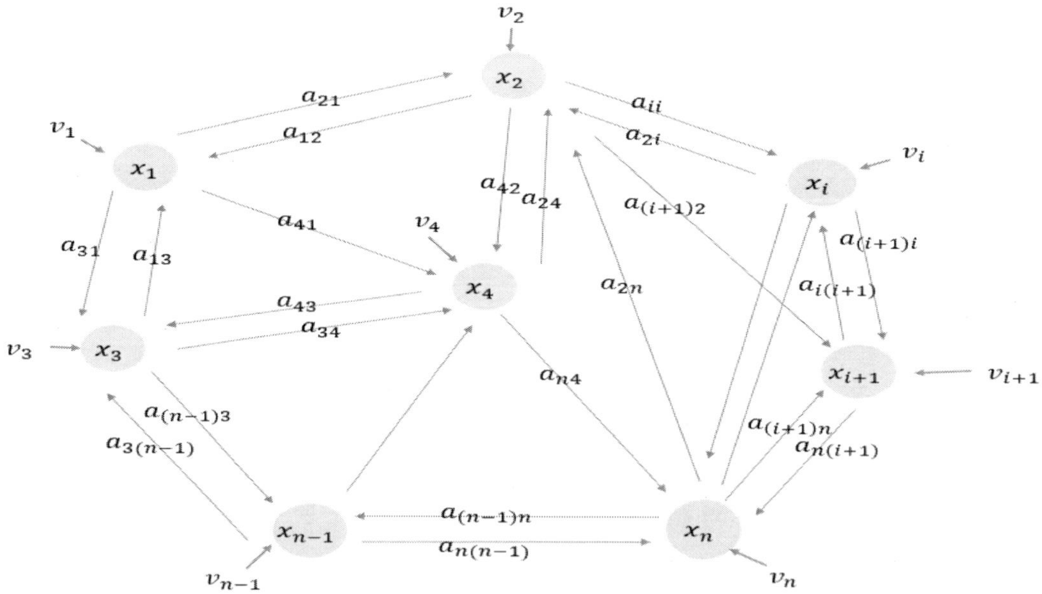

Figure 2.1. A biological system with n species.

In the steady state case, the dynamic equation in (2.1) can be modified as the following linear regression form

$$x_i = a_{i1}x_1 + \cdots + a_{in}x_n + v_i, i = 1, \ldots, n \tag{2.5}$$

In this situation, the linear state space model in (2.3) is modified as

$$x = Ax + v \tag{2.6}$$

$$(I - A)x = v \tag{2.7}$$

where I is the identity matrix. If $(I - A)$ is invertible, then the steady state of biological system can be obtained as

$$x = (I - A)^{-1} v \tag{2.8}$$

Definition 2.1

Given a square matrix A, if there exist a column vector x and a scalar λ such that

$$Ax = \lambda x \tag{2.9}$$

then λ is called an eigenvalue of A and x is the eigenvector associated with the eigenvalue λ. The eigenvalues can be obtained by solving the characteristic equation

$$\det(A - \lambda I) = 0 \tag{2.10}$$

where $\det(A-\lambda I)$ is called the characteristic polynomial. There are n eigenvalues $\lambda_1, \cdots, \lambda_n$ solving equation (2.10)

Let Λ denote the following diagonal matrix of eigenvalues

$$\Lambda = \begin{bmatrix} \lambda_1 & 0 & \cdots & 0 \\ 0 & \lambda_2 & \ddots & \vdots \\ \vdots & \ddots & \ddots & 0 \\ 0 & \cdots & 0 & \lambda_n \end{bmatrix} \tag{2.11}$$

Assume that the $n \times n$ matrix A has n independent eigenvectors x_i organized as the columns of a matrix T. In addition, let Λ be the diagonal matrix of eigenvalues of A. Then

$$T^{-1}AT = \Lambda \text{ or } A = T\Lambda T^{-1} \tag{2.12}$$

A simple example is given to explain the physical meaning of eigenvalue and eigenvector of matrix A. Suppose A is used to describe the position of a two point system in Figure 2.2, i.e., point 1 and point 2 are in a plane coordinated by a horizontal axis $[1\ 0]^T$ and vertical axis $[0\ 1]^T$, i.e.,

$$A=\begin{bmatrix} a_{11} & a_{12} \\ a_{21} & a_{22} \end{bmatrix}=\left[\begin{array}{c|c} 1 & 0 \\ 0 & 1 \end{array}\right]\begin{bmatrix} a_{11} & a_{12} \\ a_{21} & a_{22} \end{bmatrix}\left[\begin{array}{c|c} 1 & 0 \\ 0 & 1 \end{array}\right]^{-1} \tag{2.13}$$

where a_{11} and a_{12} denotes the projection of point 1 to the axis $[1\ 0]^T$ and $[0\ 1]^T$, respectively and a_{21} and a_{22} denote the projection of point 2 to $[1\ 0]^T$ and $[0\ 1]^T$, respectively (see Figure 2.2).

It is more complex to use (2.13) to describe the position of the two point system. If we want to simplify the position description of the two point system, we can use the eigenvectors t_1 and t_2 as axis to describe the position as follows

$$A=\left[t_1 \mid t_2\right]\begin{bmatrix} \lambda_1 & 0 \\ 0 & \lambda_2 \end{bmatrix}\left[t_1 \mid t_2\right]^{-1} \tag{2.14}$$

i.e., with t_1 and t_2 as orthogonal axes to describe position of the two point system, then point 1 is just at the eigenvector t_1 and point 2 just at the eigenvector t_2. Therefore λ_1 is just the distance from point 1 to the origin and λ_2 is just the distance from point 2 to the origin. Therefore we could use Λ in (2.12) to describe the position of the two-point system.

Result 2.1

The eigenvalues of a symmetrical matrix $A\in\mathbb{R}^{n\times n}$ are real and there exist orthonormal eigenvectors $t_1, \cdots, t_n$ corresponding to eigenvalues $\lambda_1, \cdots, \lambda_n$ so that

$$A=T\Lambda T^{-1}=T\Lambda T^T \tag{2.15}$$

where the columns of T are eigenvectors of A.

Result 2.2

Let A and B be any $m\times n$ matrix. Then

$$Tr\left(AB^T\right)=Tr\left(B^T A\right) \tag{2.16}$$

Result 2.3

The n eigenvalues $\lambda_1, \cdots, \lambda_n$ solving equation (2.10) also satisfy the relations

$$\begin{aligned} &\det A=\lambda_1\times\cdots\times\lambda_n \\ &Tr\left(A\right)=\lambda_1+\cdots+\lambda_n=Tr\left(\Lambda\right) \end{aligned} \tag{2.17}$$

Result 2.4 – matrix inversion lemma

$$\left(A+BC\right)^{-1} = A^{-1} - A^{-1}B\left(I+CA^{-1}B\right)^{-1}CA^{-1} \tag{2.18}$$

This lemma is useful when the ordinary least-square parameter estimation algorithm is to be transformed to the recursive parameter estimation algorithm in the sequel.

2.2. Matrix Norms

Some important examples of vector norms are

$$\begin{aligned}
&\|x\|_1 = |x_1| + \cdots + |x_n| && \text{1-norm} \\
&\|x\|_2 = \sqrt{x^T x} = \left(|x_1|^2 + \cdots + |x_n|^2\right)^{1/2} && \text{2-norm} \\
&\|x\|_p = \left(|x_1|^p + \cdots + |x_n|^p\right)^{1/p} && p\text{-norm} \\
&\|x\|_\infty = \max\left(|x_1|, \cdots, |x_n|\right) && \infty\text{-norm}
\end{aligned} \tag{2.19}$$

Let w be some positive definite symmetric matrix. A weighted 2-norm is defined as

$$\|x\|_w = \sqrt{x^T wx}, w = w^T > 0 \tag{2.20}$$

Definition 2.2
The matrix (induced) norm of a matrix A corresponding to a certain vector norm $\|x\|_p$ is

$$\|A\|_p = \sup_{x \in \mathbb{R}^n} \frac{\|Ax\|_p}{\|x\|_p} \tag{2.21}$$

Remark 2.1: (i) sup denotes the less upper bound. In general, it is considered as maximum. (ii) In the case of linear signal system $y=Ax$. $\|A\|_p=\max_{x\in\mathbb{R}^n}\|y\|_p/\|x\|_p$, i.e., $\|A\|_p$ denotes the maximum output/input ratio for all possible bounded input signals, i.e., the gain of the linear signal system from p-norm point of view.

Definition 2.3 – Singular value decomposition
Let A be an $m\times n$ matrix of generally complex-valued elements. Then there exist $m\times m$ and $n\times n$ unitary matrices U and V (i.e., $U^{-1}=U^*$ and $V^{-1}=V^*$) such that

$$A = U\Sigma V^* \tag{2.22}$$

where * denotes the conjugate transpose and Σ is an $m \times n$ matrix whose elements σ_{ij} are zero except possibly along its main diagonal. These nonnegative diagonal elements are ordered such that

$$\sigma_{11} \geq \cdots \geq \sigma_{pp} \geq 0,\ p = \min(m,n) \tag{2.23}$$

i.e.,

$$\Sigma = \begin{bmatrix} \sigma_{11} & 0 & \cdots & 0 \\ 0 & \ddots & 0 & \vdots \\ 0 & 0 & \sigma_{pp} & 0 \end{bmatrix} \in \mathbb{R}^{m\times n},\ \text{if } p = m$$
$$= \begin{bmatrix} \sigma_{11} & 0 & 0 \\ 0 & \ddots & 0 \\ \vdots & \ddots & \sigma_{pp} \\ 0 & \cdots & 0 \end{bmatrix} \in \mathbb{R}^{m\times n},\ \text{if } p = n \tag{2.24}$$

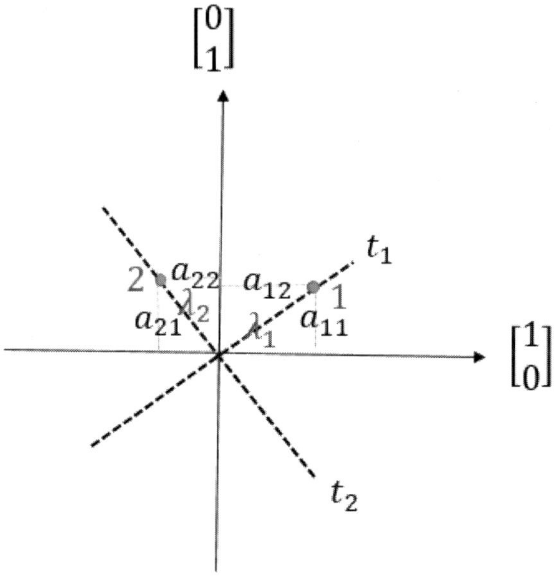

Figure 2.2 – The position of the two point system. Suppose there exists a two point system at point 1 and point 2. If we use $[1\ 0]^T$ and $[0\ 1]^T$ as axis to coordinate the plane of the two point system, then $\begin{bmatrix} a_{11} & a_{12} \\ a_{21} & a_{22} \end{bmatrix}$ is needed to describe the position of the two point system. If we use eigenvector t_1 and t_2 as axis to coordinate the plane of the two point system, then $\begin{bmatrix} \lambda_1 & 0 \\ 0 & \lambda_2 \end{bmatrix}$ is used to describe the position of the two point system because point 1 has zero projection on t_2 and point 2 has zero projection on t_1.

The diagonal elements σ_{ii} are known as the singular values of the matrix A and the nonzero singular values correspond to the positive square roots of the eigenvalue of A^*A, i.e.,

$$\sigma_{ii} = \sqrt{\lambda_i\left(A^*A\right)} \tag{2.25}$$

where $\lambda_i(A^*A)$ denotes the i-th eigenvalue of A^*A

The columns of U and V correspond to orthonormal eigenvectors of AA^* and A^*A. The matrix decomposition in (2.22) is called singular value decomposition (SVD)

Remark 2.2: The hierarchical singular values in (2.23) denote the principal components of A. Therefore the singular value decomposition of A in (2.22) is called principal component analysis. We could truncate the less components, i.e., less σ_{ii} to keep the principal components of Σ in (2.22), which is called model truncation.

Result 2.5

A useful property of the SVD is

$$\|A\|_2 = \sigma_{11} \tag{2.26}$$

i.e., the 2-norm of matrix A is its largest singular value.

Definition 2.4

The pseudo inverse (or Moore-Penrose inverse) of the matrix $A \in \mathbb{R}^{m\times n}$ is defined as

$$A^\dagger \triangleq \left(A^*A\right)^{-1} A^* \tag{2.27}$$

Clearly if $m=n=\text{rank}(A)$, then $A^\dagger=A^{-1}$. If A is a real-valued matrix, then $A^\dagger=(A^TA)^{-1}A^T$.

Result 2.6

$$A^\dagger = V\Sigma^\dagger U^* \tag{2.28}$$

where

$$\Sigma^\dagger = \begin{bmatrix} \frac{1}{\sigma_{11}} & 0 & \cdots & 0 \\ 0 & \ddots & \ddots & \vdots \\ 0 & 0 & \frac{1}{\sigma_{pp}} & 0 \end{bmatrix} \in \mathbb{R}^{n\times m}$$

Result 2.7

The singular value decomposition has a close relation to the least-squares problem in that the least square solution

$$\hat{x} = \arg\min_{x \in \mathbb{R}^n} \|Ax - b\|_2^2 = A^{\dagger} b \tag{2.29}$$

for a matrix A and a vector b. The residual sum at the above optimum is

$$\|A\hat{x} - b\|_2^2 = \left\|\left(AA^{\dagger} - I\right)b\right\|_2^2 \tag{2.30}$$

Definition 2.5 – Quadratic forms

A real-valued quadratic form is defined as

$$V(x) = x^T(t)Px(t), \; P \in \mathbb{R}^{n \times n} \tag{2.31}$$

where P is a symmetric matrix. The quadratic form and the matrix P are said to be positive definite if $x^T Px > 0$ for all non-zero x and all the eigenvalues of a positive definite matrix P are positive. The quadratic function in (2.31) can be considerd as the energy function of linear system in (2.4). In this situation, $V(x)$ in (2.31) is called Lyapunov-function of linear system in (2.4).

Definition 2.6 – Matrix differentiation

Let $x \in \mathbb{R}^n$ be a vector and let $f(x)$ be a scalar function of x. The partial derivative of f with respect to x is the row vector

$$\frac{\partial f(x)}{\partial x} = \begin{bmatrix} \frac{\partial f(x)}{\partial x_1} & \cdots & \frac{\partial f(x)}{\partial x_n} \end{bmatrix} \tag{2.32}$$

Differentiation of a vector-valued function $f{:}\mathbb{R}^n \to \mathbb{R}^m$ with respect to x gives the Jacobian matrix

$$\frac{\partial f(x)}{\partial x} = \begin{bmatrix} \frac{\partial f_1(x)}{\partial x_1} & \cdots & \frac{\partial f_1(x)}{\partial x_n} \\ \vdots & \ddots & \vdots \\ \frac{\partial f_m(x)}{\partial x_1} & \cdots & \frac{\partial f_m(x)}{\partial x_n} \end{bmatrix}, \; f(x) \in \mathbb{R}^m, \; x \in \mathbb{R}^n \tag{2.33}$$

Differentiation of a scalar function $f(A)$ with respect to matrix A gives a matrix with elements

$$\frac{\partial f(A)}{\partial A} = \begin{bmatrix} \frac{\partial f(A)}{\partial a_{11}} & \cdots & \frac{\partial f(A)}{\partial a_{1n}} \\ \vdots & \ddots & \vdots \\ \frac{\partial f(A)}{\partial a_{n1}} & \cdots & \frac{\partial f(A)}{\partial a_{nn}} \end{bmatrix} \tag{2.34}$$

Example 2.1

$$V(x) = x^T P x \tag{2.35}$$

$$\frac{\partial V(x)}{\partial x} = Px + P^T x \tag{2.36}$$

Example 2.2

$$\frac{\partial \log(\det A)}{\partial A} = A^{-T} \tag{2.37}$$

Example 2.3

$$\frac{\partial Tr\left(BA^{-1}\right)}{\partial A} = -\left(A^{-1}BA^{-1}\right)^T \tag{2.38}$$

Since the state variables of biological systems are always random or stochastic due to intrinsic random fluctuation and external disturbance, the statistical inference will be discussed before further application to the analysis and design of biological systems.

2.3. Statistical Inference

A random variable or a stochastic variable has a value which is dependent on chance and which cannot be predicted from a knowledge of the experimental condition. To describe the outcome of a random variable X it is common practice to introduce the probability distribution function

$$F(x) = P\{X \le x\},\ 0 \le F(x) \le 1,\ \forall x \in \mathbb{R} \tag{2.39}$$

where $P\{X \le x\}$ denotes the probability that $X \le x$. The derivation $f(x)$ of the distribution function $F(x)$ is called the probability density function, and the solution x_α to the equation

$$P\{X \leq x_\alpha\} = \alpha \tag{2.40}$$

is called the α–percentile of the distribution.

In cases when there is no risk of confusion, we use x to denote also the random variable. The statistical mean M_y or expectation of a variable which is function of a random variable x is defined as

$$M_y = E\{y\} = \int_{-\infty}^{\infty} y(x) f(x) dx \tag{2.41}$$

and the mean of the distribution is

$$M_x = E\{x\} = \int_{-\infty}^{\infty} x f(x) dx \tag{2.42}$$

The variance of a scalar variable $y(x)$ is defined as

$$Var(y) = E\left\{(y - M_y)^2\right\} = \int_{-\infty}^{\infty} (y(x) - M_y)^2 f(x) dx \tag{2.43}$$

In the case of vector-valued variables, it is standard to use the definition of covariance

$$\begin{aligned} Var(y) &= Cov(y) = E\left\{(y - M_y)(y - M_y)^T\right\} \\ &= \int_{-\infty}^{\infty} (y(x) - M_y)(y(x) - M_y)^T f(x) dx \end{aligned} \tag{2.44}$$

Because it also describes the statistical relations between the components of the vector y, the covariance between two variables x and y is

$$Cov(x, y) = E\left\{(x - M_x)(y - M_y)^T\right\} = E\{xy^T\} - M_x M_y^T \tag{2.45}$$

Definition 2.7 – Statistically independent variables

Two random variables x and y are said to be statistically independent if the probability density function

$$f(x, y) = f_1(x) f_2(y), \text{ for some } f_1(x) \text{ and } f_2(x) \tag{2.46}$$

Definition 2.8 – Statistical covariance and correlation

The correlation coefficient between two variables x and y is

$$\rho = \frac{Cov(x,y)}{\sigma_x \sigma_y}, \text{ where } \sigma_x^2 = Var(x) \text{ and } \sigma_y^2 = Var(y) \tag{2.47}$$

and two variables x and y are uncorrelated if

$$Cov(x,y) = 0 \tag{2.48}$$

Definition 2.9 – The p-th moment

The p-th moment of a probability distribution $F(x)$ is the mathematical expectation

$$E\{x^p\} = \int_{-\infty}^{\infty} x^p f(x)\,dx \tag{2.49}$$

The above definitions with $E\{\bullet\}$, $Var(\bullet)$, and $Cov(\bullet,\bullet)$ are based on theoretical statistical results with all possible samples. However, in practical cases, we can only calculate these statistics based on finite observable data. In this situation, we get the following definitions based on N observable data.

Definition 2.10

The sample mean of a set of N observed variables x_i for $i=1, \cdots, N$ is defined as

$$\bar{x} \triangleq \frac{1}{N}\sum_{i=1}^{N} x_i \tag{2.50}$$

and the sample (co)variance is

$$S^2 = \frac{1}{N}\sum_{i=1}^{N}(x_i - M_x)(x_i - M_x)^T \tag{2.51}$$

Or for unknown M_x we calculate the sample variance

$$S^2 = \frac{1}{N-1}\sum_{i=1}^{N}(x_i - \bar{x})(x_i - \bar{x})^T \tag{2.52}$$

i.e., the sample mean $\bar{x}$ and the sample variance S^2 are estimates of mean μ and the variance σ^2 by means of some functions of some functions of finite data.

The cross-correlation (Pearson correlation) between x and y with a lag of k time points is given as follows

$$\rho(k)=\frac{\sum_{i=1}^{N-k}(y_{i+k}-\overline{y})(x_i-\overline{x})}{\sqrt{\sum_{i=1}^{N-k}(y_{i+k}-\overline{y})^2}\sqrt{\sum_{i=1}^{N-k}(x_i-\overline{x})^2}},i=0,1,\cdots,T \tag{2.53}$$

where $\overline{y}\triangleq\frac{\sum_{i=1}^{N-k}y_{i+k}}{N-k},\overline{x}\triangleq\frac{\sum_{i=1}^{N-k}x_i}{N-k}$.

Definition 2.11– Significance level

It is often desirable to provide an interval $[\theta_L,\theta_U]$ in which the value of a parameter θ would expect to lie with some probability

$$P\{\theta_L\le\theta\le\theta_U\}=1-\alpha \tag{2.54}$$

In the case α=0.05 this interval is called a 95% confidence interval. More generally, the interval $[\theta_L,\theta_U]$ associated with a probability level 1–α is called 100(1–α) –percent confidence interval and the statistics θ_L and θ_U are the lower and upper confidence limits. In hypothesis testing α is called the significance level of a p-value test.

Definition 2.12 – Convergence in L^P, $0<p<\infty$

The random sequence $\{x_k\}$ is said to converge in L^P to x if and only if

$$\lim_{k\to\infty}E\{|x_k-x|^P\}=0 \tag{2.55}$$

Definition 2.13 – Convergence almost surely (a.s.)

The random sequence $\{x_k\}$is said to converge with probability one (w.p.1) or almost surely (a.s.) to x if and only if for every ε>0 we have

$$\lim_{n\to\infty}P\{|x_k-x|\le\varepsilon,\forall k\ge n\}=1 \tag{2.56}$$

Definition 2.14 – Convergence in probability (in pr.)

The sequence $\{x_k\}$ is said to converge in probability (in pr.) to x if and only if for every ε>0 we have

$$\lim_{k\to\infty}P\{|x_k-x|>\varepsilon\}=0 \tag{2.57}$$

A basic result of probability theory is the following implication relationship between the convergence concepts

$$\begin{array}{ccc} & x_k \xrightarrow{L^p} x & \\ & \not\Uparrow \quad \Downarrow & \\ x_k \xrightarrow{a.s.} x & \overset{\Rightarrow}{\underset{\not\Leftarrow}{}} & x_k \xrightarrow{prob} x \end{array} \tag{2.58}$$

Result 2.8 – Central limit theorem

Let $\{x_k\}$ be an random sequence of independent random variables with common distribution function $F(x)$ with finite mean μ and variance σ^2

$$\text{If } S_N = \sum_{k=1}^{N} x_k, \text{ then } x_N = \frac{S_N - N\mu}{\sigma\sqrt{N}} \xrightarrow{dist.} N(0,1) \tag{2.59}$$

i.e., x_N has a limiting normal distribution with mean 0 and variance 1 as $N \to \infty$. The central limit theorem is commonly used as an approximation theorem in parameter identification theory and is also often to justify assumptions on normal distribution of variables.

Definition 2.15 – Normal distribution

$N(\mu,\sigma^2)$ is denoted as

$$f(x) = \frac{1}{\sqrt{2\pi}\sigma} e^{-(x-\mu)^2 / 2\sigma^2} \tag{2.60}$$

for a variable x with mean μ and variance σ^2. Hence, the normal distribution is parameterized by its mean μ and its variance σ^2. $N(0,1)$ in (2.59) denotes the standard normal distribution with $\mu=0$ and $\sigma=1$. The meaning of (2.59) is that the sum of random variable S_N will be a normal distribution, which will become a standard normal distribution after normalization (i.e., x_N) in (2.59).

A multivariate normal distribution $N(\mu,\Sigma)$ with mean $\mu \in \mathbb{R}^n$ and covariance matrix $\Sigma \in \mathbb{R}^{n\times n}$ has the probability density function

$$f(x) = \frac{1}{(2\pi)^{n/2}\sqrt{\det\Sigma}} \exp\left(-\frac{1}{2}(x-\mu)^T \Sigma^{-1}(x-\mu)\right) \tag{2.61}$$

The probability density function is symmetric around the mean μ and some of the two-tail probabilities, i.e., the probability $P\{|x-\mu|>\delta\}$ where δ is chosen such that $P\{|x-\mu|>\delta\} \le \alpha$. Standard choices of α=0.001, 0.005, 0.01, 0.05, which denote different statistical confidences.

Definition 2.16 – the χ^2 distribution

By the χ^2 distribution we mean the distribution of a norm of squares of the form

$$x^2 = x_1^2 + \cdots + x_k^2 \tag{2.62}$$

of k independent, standard normal random variables [1] where k is called the degree of freedom. The mean of χ^2 distribution is $\mu=k$ and the variance $\sigma^2=2k$ for the χ^2 distribution with k degrees of freedom. The probability density function is

$$f\left(x^2\right)=\frac{1}{2^{k/2}\Gamma\left(k/2\right)}\left(x^2\right)^{\frac{k}{2}-1}e^{\frac{-x^2}{2}},x^2\geq 0 \tag{2.63}$$

where Γ(•) denotes the standard Γ function, which can be found in mathematical software and in statistical tables.

Remark 2.3: The α-percentile x_α^2 is the value for which

$$P\left\{x^2<x_\alpha^2\right\}=\alpha \tag{2.64}$$

A confidence interval at the probability level α around the mean for χ^2 distribution is $[\chi^2_{\alpha/2},\chi^2_{1-\alpha/2}]$.

Definition 2.17 – the student (t) distribution

$$f\left(x\right)=\frac{\Gamma\left(\frac{p+1}{2}\right)}{\Gamma\left(\frac{p}{2}\right)\Gamma\left(\frac{1}{2}\right)\sqrt{p}}\left(1+\frac{x^2}{p}\right)^{-\frac{p+1}{2}} \tag{2.65}$$

is the probability distribution of the respective t distribution with p degree of freedom. Here $\Gamma(z)=\int t^{z-1}e^{-t}dt$ is the gamma function.

The above χ^2 distribution and student t distribution are useful for testing statistical hypothesis in the sequel.

2.4. Statistical Hypothesis Testing

In the replicated measurement of same biological system, for example, in a treatment and a control condition, we obtain two series of experimental data, $x_1, \cdots, x_n$ and $y_1, \cdots, y_n$. The biological problem of judging difference from replicated measurement can be formulated as statistical hypothesis, the null hypothesis, H_0, and the alternatives hypothesis H_1. Statistical hypothesis testing is a statement about some parameters, e.g., mean or variance, of a probability distribution. Its application is important in contexts where the result and quality of two different experimental methods should be distinguished by means of statistical analysis of data.

Example 2.4 – Two sample location test

Assume that the elements of both samples are independently Gaussian distributed, $N(M_x,\sigma^2)$ and $N(M_y,\sigma^2)$, respectively, with equal variances. Thus we could interpret the sample data x_i as outcomes of independent random variables that are Gaussian distributed with the respective parameter (y_i likewise). We want to test the hypothesis $\mathcal{H}$ whether the sample means are equal [2], i.e.,

$$\begin{aligned} H_0 &: M_x = M_y \\ H_1 &: M_x \neq M_y \end{aligned} \tag{2.66}$$

Under the above assumption, the test statistic

$$t = \frac{\bar{x} - \bar{y}}{\sqrt{\frac{(n-1)S_x^2 + (m-1)S_y^2}{n+m-2}}\sqrt{\frac{1}{n} + \frac{1}{m}}} \tag{2.67}$$

where the sample mean $\bar{x}$ and $\bar{y}$ are defined in (2.50) and sample covariance S_x^2 and S_y^2 are defined in (2.52).

When $n=m$, (2.67) becomes

$$t = \frac{\bar{x} - \bar{y}}{\sqrt{\frac{S_x^2}{n} + \frac{S_y^2}{n}}}$$

For a calculated value of the t-statistic, $\hat{t}$, we can now judge the probability of having an even more extreme value by calculating the probability (i.e., p-value)

$$P\left\{|t| > |\hat{t}|\right\} = 2 \times P\left\{t > |\hat{t}|\right\} = 2 \times \int_{\hat{t}}^{\infty} f(x)dx \tag{2.68}$$

where $f(x)$ is the student distribution in (2.65).

Example 2.5 – A sample variance test of experimental data $x_1, \cdots, x_n$

A two-sided test is relevant for testing the hypothesis $\mathcal{H}$:

$$\begin{aligned} H_0 &: \sigma_x^2 = \sigma^2 \\ H_1 &: \sigma_x^2 \neq \sigma^2 \end{aligned} \tag{2.69}$$

where the alternative hypothesis should be rejected if

$$\chi^2_{\frac{\alpha}{2}}(n-1) \leq x^2 \leq \chi^2_{1-\frac{\alpha}{2}}(n-1) \tag{2.70}$$

where x^2 is defined in (2.62) and α is percentile, i.e., p-value in the statistical confidence. For example, suppose a sum of squares of 10 random zero-mean experimental data $x_1, \cdots, x_{10}$ is

$$x^2 = \frac{1}{9}\sum_{i=1}^{10} x_i^2 = 18.4 \tag{2.71}$$

We want to test whether the covariance of the data $\sigma^2=1$. Then the 95% confidence interval for x^2 is $[\chi_{0.025}{}^2(9),\ \chi_{0.975}{}^2(9)]=[2.70,\ 19.0]$ under the assumption of normal distribution with $\sigma^2=1$. Hence we cannot reject the hypothesis that $\sigma_x{}^2=\sigma^2=1$.

Remark 2.4:

The type I error and type II error in testing a hypothesis $\mathcal{H}$ in (2.66) or (2.69) are given in the following table

$\mathcal{H}$		Unknown true	
		H_0	H_1
Accept ed		True decision P=1−α	Type II error P=β
		Type I error P=α	True decision P=1-β

Tests of hypothesis $\mathcal{H}$ always involve risk of making false decision as shown in the above table: (i) Rejecting a true hypothesis (type I error), (ii) Accepting a false hypothesis (type II error). Further, the probability of true decision for true hypothesis is called specificity (i.e., 1–α), and the probability of true decision for false hypothesis is called the sensitivity (i.e., 1–β).

2.5. Stochastic Systems

Consider the following linear stochastic system

$$\frac{d}{dt}x(t) = Ax(t) + Mx(t)n(t) = \left(A + Mn(t)\right)x(t) \tag{2.72}$$

where $n(t)$ denotes a normalized white Gaussian white noise with unit covariance since $n(t)$ is a standard stochastic process. (2.72) is also a stochastic process due to noise $n(t)$. In general, $Mn(t)$ denotes the random parameter fluctuation of system matrix A. $n(t)$ denotes the random fluctuation source when M denotes the amplitude of the fluctuation. Therefore the term $Mx(t)n(t)$ is called the intrinsic random fluctuation of the system. The stochastic process in (2.72) can be represented by the following form

$$dx(t) = Ax(t)dt + Mx(t)dw(t) \tag{2.73}$$

where $dw(t)=n(t)dt$ and $w(t)$ is a standard Wiener process.

Since the effect of intrinsic random fluctuation is state dependent and will influence the stability of the stochastic system in (2.73), the stochastic stability of the linear stochastic system in (2.73) is discussed in the following.

Let $V(x)>0$ with $V(0)=0$ denote Lyapunov (energy like) function of linear stochastic system in (2.73). Then the stochastic system in (2.73) is stochastically stable, if the expectation of the derivative of $V(x)$ is not positive, i.e.,

$$E\left\{\frac{d}{dt}V(x)\right\}\leq 0 \tag{2.74}$$

Let us denote the Lyapunov (energy) function of the linear stochastic system in (2.73) as $V(x)=x^T(t)Px(t)$ for some symmetric positive definite matrix $P=P^T>0$. Then the change of Lyapunov function is obtained by the following Itô derivative [3, 4]

$$dV(x)=\frac{\partial V(x)}{\partial x(t)}dx(t)+\frac{1}{2}x^T(t)M^T\frac{\partial^2 V(x)}{\partial x^2}Mx(t)dt \tag{2.75}$$

The first term in the right hand side is due to the chain rule, the second term is due to the effect of the in-differentiability of Wiener process by the conventional derivative method in (2.73). Substituting (2.73) into (2.75) and with the fact $E\{dw(t)\}=0$, we get

$$\begin{aligned}E\left\{\frac{d}{dt}V(x)\right\}&=\frac{1}{2}\left[x^T(t)PAx(t)+x^T(t)A^TPx(t)+x^T(t)M^TPMx(t)\right]\\&=\frac{1}{2}x^T(t)\left[PA+A^TP+M^TPM\right]x(t)\end{aligned} \tag{2.76}$$

If the following linear matrix inequality (LMI) holds for $P>0$

$$PA+A^TP+M^TPM\leq 0 \tag{2.77}$$

then $E\{dV(x)/dt\}\leq 0$, the linear stochastic system in (2.72) and (2.73) is stochastically stable.

For the nonlinear stochastic system

$$dx(t)=A(x)dt+M(x)dw(t) \tag{2.78}$$

where $A(x)$ denotes the nonlinear interaction function of the nonlinear stochastic system, and $M(x)dw(t)$ denotes the nonlinear intrinsic fluctuation. Suppose the origin is shifted to one equilibrium point x_e of the nonlinear system, i.e., $A(0)=0$ and $M(0)=0$. Then, by the similar procedure, if the following Hamilton Jaccobi inequality (HJI) has a positive solution $V(x)>0$ with $V(0)=0$,

$$\left(\frac{\partial V(x)}{\partial x(t)}\right)^T A(x)+\frac{1}{2}M^T(x)\frac{\partial^2 V(x)}{\partial x^2}M(x)\le 0 \tag{2.79}$$

then the nonlinear stochastic system is stochastically stable, i.e., $E\{dV(x)/dt\}\le 0$.

Remark 2.5:

If $(\partial A(x)/\partial x(t))|_{x=0}=A$, $(\partial M(x)/\partial x(t))|_{x=0}=M$, i.e., the linear stochastic system in (2.73) is the local linear system of nonlinear stochastic system at x=0 in (2.78), then the HJI in (2.79) will be reduced to LMI in (2.77) if $V(x)=x^T(t)Px(t)$.

After the robust stability of stochastic system under intrinsic random fluctuation is discussed, the effect of external disturbance on the stochastic system will be discussed in the following.

If the linear stochastic system in (2.73) also suffers from environmental disturbance $v(t)$, then the stochastic system in (2.73) is modified as follows [3]

$$dx(t)=Ax(t)dt+Hv(t)dt+Mx(t)dw(t) \tag{2.80}$$

where H is the coupling matrix which denotes the influence of environmental disturbance on $x(t)$.

In this situation, by Itô formula, we get the following bt the fact $Edw(t)$=0

$$E\left\{\frac{d}{dt}V(x)\right\}=E\left\{\frac{\partial V(x)}{\partial x(t)}(Ax(t)+Hv(t))+\frac{1}{2}x^T(t)M^T\frac{\partial^2 V(x)}{\partial x^2}Mx(t)\right\} \tag{2.81}$$

Let us choose the Lyapunov function $V(x)=x^T(t)Px(t)$. Then

$$\begin{aligned}&E\left\{\frac{d}{dt}V(x)\right\}\\&=E\left\{2x^T(t)P(Ax(t)+Hv(t))+x^T(t)M^TPx(t)\right\}\\&=E\left\{x^T(t)P(Ax(t)+Hv(t))+(Ax(t)+Hv(t))^TPx(t)x^T(t)M^TPx(t)\right\}\\&=E\left\{x^T(t)(PA+A^TP+M^TPM)x(t)+x^T(t)PHv(t)+v^T(t)H^TPx(t)\right\}\end{aligned} \tag{2.82}$$

By the fact that

$$x^Ty+y^Tx\le\rho^2x^Tx+\rho^{-2}y^Ty,\ \text{for any } \rho^2 \tag{2.83}$$

we get

$$E\left\{\frac{d}{dt}V(x)\right\}\leq E\left\{x^T(t)\left(PA+A^TP+M^TPM+\rho^{-2}PHH^TP\right)x(t)+\rho^2v^T(t)v(t)\right\}$$

If the following matrix inequality holds for $P>0$

$$PA+A^TP+M^TPM+\frac{1}{\rho^2}PHH^TP\leq 0 \tag{2.84}$$

then

$$E\left\{\frac{d}{dt}V(x)\right\}\leq \rho^2 E\left\{v^T(t)v(t)\right\} \tag{2.85}$$

Obviously, the stochastically stability cannot be achieved under the environmental disturbance because the environmental disturbances always perturb the stochastic system persistently. If the stochastic system is free of environmental disturbance then $E\{dV(x)/dt\}\leq 0$ can be achieved and the stochastic system will be stochastically stable.

Some signals in biological system are point process (like vesicles in neural transmission). Therefore, Poisson jump process in biological systems will be discussed in the following. If the linear system suffers from the Poisson jump process.

$$dx(t)=Ax(t)+Mx(t)dN(t) \tag{2.86}$$

where $N(t)$ is a Poisson process with the jump intensity λ, then $dV(x)=[\partial V(x)/\partial x(t)]Ax(t)dt+\lambda[V(x+Mx)-V(x)]dt$. The second term is due to the effect of discontinuity of Poisson process on the conventional derivative.

If we choose the Lyapunov function as $V(x)=x^T(t)Px(t)$. Then [4]

$$\begin{aligned}&E\left\{\frac{d}{dt}V(x)\right\}\\&=E\left\{x^T(t)\left(A^TP+PA\right)x(t)+\lambda\left[(x^T(t)+x^T(t)M^T)P\left(x(t)+Mx(t)\right)-x^T(t)Px(t)\right]\right\}\\&=x^T\left(A^TP+PA\right)+\lambda\left[(M^TP+PM)+M^TPM\right]x(t)\end{aligned} \tag{2.87}$$

Therefore, if the following LMI holds for $P>0$

$$A^TP+PA+\lambda\left[(M^TP+PM)+M^TPM\right]\leq 0 \tag{2.88}$$

then $E\{dV(x)/dt\}\leq 0$. Then the stochastic Poisson jump system (2.86) will be stochastically stable.

2.6. Numerical Optimization

Optimization techniques are very important in systems biology. In general, there exists no closed-form solution in most of optimization problems. Therefore some numerical solutions will be introduced in this section. We will consider the optimization problem of

$$\min_{x} f(x), x = (x_1, \cdots, x_n)^T \tag{2.89}$$

where x is the vector of free variables to be found to minimize $f(x)$. Some numerical optimizations are introduced in the following [1]:

(i) Descent method :

$$x^{(i+1)} = x^{(i)} - \alpha^{(i)} \nabla f(x^{(i)}) \tag{2.90}$$

where $\alpha^{(i)}$ denotes the step size and the gradient at $x^{(i)}$ is denoted as

$$\nabla f\left(x^{(i)}\right) = \left.\frac{\partial f(x)}{\partial x}\right|_{x=x^{(i)}}$$

(ii) Newton Method

$$x^{(i+1)} = x^{(i)} - (\nabla^2 f(x^{(i)})^{-1} \nabla f(x^{(i)}) \tag{2.91}$$

where the Hessian matrix $\nabla^2 f(x^{(i)})$ is the matrix of second partial derivatives of $f(x)$ evaluated at $x^{(i)}$.

The Newton Method has the advantage of very rapid convergence when $x^{(i)}$ is close to the optimum x^* but has the disadvantage of needing time to calculate the matrix of second derivatives. Another problem is that the matrix inverse of second partial derivatives might not exist at some distance from the optimum x^*.

The above searching methods need the derivative information. Further, these methods may be attracted by local optimums. GA and EA algorithms are derivative free and may achieve the global optimum.

(iii) Genetic Algorithm (GA)

Let

$$g(x) \propto \frac{1}{f(x)} \text{ or } g(x) = -f(x) \tag{2.92}$$

then the minimization problem in (2.89) is equivalent to the following maximization problem

$$\max_{x} g(x) \tag{2.93}$$

Genetic algorithms are stochastic optimization algorithms that are originally motivated by the mechanisms of natural selection and evolutionary genetics. The fitness function $g(x)$ depends on $x=[x_1, \cdots, x_n]^T$, with each variable

$$x_i \in [x_{i\min}, x_{i\max}] \tag{2.94}$$

A common encoding method, which transforms points x from the phenotype space into the genotype space, is the binary coding. In our example the phenotype space for x_i is the interval $[x_{i\min}, x_{i\max}]$. If we decide to use l bits to represent x_i, then this corresponds to mapping the interval $[x_{i\min}, x_{i\max}]$ into the discrete genotype space $[0,2^l-1]$ consisted of 2^l binary strings. For example, in the case of $x=[x_1, x_2]^T$, the two binary coded strings of x_1 and x_2 are assumed as follows

$$x_1 : (\alpha_1\alpha_2\alpha_3)_1 = 110000011111 \tag{2.95}$$

$$x_2 : (\alpha_1\alpha_2\alpha_3)_2 = 001011001011 \tag{2.96}$$

then the phenotype $x=[x_1, x_2]^T$ is encoded as

$$x = 110000011111001011001011 \tag{2.97}$$

The strings of binary digits corresponding to the point x like (2.97) are called chromosomes. The space of chromosomes is called the genotype space. In general, chromosomes in the genotype space are abstract mathematical objects on which the so-called genetic operators work to search for the maximum of $g(x)$ in (2.93). After a stopping criterion for the algorithm is satisfied, a decoding function transforms a solution chromosome from the genotype space into phenotype space. Genetic algorithms are derivative-free optimization algorithms that use only the fitness function. They do not use the fitness function derivative. The higher the fitness of a chromosome, the better the candidate solution encoded by the chromosome. Genetic algorithm is an iterative process that at each iteration uses the information obtained from testing a current population of chromosomes to direct its search into promising areas of the search space. Thus, generating a suitable population of chromosomes at each iteration is the key to the algorithm performance. In the simplest form, creating a new population of chromosomes to be tested involves three genetic operations, reproduction, crossover, and mutation. These genetic operators work on genotypes that encode phenotypes.

The role of selection operator is to pick a population of chromosomes for further processing, called reproduction. Chromosomes are selected into a mating pool for reproduction with a probability proportion to their fitness.

The crossover operator randomly exchanges substrings between two parent chromosomes to create two offspring. The simplest crossover operator is the one-point crossover, where crossing site is randomly picked and the corresponding substrings are exchanged, for example, two chromosomes

$$\begin{array}{c:c} 110000011111 & 001011001011 \\ 101101010100 & 110000110110 \end{array} \tag{2.98}$$

are crossovered as follows

$$\begin{array}{c:c} 101101010100 & 001011001011 \\ 110000011111 & 110000110110 \end{array} \tag{2.99}$$

The mutation operator randomly flips an arbitrary bit in a chromosome. For example the chromosome

$$11000001111\underset{\uparrow}{1}001011001011 \tag{2.100}$$

might undergo a mutation in its 11th position to yield a new chromosome

$$110000011101001011001011 \tag{2.101}$$

We now provide a description of canonical genetic algorithm to solve the maximization problem in (2.93) as follows:

Canonical Genetic Algorithm (GA)

For a given optimization problem characterized by a fitness function $g(x)$ in (2.93)

Step 1: Choose a method of encoding candidate solutions into chromosomes; that is, decide upon a map between the phenotype and genotype spaces (A commonly used encoding method is to concatenate the binary strings representing each real parameter value x_i)
Step 2: Start with a randomly generated population of chromosomes.
Step 3: Evaluate the fitness of each chromosome in the population.
Step 4: Replace the current population with a new population by applying selection, crossover, and mutation operations in the current population.
Step 5: If a stopping criterion is satisfied, stop and choose the fittest chromosome from the population as a solution, else go to step 3.

Remark 2.6: The selection in step 5 is based on the principle of survival of better fitness. The fitness of i-th string, $F_i=g(x^i)$, is assigned to each individual chromosomes in the population where high F_i means better fitness. These chromosomes with large fitness would have a large number of copies in new generation. For example, in roulette wheel selection, the i-th chromosome with high fitness value F_i is given a proportionally high probability of reproduction, P_i, according to the distribution

$$P_i = \frac{F_i}{\sum F_i} \tag{2.102}$$

Once the chromosomes are reproduced or copied for possible use in the next generation, they are produced in a mating pool where they await the action of other two operators, crossover and mutation.

(iv) Evolutionary Algorithm (EA)

Canonical GAs use binary representation of elements of the parameter space. A binary representation may have some disadvantage though. In contrast to canonical GAs, the genetic operators work directly on real-valued vectors. A term used to describe an optimization algorithm for (2.93) using genetic-type operators working on the real-valued vectors is an evolutionary algorithm. Here we present a very simple evolutionary algorithm for solving optimization problem as follows :

Simple Evolutionary Algorithm (EA) [5]

Step 1: Initialize a population of candidate solutions (chromosomes) to the problem. Each chromosome is a real-valued vector of the same length as the vector x of discussion variables.

Step 2: Evaluate the fitness of each candidate solution in the population.

Step 3: Replace the current population with a new population by applying selection, crossover, and mutation operations on the current population. The roulette-wheel selection method can be used for selection. A possible implementation of the crossover operation is to use the average crossover operator. This operator takes two parents, say x^i and x^k, and returns one offspring

$$x = \frac{1}{2}(x^i + x^k)$$

Another implementation of crossover operator is a convex combination of two parents. Specifically, let $r \in (0,1)$. Then, the offspring is

$$x = rx^i + (1-r)x^k$$

As for the mutation operation, let x be a chromosome to be mutated. Then, an implementation of the mutation operation could have the form

$$x' = x + rd$$

where d is a randomly generated feasible vector $r \in (0,1)$. Note that if the feasible set is convex, then the above genetic operators produce chromosomes that are elements of the feasible set. Otherwise, conditions should be imposed to ensure that candidate solutions are the elements of the feasible set.

Step 4: If a stopping criterion is satisfied, terminate the algorithm and choose the fittest chromosome from the population as solution, else go to step 2.

Remark 2.7: In addition, both in the GA and EA, the elitist strategy can be incorporated to improve GA and EAs performance. This strategy copies the best chromosomes from the old population into the next population to prevent losing the best solution in the succeeding iteration.

CONCLUSION

Linear algebra, statistics, stochastic processes, system analysis, and optimization are useful tools for modeling, analysis, and design in systems biology.

REFERENCES

[1] Storey, JD; Xiao, WZ; Leek, JT; Tompkins, RG; Davis, RW. "Significance analysis of time course microarray experiments," *Proceedings of the National Academy of Sciences of the United States of America*, vol. 102, 12837-12842, Sep 6 2005.

[2] Johansson, R. *System modeling and identification*. Englewood Cliffs, NJ: Prentice Hall, 1993.

[3] Chen, BS; Zhang, WH. "Stochastic H(2)/H(infinity) control with state-dependent noise," *IEEE Transactions on Automatic Control*, vol. 49, 45-57, Jan 2004.

[4] Ikeda N; Watanabe, S. *Stochastic differential equations and diffusion processes*. Amsterdam: North-Holland Pub. Co., 1981.

[5] Zak, SH. *Systems and control*. New York: Oxford University Press, 2003.

Chapter 3

REVERSE ENGINEERING IN SYSTEMS BIOLOGY

ABSTRACT

Reverse engineering has been widely applied in control engineering to construct system models from system input and output data. Reverse engineering methods therefore play a role in constructing biological system models from corresponding experimental data, e.g. in the construction of GRN or PPI network dynamic models from corresponding microarray data. In this chapter, several reverse engineering methods and a system order detection method for systems biology are introduced.

INTRODUCTION

In this chapter, the least square parameter estimation, maximum likelihood, and recursive least square parameter estimation methods are discussed with an eye to their role in parameter estimation from experimental biological systems data. A system order determination method to prune false-positive components of biological system models is discussed. Finally, a nonlinear parameter estimation method for parameter estimation in nonlinear biological systems is introduced.

Reverse engineering involves identification of a system from measurement data. System identification plays a very important role in systems biology. It is employed in identifying or reconstructing biological systems or networks from different kinds of experimental data. This chapter discusses reverse engineering in systems biology using the example of identification of system matrix A in dynamic system (2.3) from measurement data $x(k)$. In general, however, we do not identify A directly. For simplicity of computation, we will identity each row of A from dynamic equations in (2.1).

The linear dynamic model (2.1) can be represented by the following linear regression model.

$$x_i(k+1) = \begin{bmatrix} x_1(k) & x_2(k) & \dots & x_n(k) \end{bmatrix} \begin{bmatrix} a_{i1} \\ a_{i2} \\ \vdots \\ a_{in} \end{bmatrix} + v_i(k)$$
$$\triangleq \phi(k) \bullet \theta_i + v_i(k) \tag{3.1}$$

with regression vector $\phi(k)$ and parameter vector θ_i.

Assume that there are n parameters in θ_i to fit the linear dynamic model (2.1) to N observations. The parameter estimation problem is to find an estimate $\hat{\theta}_i$ of the parameter vector θ_i from the following variables $\{x_i(k)\}$ and $\{\phi(k)\}$, i.e.,

$$\begin{aligned} x_i(1) &= \phi(0)\theta_i + v_i(0) \\ x_i(2) &= \phi(1)\theta_i + v_i(1) \\ &\vdots \\ x_i(N) &= \phi(N-1)\theta_i + v_i(N-1) \end{aligned} \tag{3.2}$$

In matrix notation, we obtain the vectors of observation $X_i(N)$, the regressor matrix $\Phi(N)$, and noise vector $V_i(N)$, respectively, as follows

$$X_i(N) = \begin{bmatrix} x_i(1) \\ x_i(2) \\ \vdots \\ x_i(N) \end{bmatrix}, \Phi(N) = \begin{bmatrix} \phi(0) \\ \phi(1) \\ \vdots \\ \phi(N-1) \end{bmatrix}, V_i(N) = \begin{bmatrix} v_i(0) \\ v_i(1) \\ \vdots \\ v_i(N-1) \end{bmatrix} \tag{3.3}$$

Finally, the resulting estimation model for linear regression is

$$X_i(N) = \Phi(N)\theta_i + V_i(N) \tag{3.4}$$

3.1. The Least Square Parameter Estimation Method

By the least square estimation method, we select parameter estimate $\hat{\theta}_i$ to minimize the following square error

$$V(\hat{\theta}_i) = \frac{1}{2}(X_i(N) - \Phi(N)\hat{\theta}_i)^T (X_i(N) - \Phi(N)\hat{\theta}_i) \tag{3.5}$$

The optimal estimate $\hat{\theta}_i$ to minimize the square error in (3.5) is obtained as

$$\hat{\theta}_i = (\Phi^T(N)\Phi(N))^{-1}\Phi^T(N)X_i(N) \\ = \Phi^\dagger(N)X_i(N), i = 1, 2, \cdots n \tag{3.6}$$

where $\Phi^\dagger(N)$ denotes the pseudoinverse of $\Phi(N)$

The covariance matrix of $\hat{\theta}_i$ is

$$Cov(\hat{\theta}_i) = \sigma_{v_i}^2(\Phi^T(N)\Phi(N))^{-1} \tag{3.7}$$

where $\sigma_{v_i}^2$ is the covariance of noise vector $V_i(N)$, i.e., $Cov(V_i(N))=\sigma_{v_i}^2 I$

The estimate of noise covariance $\sigma_{v_i}^2$ is

$$\hat{\sigma}_{v_i}^2 = \frac{2}{N-n}V(\hat{\theta}_i) \tag{3.8}$$

Remark 3.1:

(i) For the parameter estimation of A in the linear dynamic system (2.2) or (2.3), it is more appealing to estimate A one row by one row as the parameter estimation method in (3.6), i.e., $\hat{\theta}_i$, $i = 1, 2, \cdots n$ and finally $\hat{A}^T = \begin{bmatrix}\hat{\theta}_1 & \cdots & \hat{\theta}_n\end{bmatrix}^T$. In general, it is not suitable to estimate A directly because of its complexity. Actually, the two results are equivalent.

(ii) In the linear regression model (2.5), the static equation can also be represented by the following regression form

$$x_i = \begin{bmatrix}x_1 & x_2 & \cdots & x_n\end{bmatrix}\begin{bmatrix}a_{i1} \\ \vdots \\ \vdots \\ a_{in}\end{bmatrix} + v_i = \phi\theta_i + v_i \tag{3.9}$$

If there are N samples (for example microarray data from N patients) then we get the following regression form for each sample

$$x_{i(1)} = \phi(1)\theta_i + v_1 \\ \vdots \\ x_{i(N)} = \phi(N)\theta_i + v_N \tag{3.10}$$

which can be represented by

$$X_i(N) = \Phi(N)\theta_i + V_N \tag{3.11}$$

Therefore, the parameter estimation of θ_i in (3.11) could be obtained by the least square parameter estimation algorithm in (3.6).

(iii) The difference in parameter estimation between the dynamic model in (3.1) and static model in (3.9) is that we need time profile data to construct the regression form (3.4) when we need sample data to construct the regression form (3.11).

(iv) In order to avoid the over-fitting in the least square parameter estimation in (3.6), the number of data must be at least three times than the number of parameters to be estimated.

3.2. THE MAXIMUM LIKELIHOOD PARAMETER ESTIMATION METHOD

The regression system (3.4) can be represented by Figure 3.1.

From Figure 3.1, we want to estimate system parameter θ_i from the output $X_i(N)$. The value of θ_i should be consistent with the largest probability under output data $X_i(N)$, i.e., the estimate of θ_i should satisfy the maximum probability

$$\max_{\theta_i} P(\theta_i \mid X_i(N)) \tag{3.12}$$

By the condition probability

$$P(\theta_i \mid X_i(N)) = \frac{P(\theta_i) P(X_i(N) \mid \theta_i)}{P(X_i(N))} \tag{3.13}$$

It is more difficult to calculate $P(\theta_i|X_i(N))$ because we need three probability density functions $P(\theta_i)$, $P(X_i(N))$ and $P(X_i(N)|\theta_i)$. Since $P(X_i(N))$ is not explicit with parameter θ_i, it does not influence the optimal solution in (3.12). $P(\theta_i)$ denotes the probability density of θ_i, i.e., the priori information of θ_i. In general, we have no information of $P(\theta_i)$ and always assume $P(\theta_i)$=constant, i.e., all values of θ_i may occur with the same probability. In this situation the optimal parameter estimation method (Maximum a priori (MAP)) in (3.12) is equivalent to the following maximum likelihood method

$$\max_{\theta_i} P(X_i(N) \mid \theta_i) \tag{3.14}$$

i.e., how to select θ_i so that the probability of $X_i(N)$ is maximum. $P(X_i(N)|\theta_i)$ denotes the likelihood function, which means the output data must be consistent (likelihood) with the largest probability to exploit the occurrence of the output data. From (3.4), it is seen that the probability density function $P(X_i(N))$ of $X_i(N)$ is mainly due to the white noise vector $V_i(N)$. If we assume $V_i(N)$ is zero-mean Gaussian noise with covariance matrix $\hat{\sigma}_{v_i}^2 I$. Then

$$\begin{aligned} P(V_i(N)) &= \frac{1}{\left(2\pi\sigma_{v_i}\right)^{\frac{N-1}{2}}}\exp\left(-\frac{1}{2\sigma_{v_i}^2}V_i^T(N)V_i(N)\right) \\ &= \frac{1}{\left(2\pi\sigma_{v_i}\right)^{\frac{N-1}{2}}}\exp\left(-\frac{1}{2\sigma_{v_i}^2}\left(x_i(N)-\Phi(N)\theta_i\right)^T\left(x_i(N)-\Phi(N)\theta_i\right)\right) \end{aligned} \tag{3.15}$$

In general, the notation θ_i in the likelihood function in (3.14) is always neglected. Therefore the maximum likelihood estimation method in (3.14) is equivalent to how to select θ_i to maximize $P(V_i(N))$ in (3.15), and also maximize $\log P(V_i(N))$. The maximum likelihood estimate method (3.14) is equivalent to the following likelihood optimization method

$$\max_{\theta_i} \log L\left(\theta_i, \sigma_{v_i}^2\right) \tag{3.16}$$

where the log-likelihood function is denoted as

$$\begin{aligned} \log L\left(\theta_i, \sigma_{v_i}^2\right) &= \log P\left(V_i(N)\right) \\ &= -\frac{N-1}{2}\log\left(2\pi\sigma_{v_i}^2\right) \\ &\quad -\frac{1}{2\sigma_{v_i}^2}\left(X_i(N)-\Phi(N)\theta_i\right)^T\left(X_i(N)-\Phi(N)\theta_i\right) \end{aligned} \tag{3.17}$$

i.e., the log-likelihood method in (3.16) could simplify the parameter estimation for (3.14)Here, we expect the log- likelihood function to have the maximum value at $\theta_i = \hat{\theta}_i$ and $\sigma_{vi}^2 = \hat{\sigma}_{vi}^2$. The necessary conditions for determining the maximum estimates $\hat{\theta}_i$ and $\hat{\sigma}_{vi}^2$ must consist of

$$\begin{aligned} \frac{\partial \log L\left(\theta_i, \sigma_{v_i}^2\right)}{\partial \theta_i} &= 0 \\ \frac{\partial \log L\left(\theta_i, \sigma_{v_i}^2\right)}{\partial \sigma_{v_i}^2} &= 0 \end{aligned} \tag{3.18}$$

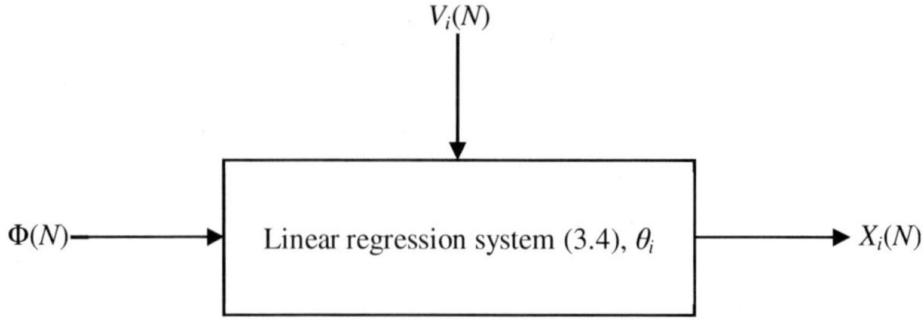

Figure 3.1. The linear regression system (3.4) can be represented by the following figure, we want to estimate system parameter θ_i from the output $X_i(N)$.

After some computational arrangements from (3.18), the estimated parameter $\hat{\theta}_i$ and $\hat{\sigma}^2_{v_i}$ are

$$\hat{\theta}_i = \left(\Phi^T(N)\Phi(N)\right)^{-1}\Phi^T(N)X_i(N) \tag{3.19}$$

$$\hat{\sigma}^2_{v_i} = \frac{1}{N-1}\left(X_i(N) - \Phi(N)\hat{\theta}_i\right)^T\left(X_i(N) - \Phi(N)\hat{\theta}_i\right) \tag{3.20}$$

$$i = 1, 2, \cdots n$$

Under the Gaussian distribution of noise $V_i(N)$ in (3.4), the maximum likelihood parameter estimate $\hat{\theta}_i$ in (3.19) is equivalent to the least square parameter estimate $\hat{\theta}_i$ in (3.6).

3.3. The Recursive Least Square Parameter Identification Algorithm

In the least square parameter estimation algorithm in (3.6) or the maximum likelihood estimation algorithm (3.19), we need accumulate a block of data matrix $\Phi(N)$ to solve the pseudoinverse $\Phi^\dagger$ in (3.6) or (3.19) to obtain $\hat{\theta}_i$. Actually, based on the matrix inversion lemma in (2.18), (3.6) or (3.19) can be transformed to an equivalent recursive least square parameter estimation algorithm as follows [1]:

$$\hat{\theta}_i(k+1) = \hat{\theta}_i(k) + P_k\phi(k)\left(x_i(k) - \phi^T(k)\hat{\theta}_i(k-1)\right)$$
$$P_k = P_{k-1} - \frac{P_{k-1}\phi(k)\phi^T(k)P_{k-1}}{1+\phi^T(k)P_{k-1}\phi(k)} \tag{3.21}$$
$\theta_i(0)$ and P_0 are given.

This recursive least square parameter estimation can use time profile data to update the parameter estimate one by one. Therefore it could be used for real-time parameter estimation. If the data number of $x_i(k)$ is small, the final parameter estimate $\hat{\theta}_i(N)$ could be used as the initial parameter estimate $\hat{\theta}_i(0)$, and then we could repeat several rounds of recursive parameter estimation algorithm in (3.21) to achieve the optimal parameter estimate.

In general, small time-profile data or sample data in the block parameter estimation algorithm (3.6) or (3.19) may lead to over fitting problem due to the singularity of $(\Phi^T(N)\Phi(N))^{-1}$. Therefore, a cubic spline method is employed for time-profile data to avoid over fitting problem. However, there is no singularity problem in recursive parameter estimation algorithm in (3.21). Therefore, in small data case, we could use several rounds of the recursive parameter estimation algorithm for precise parameter estimation.

3.4. System Model Order Determination

Unless a priori consideration in the field of application suggests that the model order (i.e., parameter number n in (2.1) or (2.2) in our case) is the correct one, the correct model is often not known a priori. It makes sense to postulate several different model orders about system or network before system identification. When only experimental data or measurement data is available, as shown in the covariance of least square estimation error in (3.8), the loss function decrease monotonically with increasing model order. Thus the estimation error or loss function in (3.7) or (3.8) alone might not be sufficient to indicate when to terminate the search for the correct model. One attempt to include both the estimated variance and model complexity in one statistic is the Akaike information criterion (AIC) as follows [1]

$$AIC(n) = \log\left(\frac{1}{N}\right)\left(X_i(N) - \hat{X}_i(N)\right)^T\left(X_i(N) - \hat{X}_i(N)\right) + \frac{2n}{N} \tag{3.22}$$

where $X_i(N)$ denotes the estimated expression profile of linear dynamic model in (3.4) i.e., $X_i(N)=\Phi(N)\hat{\theta}_i$. AIC in (3.22) decreases as the residual variance $\sigma^2_{\hat{\theta}_i}$ in (3.7) or the estimated covariance (3.8) decreases and increases as the number of the number of parameters n increases. AIC should be with a minimum at the correct model order or parameter number. Therefore the true model n could minimize the $AIC(n)$ in (3.22).

In the reconstruction of gene regulatory network or protein-protein interaction network from the corresponding microarray data, we always construct the candidate gene regulatory network and protein-protein interaction network by bio-information method and data mining method at the beginning. There exist many false positive regulations or interactions in these candidate networks. Therefore, we always use the parameter estimation methods to estimate these regulation abilities or interaction abilities (network parameters). Then, based on model order detection method, we use the AIC method to prune some false positive regulations in candidate gene regulatory network or some false positive protein interactions in candidate protein-protein interaction network to obtain the more correct biological network from microarray data.

3.5. Nonlinear Parameter Estimation Methods

Consider the following nonlinear parameter system

$$x(k+1)=f\left(x(k),\theta\right)+v(k) \tag{3.23}$$

Let us denote the square error equation as follows

$$V(\theta)=\sum_{k=1}^{N}\left(x(k+1)-f\left(x(k),\theta\right)\right)^{T}\left(x(k+1)-f\left(x(k),\theta\right)\right) \tag{3.24}$$

Then we need to select the optimal θ^* to minimize $V(\theta)$ to achieve the least square parameter estimation by the descent recursive method as

$$\hat{\theta}(k+1)=\hat{\theta}(k)-\alpha(k)\nabla V\left(\hat{\theta}\right),\theta(0)=\theta_0 \tag{3.25}$$

and by Newton recursive method as

$$\hat{\theta}(k+1)=\hat{\theta}(k)-\left(\nabla^2 V\left(\hat{\theta}(i)\right)\right)^{-1}\nabla V\left(\hat{\theta}(i)\right),\theta(0)=\theta_0 \tag{3.26}$$

In general, the initial $\hat{\theta}(0)$ of the descent method in (3.25) and the Newton method in (3.26) should be chosen near the optimal value θ^* or they may converge to other local optimal values.

In order to obtain the global optimal parameter estimation, the GA and EA algorithm in chapter 2 will be employed to solve the nonlinear parameter estimation problem. Let the fitness function $G(\theta)$ to replace $V(\theta)$ in (3.24) as

$$G(\theta)\propto\frac{1}{V(\theta)} \text{ or } G(\theta)=-V(\theta) \tag{3.27}$$

Then the optimal parameter estimation for nonlinear parameter estimation problem in (3.23) and (3.24) is to employ GA or EA algorithm to solve the following maximum fitness problem

$$\theta^* = \arg \max_{\theta \in [\theta_l, \theta_u]} G(\theta) \tag{3.28}$$

Therefore, the GA and EA algorithm in the above chapter will be employed to solve the optimal parameter searching problem. In general GA and EA algorithm could achieve θ^* globally.

For example, a simple gene regulatory system with parameters to be estimated is given as follows:

$$\begin{aligned} x_1(k+1) &= \frac{x_1(k)}{(c_1 + x_2(k))^{n_1}} - \lambda_1 x_1(k) + v_1(k) \\ x_2(k+1) &= \frac{x_2(k)}{(c_2 + x_1(k))^{n_2}} - \lambda_2 x_2(k) + v_2(k) \end{aligned} \tag{3.29}$$

Suppose we want to estimate parameters c_1, c_2, n_1, n_2, λ_1, and λ_2 from the output measurement $x_1(k)$ and $x_2(k)$, k=1, ⋯, N. Let us denote $\theta=(c_1, c_2, n_1, n_2, \lambda_1, \lambda_2)$, then it is a nonlinear parameter estimation problem and the EA and GA algorithm could be employed to treat the maximum fitness problem in (3.28).

CONCLUSION

Reverse engineering is employed to construct accurate models of different biological system levels from experimental data. Systems biologists can analyze system characteristics to gain more insight into biological systems, and synthetic biologists and metabolic engineers can design desired biological systems or pathways based on these types of accurate system models.

REFERENCES

[1] Johansson, R. *System modeling and identification*. Englewood Cliffs, NJ: Prentice Hall, 1993.

Chapter 4

Systems Biology Approach to GRNs and PPI Networks

Abstract

GRNs and PPI networks are two core biological networks that integrate cellular functions and mechanisms. High throughput data, microarray data and large databases have recently become so enriched that GR and PPI networks of a specific biological condition can be studied from the global cell point of view. In this chapter, a candidate GRN is constructed using high throughput datasets. Then, a dynamic regulatory model of each gene in the candidate GRN is identified using reverse engineering as described in Chapter 3 by employing microarray data from a specific biological condition. Some false positive regulators and regulators with less characterized abilities are pruned using model order determination (AIC) and p-values to obtain the real GRN under the specific biological condition. The construction of a PPI network is also a two-step process. The first step involves constructing a candidate PPI network by mining data from literature and databases such as BioGRID, SGD, and CGD. In the second step, false positive PPIs of the candidate PPI network are pruned using PPI model and microarray data of a specific biological condition, resulting in the real PPI network.

Introduction

In this chapter, we first construct the GRN of a yeast cell cycle using corresponding microarray data and ChIP data of transcription factors. Then, a PPI network of the apoptosis network in a cancer cell is constructed from corresponding microarray data. The construction of an integrated GR and PPI cellular network is also performed for the case of hyperosmotic stress in *S. cerevisiae*. An interspecies PPI network between host and pathogen is constructed during zebrafish infection with *C. albicans* using simultaneous microarray data. Finally, a stochastic spatial-temporal dynamic model for gene/protein interaction networks in the early development of fruit flies is introduced.

4.1. Construction of GRNs via Microarray Data

The construction of a GRN using a systems biology approach can be divided into two steps. The first step is construction of a candidate (or putative) GRN from two high-throughput data sets, namely, a ChIP network and a mutant network. The ChIP data are based on experimental data published by Harbison et al. [1], who also used a genomic tiling array to identify the genomic region bound by transcription factors (TFs). The mutant data are the gene expression data matrix published by Hughes et al. [2] with different gene deletion mutants. In general, the candidate GRN is suitable for all possible biological conditions. Therefore, the GRN for a specific biological condition needs to be confirmed using microarray gene expression data of the specific biological condition; that is, the real GRN is derived by pruning the candidate GRN with specific microarray data. Let $x_i(t)$ denote the gene expression of the i-th target gene in the candidate GRN. The dynamic equation for the regulation of the i-th gene in the candidate GRN is then modeled as [3, 4]

$$x_i(t+1) = x_i(t) + \sum_{j=1}^{N} a_{ij} x_j(t) - \lambda_i x_i(t) + k_i + n_i(t) \tag{4.1}$$

where $x_i(t)$ represents the mRNA expression level of target gene i at time t, $x_j(t)$ represents the regulation function of the j-th TF binding to the target gene i, a_{ij} denotes the regulatory ability from the j-th regulatory gene to the i-th target gene (the positive sign indicates activation and the negative sign indicates repression), λ_i indicates the degradation effect of the present time point on the next time point $t + 1$, k_i represents the basal level, and $n_i(t)$ is the stochastic noise due to model uncertainty and fluctuation of microarray data of the target gene. Expression of the i-th gene in (4.1) can be represented by the following regression equation [5]:

$$\begin{aligned} x_i(t+1) &= \begin{bmatrix} x_1(t) & \cdots & x_N(t) & 1 \end{bmatrix} \cdot \begin{bmatrix} a_1 \\ \vdots \\ a_{i-1} \\ (1-\lambda_i) \\ a_{i+1} \\ \vdots \\ a_N \\ k_i \end{bmatrix} + n_i(t) \\ &= \phi_i(t)\theta_i + n_i(t) \end{aligned} \tag{4.2}$$

where ϕ_i denotes the regression vector, which can be obtained from microarray data. θ_i is the regulatory parameter vector of target gene i, which is to be estimated. The regulatory parameter can be estimated from microarray data of the corresponding gene through the least-squares or maximum-likelihood parameter estimation in (3.6) or (3.19) [5] (known as the reverse-engineering method in chapter 3). After regulator parameters in θ_i are estimated, the system order (i.e., the number of regulatory genes) is determined by model detection method in (3.22) (e.g., Akaike's information criterion (AIC) in chapter 3), and the *P*-value statistical method is employed to determine the significant regulatory genes for target gene *i*. This is done by pruning false-positive regulations in the candidate GRN. That is, some a_{ij} is pruned because of false positive deletion. Based on the dynamic model in (4.1), the true GRN can then be constructed one target gene at a time through microarray data. Using similar methods, GRNs for yeast cell cycles [6-8], cancer cell cycles [9], stress response [10], and inflammation [11] can be constructed. The GRN for yeast cell cycle is shown in Figure 4.1 by dynamic gene regulatory network in (4.1) via microarray data [6]. The inflammatory GRN in immune system is shown in Figure 4.2 by microarray data [11].

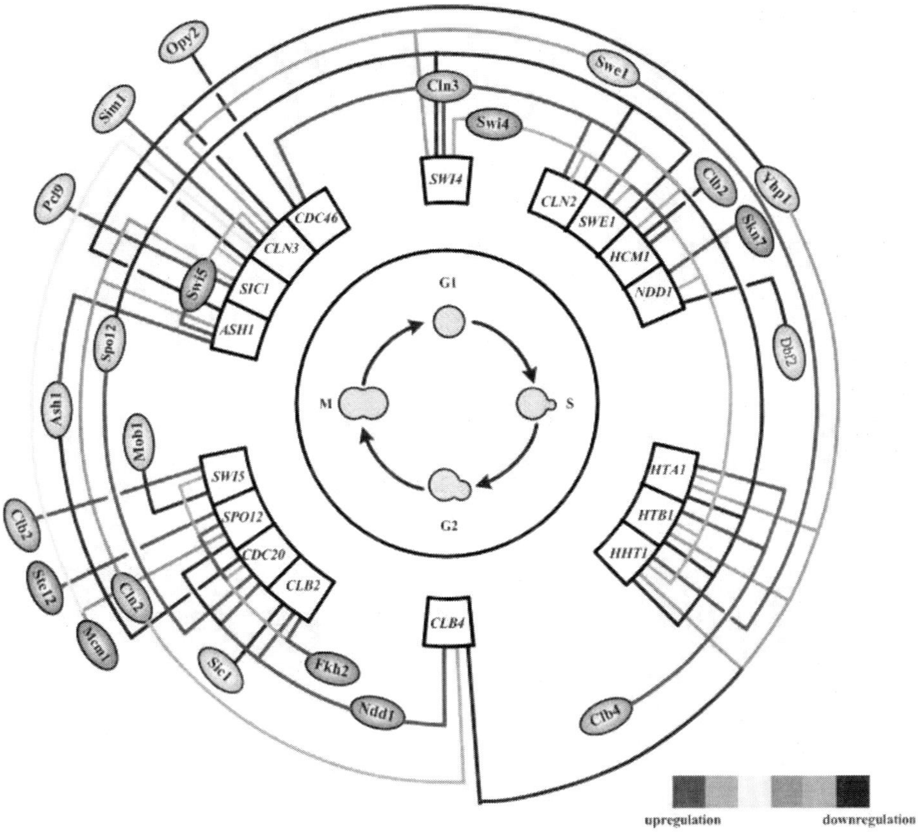

Figure 4.1. A transcriptional regulatory network of seventeen selected target genes involved in the yeast cell cycle. Each white box represents a target gene and the target gene is allocated in the cell cycle according to the time of its expression peak. Regulators are indicated by ovals of different colors. Blue ovals denote transcription factors, green ovals denote cyclins and orange ovals denote cell cycle control proteins. The target gene and regulators are connected by solid lines. The regulatory abilities of the regulators for specific target are characterized by colored lines; red for upregulation and blue for downregulation as indicated [6].

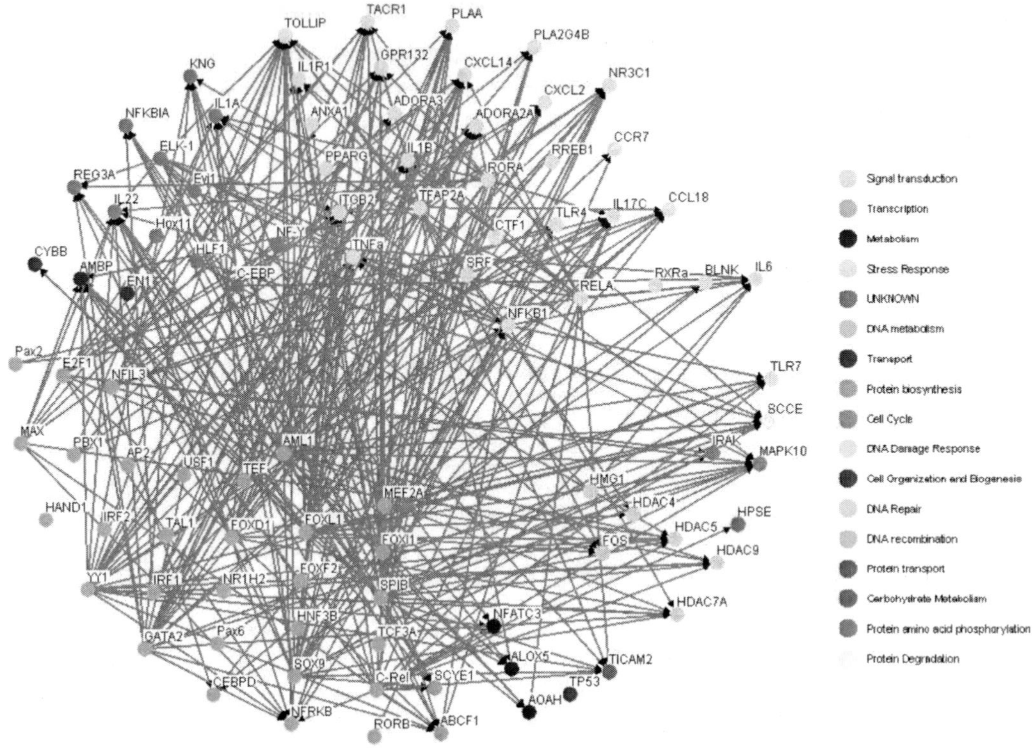

Figure 4.2. The inflammatory transcriptional gene network in immune system with LPS for Gram-negative bacteria. The inflammatory gene network with LPS containing [11].

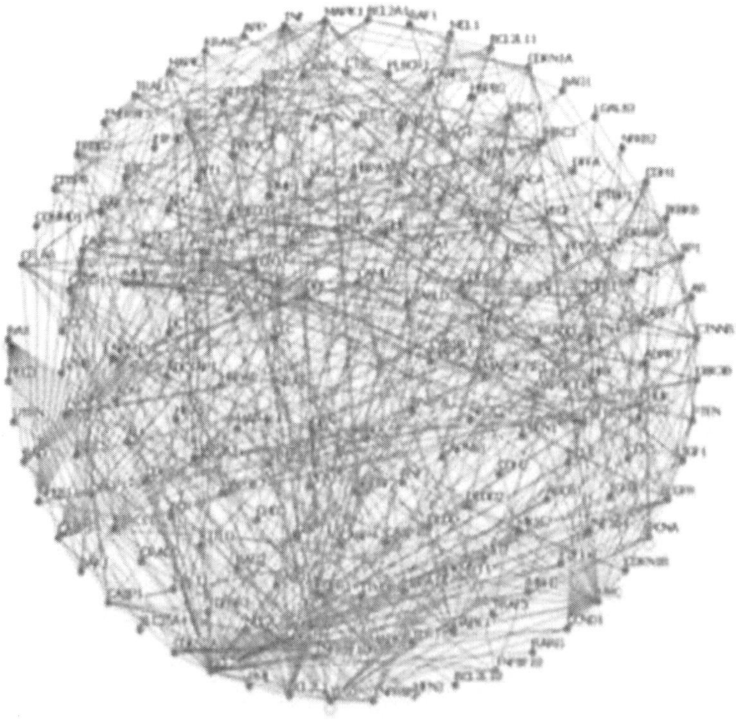

(A)

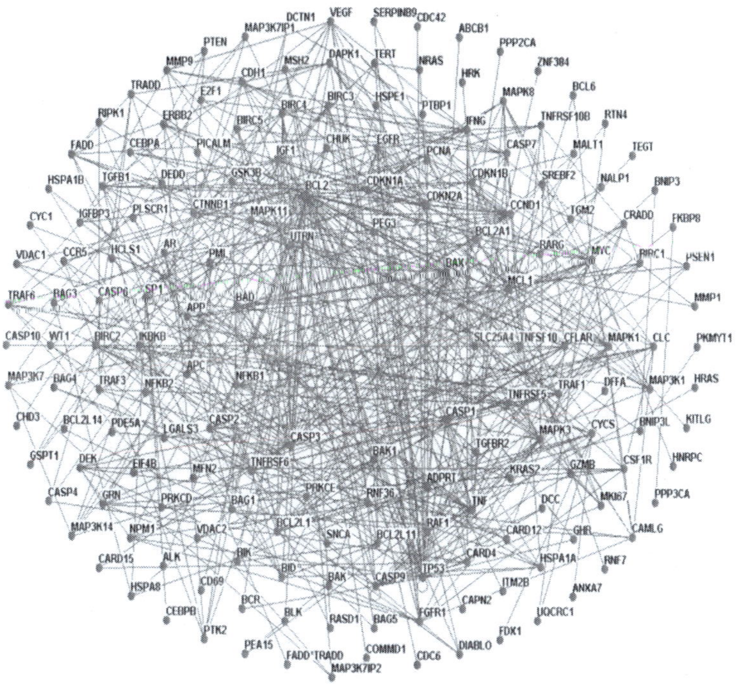

(B)

Figure 4.3. Global protein-protein interactions of apoptosis in cancerous and normal cells. (A) Apoptotic protein-protein interaction network in HeLa cells, showing 183 nodes and 552 edges. (B) Apoptotic protein-protein interaction network in normal primary lung fibroblasts, showing 175 nodes and 547 edges. Each interaction was calculated twice and only interactions with two '1' scores after AIC evaluation was considered 'true' interactions. All protein-protein interaction networks in this chapter were constructed with Osprey version 1.2.0.

4.2. CONSTRUCTION OF PPI NETWORKS

The construction of PPI network via a systems biology approach is also a two-step process. The first step is constructing a candidate PPI network via data mining from literature and databases such as BioGRID, SGD, and GO [12, 13]. As this is only a candidate network based on many biological conditions, the second step is pruning false positive PPIs by using a protein expression microarray of a specific biological condition. For a target protein i in the candidate PPI network, the dynamic model of protein activity is [14, 15]

$$y_i(t+1) = y_i(t) + \sum_{j=1}^{M} b_{ij} y_i(t) y_j(t) - \beta_i y_i(t) + h_i + v_i(t) \tag{4.3}$$

where $y_i(t)$ represents the protein activity level of the target protein i at time t, b_{ij} denotes the interaction ability of the j-th interactive protein with the i-th protein, $y_j(t)$ represents the activity level of the j-th protein interacting with the target protein i, β_i denotes the degradation effect of the protein, h_i represents the basal activity level, and $v_i(t)$ is the stochastic noise. The rate of PPI is proportional to the product of the concentrations of both proteins [16], that is, it is proportional to the probability of molecular collisions between them. The physical interpretation of equation (4.3) is that the activity of target protein i at time t+1 is the combination of present protein activity, regulatory interactions with M interactive proteins in the cell, levels of basal protein from other sources, stochastic noise, and the protein degradation of the present state. The PPI dynamic equation of target protein i in the candidate PPI network can be represented by the following regression equation [5]:

$$y_i(t+1) = \left[y_i(t)y_1(t) \quad \cdots \quad y_i(t)y_{i-1}(t) \quad y_i(t) \quad \cdots \quad y_i(t)y_M(t) \quad 1 \right] \cdot \begin{bmatrix} b_{i1} \\ \vdots \\ b_{ii-1} \\ 1-\beta_i \\ \vdots \\ b_{iM} \\ h_i \end{bmatrix} + v_i \tag{4.4}$$

$$= \Phi_i(t)\theta_i + v_i$$

The interaction parameter θ_i can be estimated from protein profile microarray data by least-squares or maximum-likelihood parameter estimation in chapter 3 [5] (if protein profile microarray data are unavailable, then mRNA microarray data could be used to replace them, with some modification [14, 15]). By using AIC in chapter 3 to prune false positive interactions, the real PPI network can then be constructed one target protein at a time by following the above two-step procedure. Some dynamic metabolic pathways [15] and PPI networks of cancer [17] and inflammation [11] have recently been constructed by using the microarray data and AIC method. Host and pathogen interspecies PPI network could be obtained by the cross-correlation in (2.53) between surface proteins of pathogen and PPI network of host [18] or the direct dynamic interaction method [19]. Comparison of PPI networks between healthy and cancer cells can provide network-based biomarkers for molecular investigation and diagnosis of cancer [20]. The construction of PPI networks of apoptosis in cancerous and normal cells is shown in Figure 4.3 [17]. When cancer network is compared with normal network [17], the "gain-of-function" network and "loss-of-function" network are given in Figure 4.4.

4.3. Construction of Integrated GRN and PPI Cellular Networks

Living organisms have evolved complex mechanisms to respond to changes in environmental conditions. This is the case even in unicellular organisms like the yeast *Saccharomyces cerevisiae*. Such environmental changes, commonly termed as "stress", are harmful or even lethal to the survival of the cells, especially of microorganisms whose environment is very changeable. Cellular responses to stresses require complex cellular networks of sensing and signal transduction pathways, as well as metabolic pathways to adapt cell growth and proliferation for adjustments of gene expression programs, metabolic activities, and other features of the cell (see Figure 4.5).

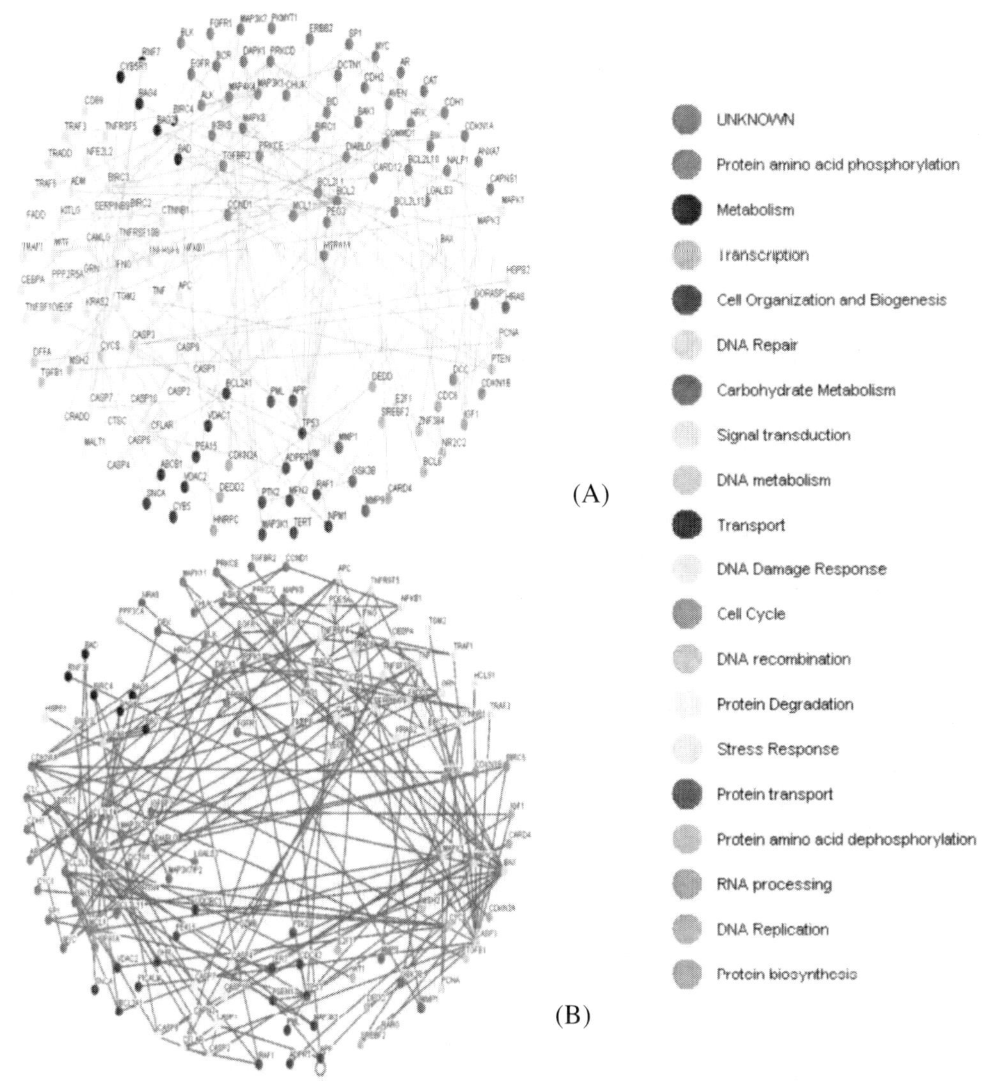

Figure 4.4. Cancer-perturbed protein-protein interactions in the apoptosis network. (A) 'Gain-of-function' network, showing 140 nodes and 157 edges. (B) 'Loss-of-function' network, showing 126 nodes and 162 edges. Colors of nodes represent Gene Ontology annotations.

These regulatory systems and signal and metabolic pathways can therefore be suitably described by an integrated cellular network of transcription regulation and PPIs. The flowchart for constructing the integrated cellular network is shown in Figure 4.6. The same two steps already described apply (construction of a candidate cellular network and pruning of false positives via specific microarrays of gene expression and protein expression data). Several kinds of omics data and database information were integrated, including data on microarray gene expression and protein expression, regulatory associations between TFs and genes, and PPI. From these data, the candidate integrated GRN and PPI networks are retrieved. In the second step, based on the dynamic model of integrated gene/protein interactions in the candidate cellular network, we find that [21]

$$x_i(t+1) = x_i(t) + \sum_{j=1}^{N} a_{ij} z_j(t) - \lambda_i x_i(t) + k_i + n_i(t)$$
$$y_n(t+1) = y_n(t) + \sum_{m=1}^{M} b_{nm} y_n(t) y_m(t) + \alpha_n x_n(t) - \beta_n y_n(t) + h_n + v_n(t) \tag{4.5}$$

where the regulation function $z_j(t)$ is modeled as the sigmoid regulation function of $y_j(t)$, that is, for the protein activity profiles of transcription factor j [21],

$$z_j(t) = f_j\left(y_j(t)\right) = \frac{1}{1+\exp\left\{-\left(y_j(t)-u_j\right)/\sigma_j\right\}} \tag{4.6}$$

where $f_j(y_j(t))$ denotes the sigmoid function, u_j and σ_j represent the mean and deviation of protein activity level of TF j, and α_n denotes the translation effect from mRNA $x_n(t)$ to protein $y_n(t)$.

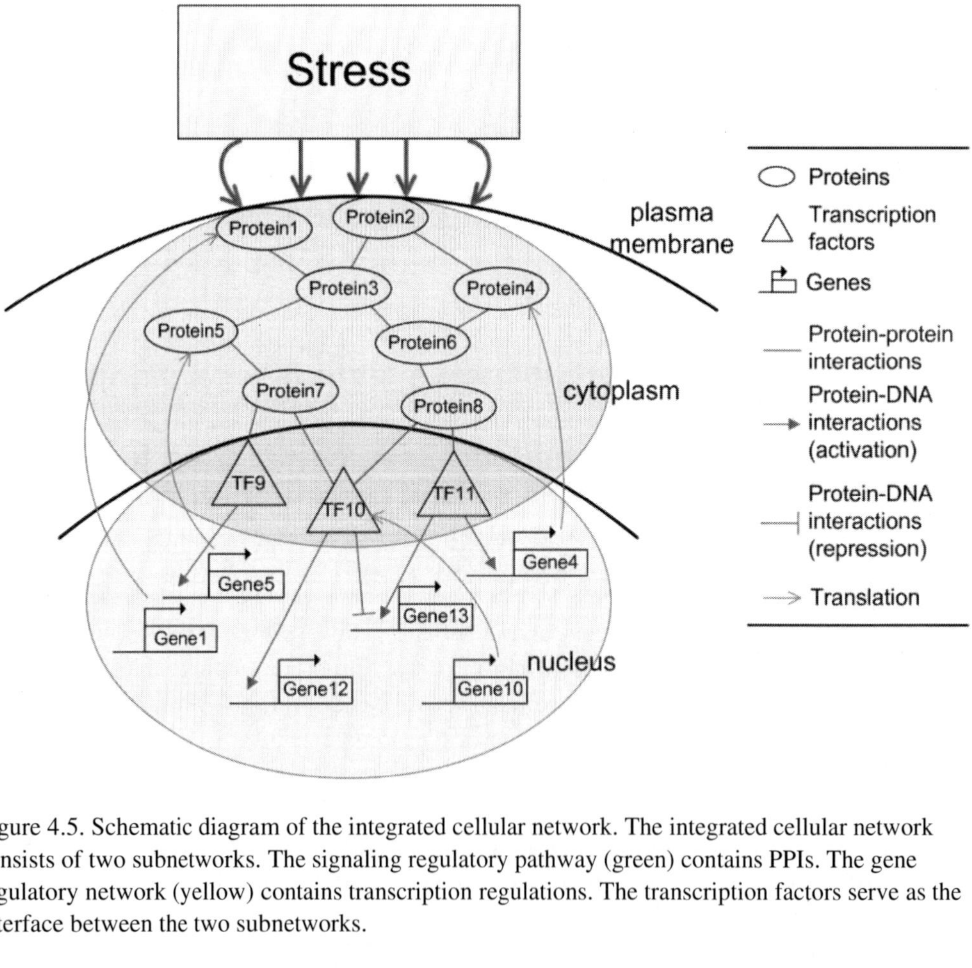

Figure 4.5. Schematic diagram of the integrated cellular network. The integrated cellular network consists of two subnetworks. The signaling regulatory pathway (green) contains PPIs. The gene regulatory network (yellow) contains transcription regulations. The transcription factors serve as the interface between the two subnetworks.

Interaction and coupling among genes and proteins in the integrated cellular network in (4.5) are described as follows. If some TFs $y_j(t)$ at the end of the signal regulatory pathway regulate their target genes through the regulation function $z_j(t) = f_j(y_j(t))$ in (4.6), then the regulatory genes influence their corresponding proteins in signal and metabolic pathways through translation effect $\alpha_n x_n(t)$. The interplay between genes and proteins can be seen from their coupling dynamic equations in (4.5) and Figure 4.5. Here, the TFs serve as the interface between the signaling regulatory pathway and gene regulatory network. In other words, the interplay between transcription regulation and PPIs constitutes the framework of the integrated cellular network.

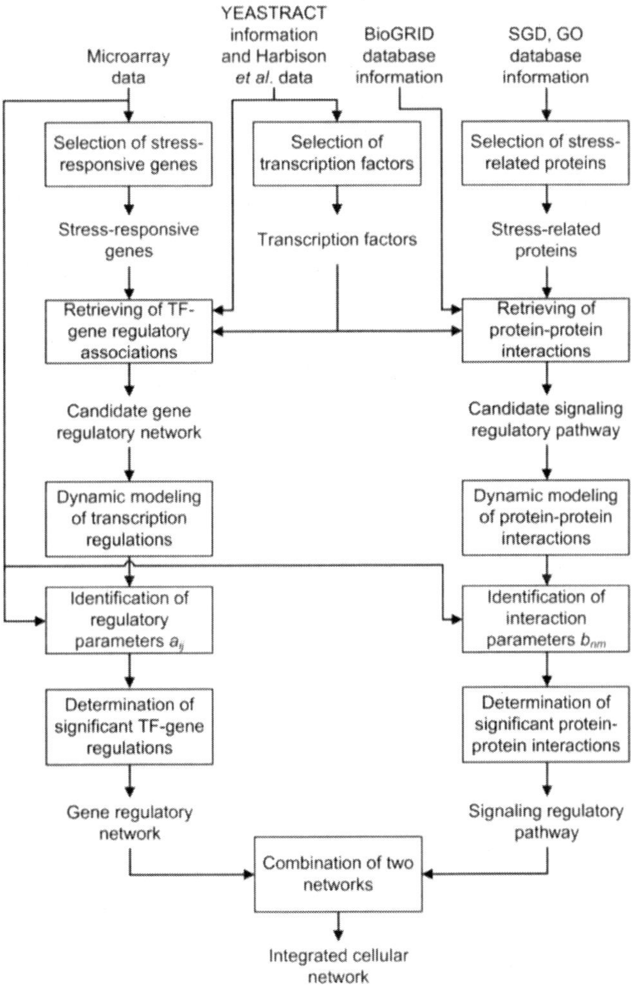

Figure 4.6. Schematic diagram of the integrated cellular network [21]. The construction of integrated GRN and PPI cellular networks consists of two steps. The first step is to construct the candidate GRN and PPI networks by data mining. The second step is to construct real cellular GRN and PPI networks by dynamic modeling of microarray and protein expression data, respectively, via parameter identification methods (i.e., reverse-engineering methods). System order detection is also employed to prune false positives in the candidate cellular network to obtain a GRN and PPI integrated network. The integrated cellular network of *S. cerevisiae* under hyperosmotic stress is shown in Figure 4.7.

Based on the dynamic coupling equation in (4.5), the candidate integrated cellular network can be linked through the regulatory parameter a_{ij} between genes and their possible regulatory TFs and through the translation parameter α_n for gene expression to protein expression. The candidate signaling or metabolic pathways can be linked through the interaction parameter b_{nm} between possible interaction proteins. Since omics data on the candidate gene regulatory network and candidate signaling or metabolic pathway only indicate possible TF-gene regulation and protein interactions, they should be confirmed using microarray data of gene and protein expressions. In particular, values of a_{ij} and b_{nm} in (4.5) should be identified and validated by least-squares estimation via microarray data in a specific biological condition or phenotype [21].

Significant regulations and interactions between genes and proteins were detected using model selection methods such as AIC and statistical assessments like such as Student's t-test in chapter 3 [5]. Based on the interface between gene regulatory and signal/metabolic networks (i.e., transcription factors) in a specific biological condition, the two networks are coupled together to form the integrated cellular network. The integrated cellular network for *S. cerevisiae* under hyperosmotic stress is shown in Figure 4.7. After the construction of GRN and PPI network from microarray data and bioinformatic method, various system characteristics of the biological network are estimated or measured using systems biology methods in the following sections so that these systematic methodologies can be applied to the systematic design of synthetic biology and metabolic engineering.

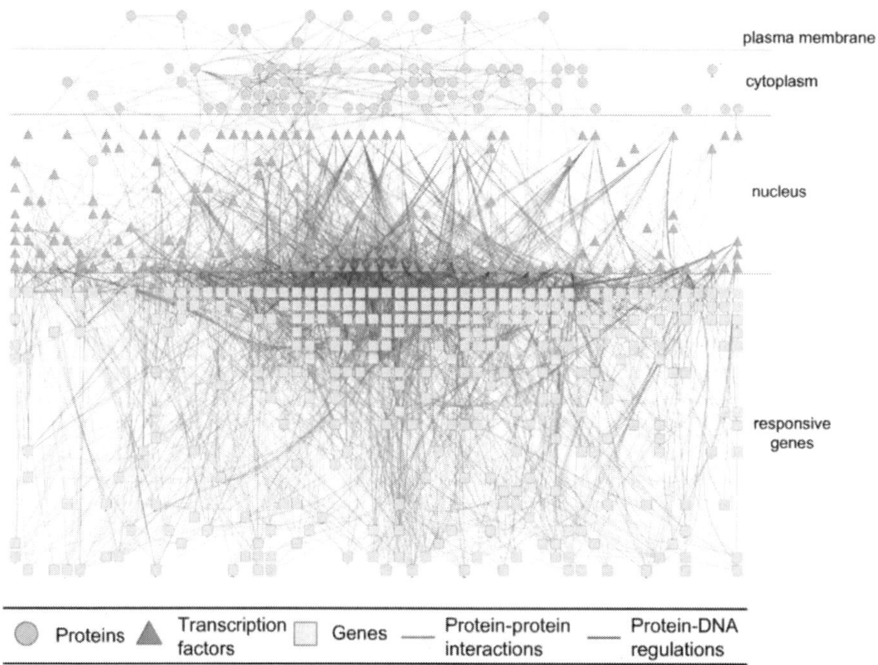

Figure 4.7. The *S. cerevisiae* integrated cellular network under hyperosmotic stress [21]. By following the schematic diagram of an integrated GRN and PPI cellular network in Figure 4.6, the integrated cellular network of signaling regulatory pathway and GRN for hyperosmotic stress in *S. cerevisiae* is identified by dynamic modeling and data mining. Receptor proteins in the plasma membrane, signal regulatory pathways in the cytoplasm, and transcription factors and GRNs in the nucleus are used to construct an integrated cellular network for *S. cerevisiae* under hyperosmotic stress.

4.4. INTERSPECIES PPI NETWORK BETWEEN HOST AND PATHOGEN

The infection-related PPI networks of two species and the intercellular network between host and pathogen are crucial to investigate pathogenicity of pathogen and immune response of host and may help improve medical therapies and facilitate the development of new antifungal drugs. In [18], we construct the PPI networks of host and pathogen based on PPI dynamic model in (4.3) by their microarray data in two stages (adhesion and invasion) of infection process, respectively. Then we used the GO databases to select cell surface proteins, which may interact with any protein of zebrafish in the infection process. The PPIs between host and pathogen can be constructed via Pearson cross-correlations (2.53) of their time series microarray data. Therefore we could construct the integrated intercellular PPI network of *C. albicans* and zebrafish during *C. albicans* infection of zebrafish as shown in Figure 4.8.

With the *C. albicans* microarray data spanning 0.5–4 hours, we constructed a network called the 'adhesive stage network', which represents *C. albicans* cells in the adhesion stage. Since cubic spline interpolation requires at least four data points to solve a cubic polynomial [22], we included the 4-hour data point to construct this network. With the *C. albicans* microarray data spanning 2–12 hours, we constructed another network called the 'hyphal stage network', which represents *C. albicans* cells transitioning to the hyphal form. Similarly, we collected two groups of data at different stages of infection to construct two separate PPI networks for zebrafish as well: one for microarray data from 0.5 to 4 hours, and another for data from 2 to 12 hours, named the zebrafish stage 1 network and zebrafish stage 2 network, respectively. By estimating the system parameters using the time series microarray data and selecting model order using the *AIC* measurement [5, 23], the likelihood of false positive interactions in the potential PPI network for the infection process was reduced. Network refinement yielded 550 proteins with 2725 PPIs in the adhesive stage network and 555 proteins with 3171 PPIs in the hyphal stage network: these two networks could then be combined into the *C. albicans* dynamic hyphal PPI network for the infection process. Similar refinements in the zebrafish data returned 1248 proteins with 2344 PPIs in the zebrafish stage 1 network and 1265 proteins with 2379 PPIs in the zebrafish stage 2 network, and these two networks could then be combined into the zebrafish dynamic PPI network for the defensive process. The *C. albicans* dynamic hyphal PPI network, the zebrafish dynamic PPI network, and the host-pathogen intercellular PPI network could be merged into an integrated infection intercellular PPI network.

The global system view of the *C. albicans*- and zebrafish-integrated infection intercellular PPI network is illustrated in Figure 4.8. The entire integrated infection intercellular network can be divided into eight levels according to the location of protein action (i.e., nucleus, intracellular, cell surface, or extracellular) and species (i.e., *C. albicans* or zebrafish), and is composed of three subnetworks. The upper subnetwork is the dynamic hyphal PPI network of *C. albicans*. The middle subnetwork shows the host-pathogen intercellular interaction network. For simplicity, only the top five correlated interactions of the *C. albicans* cell surface proteins are listed. The bottom subnetwork is the dynamic defensive protein interaction network of zebrafish.

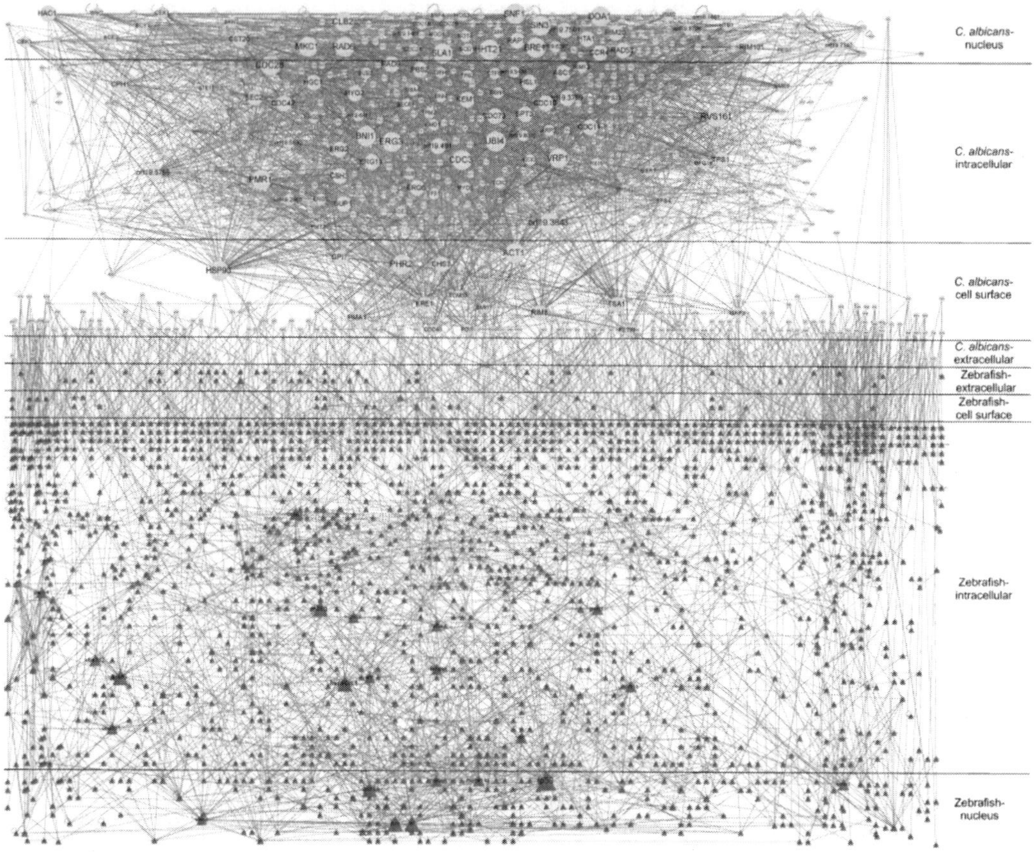

Figure 4.8. *C. albicans* and zebrafish integrated intercellular dynamic PPI network during *C. albicans* infection of zebrafish [18]. The infectious intercellular network is composed of three subnetworks. The upper subnetwork is the dynamic hyphal PPI network of *C. albicans*. The middle subnetwork shows the host-pathogen intercellular interaction network. For simplicity, only the top five correlated interactions of the *C. albicans* cell surface proteins are listed. The bottom subnetwork is the dynamic defensive protein interaction network of zebrafish. This infectious intercellular PPI network contains lines and nodes of three different colors. The red lines denote PPIs that did not appear in the stage 1 network but did in the stage 2 network. The green lines denote PPIs that appeared in the stage 1 network but did not in the stage 2 network. The blue lines denote PPIs that appeared in both the stage 1 and 2 networks. The node size denotes connectivity degree. The drawing of the PPI network was created using Cytoscape.

In order to verify the accuracy of our dynamic hyphal growth protein interaction network, we investigated whether this network contains previously identified pathways related to hyphal growth [24]. An inspection of the integrated infectious intercellular network seen in Figure 4.8 confirmed that our *C. albicans* dynamic hyphal PPI network includes the MAP kinase cascade, cyclic AMP/PKA pathway, and other hyphae-associated pathways. We isolated these pathways from Figure 4.8, and then constructed a new hyphae-related subnetwork as Figure 4.9. Our new hyphae-related subnetwork contains almost all of the proteins and interactions of the already known hyphae-related pathways. The GTP binding protein Ras1 was not contained in our dynamic hyphal growth PPI network because its p-value is greater than 0.01 in the original protein pool selection step for *C. albicans*; however, Ras2, which is in the same family as Ras1, appeared in the new subnetwork. Ras2 is similar to *S. cerevisiae* Ras2p, which can activate adenylate cyclase and is involved in *S. cerevisiae* pseudohyphal growth, and Ras2 mutants could have altered filamentous growth patterns [25]. Ras2 also stimulates Cyr1 (Cdc35), which in turn acts as an intracellular second messenger during morphological switching. Ras2 also stimulates Cdc42 through Ras-related protein (Rsr1), which is involved in budding, cell morphogenesis, and hyphal development processes [26]. In addition, an interaction between Cdc42 and Wal1 is not completely known. However, we can see that Cdc42 links to Wal1 via Myo2 and Rho3 in our new hyphae-related subnetwork. Myo2 is required for polarized cell growth and dimorphic switching in *C. albicans* and is also involved in hyphal development [27]. Rho3 is required for polarized cell growth and cell separation and also involved in hyphal development [28]. Fortunately, the previously uncertain pathway between Cdc42 and Wal1 was more completely elucidated in our dynamic hyphal growth PPI network.

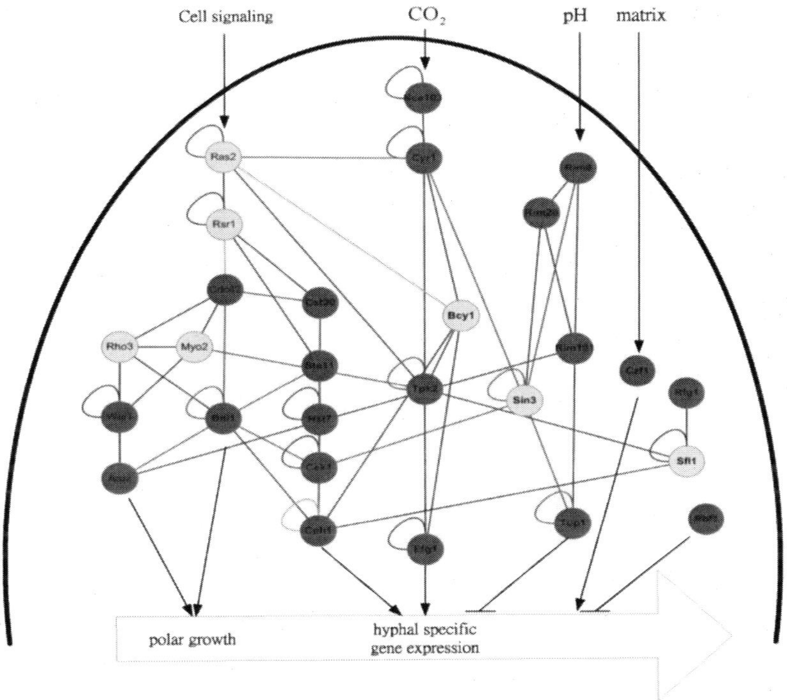

Figure 4.9. Signaling cascades involved in the dynamic hyphal growth protein interaction subnetwork of *C. albicans* at different stages of infection [18]. The MAP kinase signaling pathway, cyclic AMP signaling pathway, polarized cell growth pathway, and Rim101 signaling pathway were identified in our dynamic hyphal growth PPI network of *C. albicans* as occurring during *C. albicans* infection of zebrafish. These signal transduction pathways are involved in the dynamic regulation of *C. albicans* morphological transitions. The red lines indicate PPIs that did not appear in the adhesive stage network but did in the hyphal stage network. The green lines indicate PPIs that appeared in the adhesive stage network but did not appear in the hyphal stage network. The blue lines indicate PPIs that appeared in both the adhesive and hyphal stage networks.

Aside from these three well-known hyphae-related signaling pathways–i.e., the MAP kinase and cyclic AMP signaling pathways and the polarized cell growth pathway–the pH-dependent Rim101 pathway was also identified from the dynamic integrated infection intercellular network. In this pathway, Nrg1 was not identified for inclusion in Figure 4.9, due to its p-value being greater than 0.01. However, Tup1, which has the same function as Nrg1, fits into the pathways shown in Figure 4.9, and hence the pH-dependent pathway would seemingly be uninterrupted. In conclusion, 20 out of 22 proteins from already known pathways are included in our *C. albicans* dynamic hyphal PPI network, in which the major hyphae-related pathways are all visible. These results verify the high accuracy of our infection intercellular PPI network. Moreover, our hyphal growth protein interaction subnetwork provides a dynamic and more complete hyphal network. These proteins are all related to hyphal growth or filamentous growth, and the figure reflects true pathways of these proteins.

From Figure 4.8, the hyphal growth PPI subnetwork is shown in Figure 4.9. This could help us to identify infection and defense-related genes via dynamic host-pathogen interaction network to investigate the infection and defense mechanism between host and pathogen. Another method is to model the interspecies PPIs between host and pathogen as the following multivariate linear dynamic model [19]

$$x_i(t+1) = x_i(t) + \sum_{j=1}^{J_i} a_{ij} x_j(t) + \sum_{k=1}^{K_i} b_{ik} y_k(t) - \lambda_i x_i(t) + h_i + \varepsilon_i(t) \tag{4.7}$$

where $x_i(t)$ denotes the expression level of target protein i, a_{ij} denotes the interactivity between the j-th intracellular interactive protein and the i-th target protein, b_{ik} denotes the interactivity between the k interspecies interactive protein to the i-th protein, $y_k(t)$ represents the expression level of the k-th interspecies protein to the i-th target protein, J_i and K_i denote the number of intracellular and interspecies proteins interacting with target protein. In other word, only those proteins interacting with target protein i in the candidate interspecies PPI network are described by the multivariate linear equation in (4.7). By the least square parameter identification method and AIC order detection for dynamic interspecies model in (4.7) by microarray data to prune the false positive PPIs in candidate PPI network one protein by one protein, we could obtain the real interspecies PPI network between *C. albicans* and zebrafish in the infection process in Figure 4.10. The interspecies PPI network can get insight into the invasion and defense mechanism between host and pathogen in the infection process.

Novel interspecies interactions highlight the association between *C. albicans* pathogenesis and the zebrafish redox process On the basis of the proposed computational framework, several novel interspecies interactions were identified in the constructed host-pathogen interspecies PPI network. Since the interspecies network construction method was proposed to elucidate pathogenic and defensive mechanisms during the infection process, the interspecies subnetwork for *C. albicans* virulence proteins, namely, proteins annotated with GO term pathogenesis, were further investigated. From these identified interspecies PPIs, 24 zebrafish proteins were found to interact with *C. albicans* virulence proteins (Figure 4.11). In addition, oxidation-reduction process was the only significant GO term shared among these 24 proteins ($p < 0.01$, Fisher's exact test), highlighting the association between *C. albicans* pathogenesis and the zebrafish redox process. Six zebrafish proteins in the pathogenesis subnetwork, that is, Cyb5r2, Cyp51, Kmo, Nsdhl, Sc5d, and zgc:77112, were annotated with oxidation-reduction process (Figure 4.11) and all of their gene expressions were repressed over time, except for zgc:77112. Host defense against *C. albicans* infection relies mainly on phagocytes of the innate immune system, and one important host response generated by phagocytes is the production of reactive oxygen species (ROS). Free oxygen radicals produced by the oxidation-reduction process are highly toxic to pathogens and are utilized for pathogen clearance. In addition, ROS has been demonstrated to act as secondary signaling molecules contributing to signaling cascades related to inflammation, apoptosis, and immune responses [29]. For example, in certain cell lines, activation of the proinflammatory transcription factor NF-κB is dependent on ROS [30]. Therefore, immune cells depend on ROS to not only kill phagocytosed pathogens directly, but also to mediate inflammatory and immune signaling pathways [29]. As a result, a wide variety of pathogens have developed various molecular strategies to prevent host ROS generation, including *Helicobacter pylori*, *Legionella pneumophila*, and *Aspergillus fumigatus* [31-33]. Through these ROS inhibition mechanisms, these pathogenic organisms evade host immune responses. Similarly, *C. albicans* has the ability to suppress ROS production in host immune cells [34]. The precise mechanism utilized by *C. albicans* remains unclear; however, *C. albicans* catalase and surface superoxide dismutase have been implicated in counteracting the oxidative burst from phagocytes [35, 36]. Since ROS also carry out important signaling functions as stated above, suppression of ROS production by *C. albicans* may not only result in evasion of phagocytic killing, but also in significant modulation of anti-*Candida* inflammatory responses, which directly benefits the pathogen further. As a result, suppression of ROS production may represent an important immune evasion mechanism for *C. albicans*.

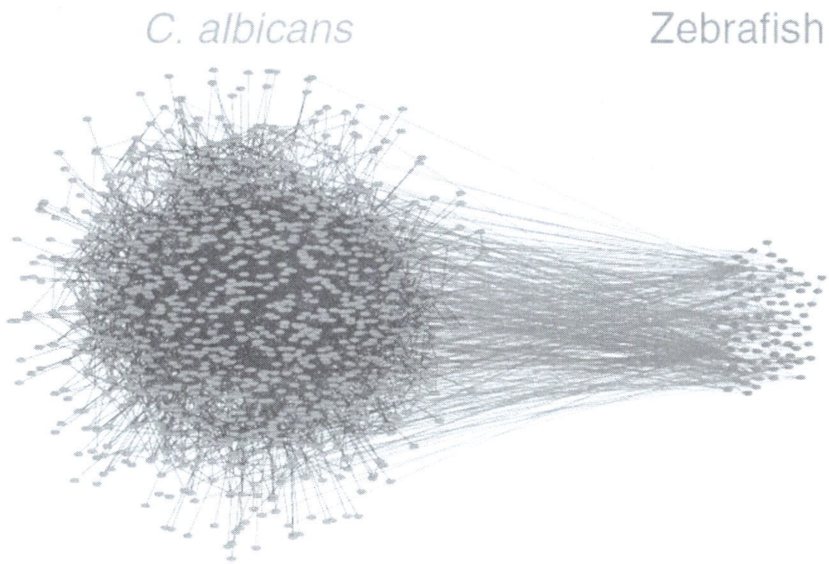

Figure 4.10. The constructed *C. albicans*-zebrafish interspecies protein-protein interaction network [19]. There were 371 interspecies interactions, 3,504 *C. albicans* intracellular interactions, and 35 zebrafish intracellular interactions among 1,127 *C. albicans* proteins and 87 zebrafish proteins in the constructed interspecies network.

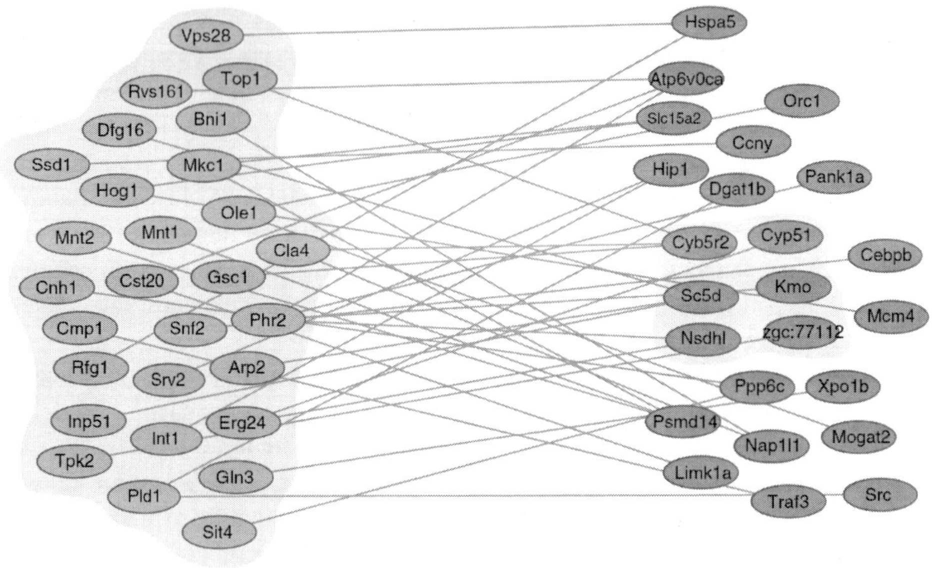

Figure 4.11. The interspecies subnetwork for *C. albicans* pathogenesis proteins [19]. This figure indicates the *C. albicans* pathogenesis subnetwork extracted from the interspecies PPI network in Figure 4.10. The 24 zebrafish proteins interacting with *C. albicans* pathogenesis proteins (purple shadow) were found to be statistically enriched with proteins annotated with oxidation-reduction process (yellow shadow) ($p < 0.01$), highlighting the association between *C. albicans* pathogenesis and the zebrafish redox process. The intracellular protein interactions for both *C. albicans* and zebrafish were omitted for simplicity.

On the other hand, infection by some pathogens such as *Entamoeba histolytica* and Japanese encephalitis virus results in enhanced ROS formation [37, 38]. These pathogens have been reported to utilize enhanced ROS generation to induce host cell death, thus allowing themselves to escape the cell. This mechanism is likely to contribute to the spread of the pathogens [29]. Recently, it has been demonstrated that *Sclerotinia sclerotiorum*, a fungal pathogen that infects virtually all dicotyledonous plants, can both suppress and induce host ROS formation during infection via the secretion of oxalic acid [39]. During the initial stages of infection, *S. sclerotiorum* dampens the oxidative burst of the plant and generates a reducing environment in host cells. Once infection is established, the pathogen induces the generation of plant ROS (oxidizing conditions) leading to programmed cell death of the host, which directly benefits the pathogen [39]. In this manner, *Sclerotinia* uses a novel strategy involving regulation of host redox status to establish infection. Although there is no evidence that *C. albicans* is capable of inducing ROS production in host cells to date (Wellington et al. demonstrated that *C. albicans* suppresses production of ROS in phagocytes within 180 minutes of infection [34]), we postulated that *C. albicans* may also modulate ROS levels to subvert immune defenses in the same way as *S. sclerotiorum*, that is, by suppressing host ROS production in the initial stages of infection and inducing ROS in the later stages. Further experiments measuring ROS at different infection stages are needed to validate this hypothesis. Taken together, novel interspecies interactions identified in this chapter highlight the association between *C. albicans* pathogenesis and the zebrafish redox process. The redox status in both the host and pathogen can be a critical factor that determines the outcome of the battle between the host and pathogen. From the perspective of the pathogen, it is essential to be in control of the redox environment and the cell death pathways of the host in order to subvert immune defenses by the host and support self-survival. In contrast, during an immune response, the host seeks to control the redox environment and the cell death pathways to the detriment of the pathogen.

In the study of host-pathogen PPI interaction network, a computational framework which integrated multiple omics data was proposed to construct an interspecies PPI network for characterization of host-pathogen interactions. The proposed method was shown to be useful, with results highlighting an association between *C. albicans* pathogenesis and the zebrafish redox process during *C. albicans* infection in zebrafish. Results further indicated that redox status is critical during the battle between the host and pathogen, which could determine the outcome of infection. While the pathogen can control the redox environment to subvert immune defenses and support self-survival, in contrast, the host controls the redox environment to the detriment of the pathogen during an immune response. With continued accumulation of interspecies transcriptomics data, the proposed method could be helpful in the development of network medicine for infectious diseases from an interspecies network perspective.

4.5. Stochastic Spatial-Temporal Dynamic Model for Gene/Protein Interaction Network in Early Drosophila Development

An early embryonic stage in Drosophila embryogenesis, i.e., the syncytial blastoderm stage, is completed two hours after the onset of fertilization and periodic segments are then characterized. Before the determination of periodic segments, the embryo is not yet separated by membranes, and macromolecules such as transcription factors (TFs) can diffuse freely and regulate downstream target genes in neighboring nucleus. Hence, at the syncytial blastoderm stage diffusion mechanism is fast enough to vary the concentrations of TFs in transcription regulations. Through a series of high/low affinity bindings of transcription regulations, downstream genes are dictated to express in their corresponding space of an embryo. Therefore, we assume that the transcription regulation and diffusion mechanism may play a cooperative role in characterizing embryonic segments.

Although some topics about protein diffusion have been well studied [40, 41], gradient dynamics of concentrations of transcription factors is still hard to be analyzed without quantitative inference under dynamic modeling. For example, critical boundaries settled by protein concentration gradient in dynamic models of early embryogenesis have allowed investigators to re-examine quantitatively concentration gradient dynamics [42]. Jaeger and his colleagues have used mRNA spatial-temporal data and dynamic model to characterize the establishment of gap domains [43]. Therefore, in order to analyze the diffusion mechanisms of transcription factors at different domains of *Drosophila* embryo, a spatial-temporal model is needed. In recent studies, early embryogenesis in Drosophila includes at least 31 genes in subdividing the embryonic patterns into 14 segmental primordia along the anterior-posterior (A-P) axis [44]. In the past several decades, the spatial-temporal expressions of the early development-related genes (*bicoid* (*bcd*), *caudal* (*cad*), *hunchback* (*hb*), *giant* (*gt*), *knirps* (*kni*), *Krüppel* (*Kr*), *tailless* (*tll*), *even-skipped* (*eve*), *fushitarazu* (*ftz*), *hairy*, *odd-skipped* (*odd*), *paired* (*prd*), *runt* and *sloppy-paired* (*slp*)) have been provided and studied during the early developmental stages of Drosophila melanogaster. The 14 early development-related genes can be roughly divided into three classes, i.e., maternal genes, gap genes and pair-rule genes, which have been regarded as hierarchical transcription regulations with positive auto-regulations to generate and refine the constitutions of segments [45, 46]. At the beginning of early embryogenesis, gap genes are regulated by high-level expressions of maternal TFs to initiate an early embryo development. Gene expression boundaries are determined by thresholds of protein concentration, while gene expression borders are refined by auto-regulation and repression [47, 48].

Three classified genes (i.e., maternal genes, gap genes and pair-rule genes) into which the 14 early development-related genes can be divided are described in detail in the following. The maternal genes, i.e., *bcd*, *cad* and *hb*, diffuse and regulate gap genes with different expression levels in each spatial region along the A-P axis of the Drosophila embryo. The gap genes, i.e., *gt*, *hb*, *kni*, *Kr* and *tll*, define roughly the differences between two neighboring stripes by protein diffusion. The pair-rule genes, i.e., *eve*, *ftz*, *hairy*, *odd*, *prd*, *runt* and *slp*, define periodic patterns of the embryo by transcription regulation and protein diffusion. Two of these pair-rule genes, i.e., *eve* and *ftz*, are involved in defining even and odd segments of the 14 segmental primordia along the A-P axis [49, 50]. The odd and even segments of concern are the seven eve stripes and seven *ftz* stripes, respectively. Moreover, at the blastoderm stage, along the D-V axis, three main regions, i.e., non-neural ectoderm (prospective epidermis), neurectoderm (prospective nervous system and larval ventral epidermis) and mesoderm (prospective muscle and connective tissue) are also divided [51]. The genes, which determine the three primary regions along the D-V axis, are different from these 14 early development-related genes which determine periodic segments along the A-P axis. In this chapter, for the convenience of analysis and system identification we will define spatial regions in the two-dimensional (2-D) space of the embryo along the A-P and D-V axes according to the above information. However, we only analyze the A-P formation of embryo after system modeling of transcriptional regulatory net- work, and the D-V formation can be analyzed by a similar procedure.

At the early developmental stages of Drosophila, the three-dimensional (3-D) spatial-temporal expression data of 14 early development-related proteins (http://flyex.ams.sunysb.edu/flyex/) [52-55], genome-wide mRNA time-course expression data [56] and mRNA 3-D spatial-temporal expression data (http://flyex.ams.sunysb.edu/lab/gaps.html) [43, 45] have been published and can be used for a system dynamic modeling of early Drosophila development. Interestingly, by comparing the normalized protein spatial-temporal expression data with mRNA spatial-temporal expression data, the trends of gene expressions along the A-P axis are found [42]. In this chapter, we incorporate the mRNA 3-D data with protein 3-D data to construct the gene/protein interaction network for the transcription regulations and diffusion mechanisms of early embryogenesis via our stochastic 3-D dynamic model. However, there are some expression data within the 14 early development-related genes are unavailable in mRNA 3-D spatial-temporal expression database. In recent studies, Neural-Network (NN) model, which could be trained to optimize its internal network to learn the behaviors of complex systems, has been used to not only infer gene network regulatory relationships based on genome-wide microarray data [57] but also build the relationship between input and output information by using a back-propagation algorithm to learn from the training data [58-61]. Therefore, for the unavailable mRNA data, we will use the back-propagation NN training method to obtain the mimic mRNA data according to the available protein and mRNA 3-D data.

In recent years, since the development of experimental techniques has increased the quality and amount of available mRNA and protein expressions, many approaches, e.g. fuzzy logic [62, 63], recurrent neural networks [64-66], Bayesian networks [67, 68], Boolean networks [69, 70], and differential equations [15, 71, 72], have been widely exploited to unravel regulation networks from the perspective of systems biological. For the well available protein spatial-temporal data in early Drosophila development, nonlinear 2-D dynamic models have been employed to analyze the transcription regulation properties and effect of gap genes on eve stripe formation [42, 45, 55, 73-76]. However, more efforts are needed to incorporate these pathways and gene networks with a spatial-temporal gene/protein interaction network to interpret the dynamic system behavior in early Drosophila development since not only protein but also mRNA 3-D spatial-temporal data are both available for dynamic interplay of genes and proteins at different compartments of Drosophila embryo in early embryogenesis. The mechanisms of early Drosophila development in the whole embryo can be unraveled clearly if the dynamic interactions of genes and proteins are considered at different compartments in early embryogenesis. Therefore, in this chapter, we propose a stochastic 3-D dynamic model for constructing the gene/protein interaction network of early Drosophila development.

In this section, we focus on the topic of investigating the possible mechanisms for the eve stripe formation of Drosophila embryo. In this biological development approach, it is assumed that transcription regulations consist of cis-effect and trans-effect. Since edges, i.e., transcription regulations, in a gene regulatory network must be constantly selected in order to survive randomization forces, trans-effects, which are the binding affinities of specific transcription factors to cis-regulatory regions in the promoter of the target gene, would be varied rapidly while cis-effects, which are regulated directly by physical attachment of TF's binding cis-regulatory regions, are relatively fixed [77]. Thus, we assume that regulation abilities, i.e., trans-effects, should vary with different spatial regions of the embryo, which results from different binding affinities of diffusible TFs. Based on the constructed stochastic 3-D Embryo Space-Time model (stochastic 3-DEST model), we analyze the transcription regulations and diffusion mechanisms for gene/protein interaction network. The stochastic 3-DEST model with 28 state variables is employed to represent the transcription/translation regulation process between 14 mRNA genes and the corresponding TFs in early embryogenesis. Moreover, because we consider both the environmental noises and the intrinsic noises in mRNA and protein data, stochastic partial differential equations (PDEs) are employed for the transcriptional and translational regulatory model of early embryogenesis. In order to understand the roles of TFs in each spatial region, according to the signs of diffusion parameters of the stochastic 3-DEST model, a TF can be considered as a donor (>0) or an acceptor (<0) in each spatial region to balance instant concentrations of the whole embryo. Hence, the TF in a spatial region that diffuses to (from) the neighboring spatial regions, is called a donor (acceptor). In addition, from previous studies we know that transcription regulations can be inferred by a dynamic model via microarray data [6, 15, 74]. However, how to sieve out the insignificant transcription regulations from the whole gene/protein interaction network is still a problem. For this reason, according to the stochastic 3-DEST model, the Akaike Information Criterion (AIC) in chapter 3 [5] for model order detection combined with the maximum likelihood (ML) for parameter estimation in system identification is used in this chapter to detect significant upstream regulators and to prune insignificant transcription regulations for refining the gene/protein interaction network of early Drosophila development. From the identified stochastic 3-DEST model, we can not only find the significant transcription regulations of the corresponding TFs, which control the anterior/posterior border formation of eve stripes, but also validate these results with wet experiments. In order to validate the identified effect of transcription regulation and diffusion on early Drosophila development, the wet experiments, i.e., gene mutations [46, 48, 49, 78-82], regulatory module classification [83] and cis-regulatory module detection [84], have been employed to trace back the direct or indirect transcription regulations and protein diffusions in early Drosophila development. From the perspective of the network motifs of the identified gene/protein interaction network in the embryo, we find that transcription regulations and protein diffusion mechanisms may play a cooperative role in the formation of eve stripes in early Drosophila development.

4.5.1. System Modeling and Identification for Gene/Protein Interaction Network

To identify the dynamic behavior of the early development-related genes, the procedure of system identification in early embryogenesis is divided into four steps. First, utilizing fully the well-published spatial-temporal data and the prior knowledge of early embryogenesis, we construct a stochastic 3-DEST model to identify the molecular dynamics of gene/protein interaction network in early embryogenesis. Second, for system modeling, we use Eve's spatial expression at the cleavage cycle 14A temporal class 8 (c14A8) of the nuclear cleavage to settle stripe boundaries and region boundaries of each stripe for dividing the embryo into seven eve stripes along the A-P axis and into three spatial regions (i.e., anterior part, middle part and posterior part) along the D-V axis, respectively. Third, for the early development-related genes, since a part of the mRNA spatial-temporal data are unavailable, we incorporate the available mRNA and protein spatial-temporal expression data with the back-propagating NN training method to train and simulate the mimic data for the unavailable mRNA spatial-temporal expression data (see 4.7.1). Fourth, we identify the model parameters and select the significant regulatory parameters for the stochastic 3-DEST model to construct the transcriptional regulatory network in every spatial region by the ML estimation method and the AIC backward elimination method, respectively. Finally, the transcriptional regulatory networks in every spatial region are connected together to construct the entire spatial-temporal gene/protein interaction network for early Drosophila development.

Remark 4.1: If the information of cooperation bindings is richer in future, the transcriptional regulations due to cooperation binding can be easily extended to the regulation candidates of the 3-DEST model, which can improve the proposed model of gene/protein network but with increased computation burden when using the AIC method in early embryogenesis.

4.5.2. Stochastic PDEs Model in Eve Stripe Formation

In previous studies, dynamic models with protein synthesis, protein diffusion and protein decay have been utilized in the description of the mechanism of embryonic development [42, 43, 45, 73-76]. To analyze the dynamic interplay of genes and proteins in early embryogenesis, six stochastic molecular dynamics are incorporated in the 3-DEST model, i.e., (1) protein synthesis, (2) protein decay, (3) mRNA decay, (4) protein diffusion, (5) transcription regulations, and (6) auto-regulation. In addition, in order to differentiate mRNA expressions from protein expressions, we define two state variables X_i and Y_i to represent the 3-D spatial-temporal mRNA profiles of the i-th target gene and its corresponding TFs, respectively. According to the transcription regulation model proposed in previous studies [6, 15, 45, 74], the stochastic 3-DEST model for the i-th target gene and their upstream regulatory TFs in the gene/protein interaction network of *Drosophila* development is proposed as follows [85]:

$$\begin{aligned}\frac{\partial X_i(t,x,y)}{\partial t} &= K_i(x,y)-\alpha_i(x,y)X_i(t,x,y)+\sum_{j=1}^{14}\beta_{ij}f\left(Y_j\left(t-\tau_j,x,y\right)\right)\\ &\quad +v_i(t,x,y)\\ \frac{\partial Y_j(t,x,y)}{\partial t} &= \varpi_j(x,y)+\alpha_j(x,y)X_j(t,x,y)-\lambda_j(x,y)Y_j(t,x,y)\\ &\quad +\gamma_j(x,y)\nabla^2 Y_j(t,x,y)+\zeta_j(t,x,y)\end{aligned} \tag{4.8}$$

where $i,j=1,\ldots,14$, $X_i(t,x,y)$ represents the mRNA expression of the i-th target gene, $Y_j(t,x,y)$ denotes the expression of the j-th TF of the target gene, and $f(Y_j(t,x,y))$ defined as $f(Y)=Y^n/(P^n+Y^n)$ is a sigmoid function to denote the regulatory bindings of TFs on the promoters of targets [77, 86, 87]. Here, P is defined as the means of protein expressions, which imply cis-effects of transcription regulations. The term $\sum_{j=1}^{14}\beta_{ij}(x,y)f(Y_j(t-\tau,x,y))$ denotes the transcription regulation, i.e., trans-effect, on the i-th target gene from its TFs. $\alpha_i(x,y)$ stands for mRNA decay rate for the i-th gene and is equal to the synthesis rate of the i-th protein, and $\lambda_j(x, y)$ stands for protein decay rate. $K_i(x,y)$ and $\varpi_j(x,y)$ are basal level of mRNA and protein generation, respectively, and they satisfy $K_i(x,y)$, $\varpi_j(x,y)\geq 0$. The diffusion operator $\nabla^2=\partial^2/\partial x^2+\partial^2/\partial y^2$ is the Laplacian operator in 2-D to denote the diffusion of protein at the location (x, y). In (4.8), mRNA expressions are transcriptionally regulated byTFs (i.e., $\sum_{j=1}^{14}\beta_{ij}(x,y)f(Y_j(t-\tau,x,y))$ and translated for protein synthesis $\alpha_i(x,y)X_i(t,x,y)$ in the downstream translation process. In the second equation of (4.8), the j-th TF, $Y_j(t,x,y)$, is assumed to be produced in the translation process by the corresponding mRNA $\alpha_i(x,y)X_i(t,x,y)$ from the upstream transcription process and decayed by degradation $\lambda_j(x,y)Y_j(t,x,y)$ and diffusion $\gamma_j(x,t)\nabla^2 Y_j(t,x,y)$ [88]. Diffusion coefficients of the j-th TF are represented by $\gamma_j(x,t)$. $\beta_{ij}(x,y)$ denotes the regulatory ability of the j-th TF (or regulatory protein), Y_j, on the promoter region of the target gene X_i. $\beta_{ij}(x,y)>0$ stands for the i-th target gene activated by the j-th TF (prospective activator) or not repressed by the j-th TF (prospective repressor) while $\beta_{ij}(x,y)<0$

stands for the i-th target gene not activated by the j-th TF (prospective activator) or repressed by the j-th TF (prospective repressor) [77]. Therefore, the gene/protein interaction network of early *Drosophila* development is constructed by linking up all target genes through the regulations of their upstream TFs, $\sum_{j=1}^{14} \beta_{ij}(x,y)f(Y_j(t-\tau,x,y))$ in (4.8). Moreover, the productions of Y_j in (4.8) are synthesized by the corresponding mRNA X_j and diffused from Y_j in the neighborhood. Model uncertainty, fluctuations of the basal levels and measurement noises in the mRNA (transcription) dynamics and protein (translation) dynamics are denoted by stochastic noise $v_i(t,x,y)$ and $\zeta_j(t,x,y)$, respectively. x and y denote the location of the embryo in the 2-D space, i.e., the coordination in the x-axis and y-axis.

Remark 4.2: The dynamic model in (4.8) is to interpret the transcription/translation regulation processes of 14 genes in early embryogenesis. The first Equation of (4.8) describes the transcription regulation of the i-th gene; and the mRNA productive rate is mainly due to the transcription regulations of 14 proteins (i.e., TFs), the influence of basal level and degradation of mRNA. The noise $v_i(t,x,y)$ denotes the fluctuation of basal level, measurement noise and modeling residue. Since the expression levels of TFs can be altered with different spatial regions of the whole embryo by diffusion mechanism, the relationship of transcription regulation between one TF and its target gene is also different in different spatial regions. The second equation of (4.8) describes protein production in the translational diffusion process at the location (x,y). The protein productive rate is mainly influenced by the translation of mRNA, diffusion from the neighboring space, and degradation rate of the protein. The noise $\zeta_j(t,x,y)$ is due to the fluctuation of the basal level of protein, measurement noise and modeling error. The model in (4.8) describes the interplay of gene/protein interactions at the location (x,y). The parameters of the stochastic spatial-temporal dynamic model in (4.8) can be estimated by the spatial-temporal profile of mRNA data and protein data in each spatial region. The regulatory gene/protein network can be linked gene by gene through the transcription regulations $\sum_{j=1}^{14} \beta_{ij}(x,y)f(Y_j(t-\tau_j,x,y))$ to other regulatory TFs iteratively.

In the proposed stochastic dynamic model in (4.8), the interplay of six stochastic processes, i.e., protein diffusion, transcription regulations and auto-regulation, mimics the dynamics in early embryogenesis. We are the first to combine mRNA dynamic equations with protein dynamic equations to mimic the dynamic interaction network of target genes and their regulatory proteins via 3-D mRNA and protein data at different compartments in early *Drosophila* development. Our main purpose is to infer the possible mechanisms of eve stripe formation by investigating the estimated parameters K_i, ϖ_j, α_j, β_{ij}, λ_j, and γ_j, i=1,…,14, of the system dynamic model in (4.8)via mRNA and protein data. Since it is hard to solve directly the identification problem of the continuous 3-DEST model in (4.8), we discretize the continuous 3-DEST model in (4.8) [89] and the location (x, y) on the continuous plane is transformed into the location (l, m) on the discrete plane. The discrete 3-DEST model is shown as follows:

$$X_i(k+1,l,m) = (1-a_{i,l,m})X_i(k,l,m) + \sum_{j=1}^{14} b_{ij,l,m}Y_j(k,l,m) + \varepsilon_i(k,l,m)$$

$$\begin{aligned} Y_i(k+1,l,m) &= a_{i,l,m}X_i(k,l,m) + c_{i,l,m}Y_i(k,l,m) \\ &+\rho_{i,l,m}\frac{Y_i(k,l,m) - Y_i(k,l+1,m) - Y_i(k,l,m+1) + Y_i(k,l+1,m+1)}{h_x h_y} \\ &+\delta_i(k,l,m) \\ i,j &= 1,\ldots,14, l = 1,2,3, m = 1,\ldots,21 \end{aligned} \tag{4.9}$$

With the stability constraints $d_{i,l,m}\geq 0$, $w_{j,l,m}\geq 0$, $|1-\alpha_{j,l,m}|\leq 1$, $|c_{j,l,m}-4\rho_{j,l,m}(h_x^{-2}+h_y^{-2})|\leq 1$, if $\rho_{j,l,m}<0$; $|c_{j,l,m}|\leq 1$, if $\rho_{j,l,m}\geq 0$ (see 4.7.2), where k denotes the k-th time point, l and m denote the location (l,m) on the discrete plane, and h is the distance between two locations along two axes, i.e., A-P axis(h_x) and D-V axis(h_y). The parameters are defined as follows: $d_{i,l,m}=K_i(x_l,y_m)\Delta t$, $w_{j,l,m}=\varpi_j(x_l,y_m)\Delta t$, $a_{j,l,m}=\alpha_j(x_l,y_m)\Delta t$, $b_{ij,l,m}=\beta_{ij}(x_l,y_m)\Delta t$, $c_{j,l,m}=1-\lambda_j(x_l,y_m)\Delta t$, and $\rho_{j,l,m}=\gamma_j(x_l,y_m)\Delta t$, where $\Delta t\approx 2.568$ minutes. Then, by using the discrete 3-DEST model in (4.9) and mRNA and protein data, the parameters $K_i(x_l,y_m)$, $\varpi_j(x_l,y_m)$, $\alpha_i(x_l,y_m)$, $\beta_{ij}(x_l,y_m)$, $\lambda_j(x_l,y_m)$, and $\gamma_j(x_l,y_m)$ in (4.9) can be estimated by the system identification method in a spatial region one by one, which will be described in the sequel. Therefore, before the system identification of discrete 3-DEST model in (4.9), we need to define the 2-D spatial regions of *Drosophila* embryo in the following section.

4.5.3. Specification of 2-D Spatial Regions of *Drosophila* Embryo

To identify the discrete 3-DEST model in (4.9), we have to define (l,m) as the center of the spatial regions of the embryo by specifying the boundaries of the spatial regions. Along the A-P axis, two boundaries of the i-th eve stripe are denoted by Bi and $Bi+1$, respectively. The boundaries of seven eve stripes along the A-P axis are denoted by $\{B1, B2, \ldots, B8\}$ (Figure 4.12A). Each of the eve stripes along the A-P axis is separated into three parts, and the boundaries of the middle part of the eve stripe i are denoted as Bia and Bip (Figure 4.12B). Therefore, there are totally 21 spatial regions (i.e., $m=1, \ldots, 21$ in (4.9)) within seven eve stripes along the A-P axis, and the 22 boundaries of the 21 spatial regions are specified as follows: $\{B1, B1a, B1p, B2, B2a, B2p, \ldots, B7, B7a, B7p, B8\}$ ={25%, 29%, 32.5%, 35%, 38.26%, 40.44%, 42.17%, 47%, 49%, 50%, 54.5%, 55.5%, 57.5%, 62%, 64%, 67%, 69%, 72%, 75%, 79%, 81%, 85%}.

Additionally, three spatial regions (i.e., l =1, 2, 3 in (4.9)) along the D-V axis are defined with their boundaries $\{bh1, bh2, bh3, bh4\}$ ={8.54%, 33.67%, 61.40%, 82.25%} (Figure 4.12B). For the convenience of illustration, we define a symbol, $R_{stripe,lk}$ to be a spatial region of the location (l,k) in the *stripe*-th eve stripe. The transformation from (l,m) in the whole embryo to (l,k) in the *stripe*-th eve stripe, i.e., $R_{stripe,lk}$ is given by $m=k+3\times(stripe-1)$. For example, (l,k) =(3,3) in the second eve stripe, i.e., the spatial region R corresponds to (l,m) =(3,6) in the whole embryo, with l=3 and m=3+3×(2–1)=6 (Figure 4.12B). After the determination of the spatial regions, expression levels of protein and mRNA are interpolated to the determined spatial regions, which will be used for model identification of (4.8).

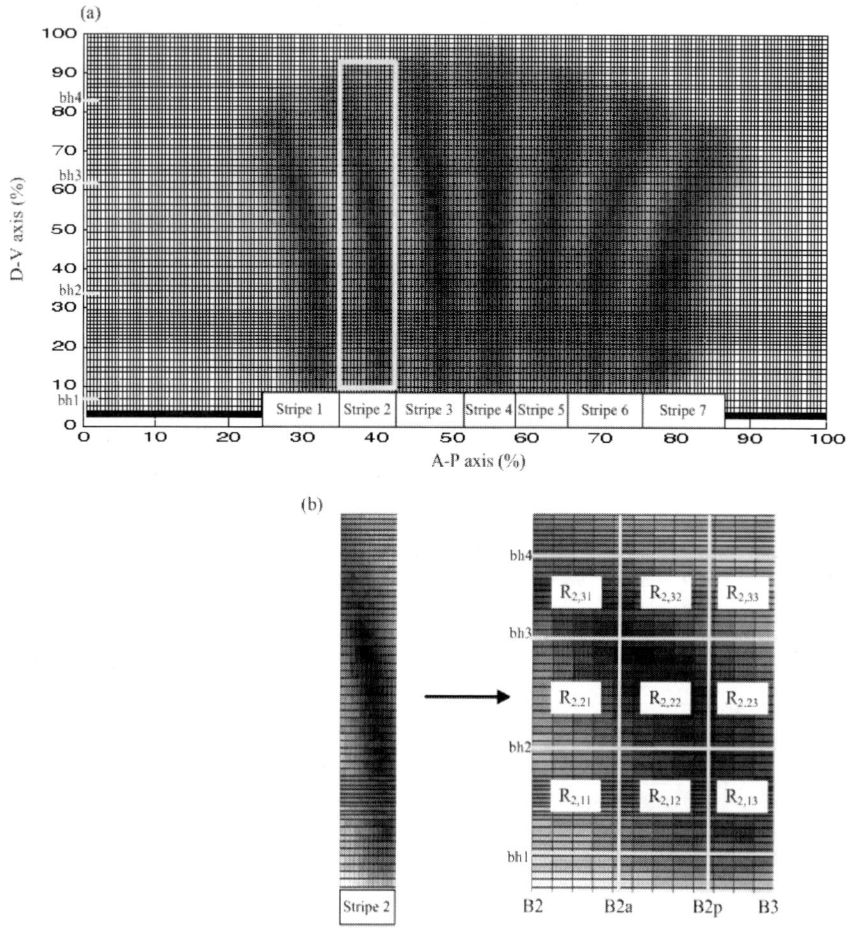

Figure 4.12. Determination of eve stripe boundaries at c14A8. a) Along the A-P axis and D-V axis, seven eve stripe boundaries $\{B1, B2, \ldots, B8\}$ and three spatial region boundaries $\{bh1, bh2, bh3, bh4\}$ are defined, respectively. b) The yellow square frame as shown in (a) is enlarged for the second eve stripe. Nine spatial regions with symbol $R_{stripe,lk}$ are defined in each stripe [85].

4.5.4. System Identification for Stochastic 3-DEST Gene/Protein Interaction Networks in Different Spatial Regions of *Drosophila* Embryo

When the data points $\{X_i(k,l,m), Y_j(k,l,m)\}$ for $i,j \in \{1,\ldots,14\}$, $k \in \{1,\ldots.N\}$, $l=\{1,2,3\}$, $m=\{1,\ldots,21\}$ are ready, the parameters of stochastic 3-DEST model can be estimated using (4.9) for gene/protein interaction networks in each spatial region of *Drosophila* embryo. For the convenience of parameter estimation, (4.9) with N data points can be translated into the following linear regression matrix form [85]:

$$Y_{l,m} = \Phi_{l,m}\Theta_{l,m} + E_{l,m}, l = \{1,2,3\}, m = \{1,\ldots,21\} \tag{4.10}$$

where

$$Y_{l,m}=\begin{bmatrix}\chi_{1,l,m}\\ \chi_{2,l,m}\\ \vdots\\ \chi_{14,l,m}\\ \psi_{1,l,m}\\ \psi_{2,l,m}\\ \vdots\\ \psi_{14,l,m}\end{bmatrix},\ \Phi_{l,m}=\begin{bmatrix}H_{l,m} & 0 & -Z_{l,m} & 0 & 0 & \Sigma\Upsilon_{l,m}\\ 0 & H_{l,m} & Z_{l,m} & \Upsilon_{l,m} & \Delta\Upsilon_{l,m} & 0\end{bmatrix},\ \Theta_{l,m}=\begin{bmatrix}D_{l,m}\\ \Omega_{l,m}\\ M_{l,m}\\ N_{l,m}\\ \Xi_{l,m}\\ O_{l,m}\end{bmatrix},$$

$$E_{l,m}=\begin{bmatrix}\xi_{1,l,m}\\ \xi_{2,l,m}\\ \vdots\\ \xi_{14,l,m}\\ \varsigma_{1,l,m}\\ \varsigma_{2,l,m}\\ \vdots\\ \varsigma_{14,l,m}\end{bmatrix},\ \chi_{i,l,m}=\begin{bmatrix}X_i(2,l,m)-X_i(1,l,m)\\ X_i(3,l,m)-X_i(2,l,m)\\ \vdots\\ X_i(N,l,m)-X_i(N-1,l,m)\end{bmatrix},\qquad \psi_{i,l,m}=\begin{bmatrix}Y_i(2,l,m)\\ Y_i(3,l,m)\\ \vdots\\ Y_i(N-1,l,m)\end{bmatrix},$$

$H_{l,m}=blockdiag(1_{N-1},1_{N-1},\ldots,1_{N-1})$,

$$Z_{l,m}=\begin{bmatrix}X_1(1,l,m) & 0 & \cdots & 0\\ X_1(2,l,m) & \vdots & \vdots & \vdots\\ \vdots & \vdots & \vdots & \vdots\\ X_1(N-1,l,m) & 0 & \vdots & \vdots\\ 0 & X_2(1,l,m) & \vdots & \vdots\\ \vdots & X_2(2,l,m) & \vdots & \vdots\\ \vdots & \vdots & \vdots & \vdots\\ \vdots & X_2(N-1,l,m) & 0 & \vdots\\ \vdots & 0 & \ddots & 0\\ \vdots & \vdots & 0 & X_{14}(1,l,m)\\ \vdots & \vdots & \vdots & X_{14}(2,l,m)\\ \vdots & \vdots & \vdots & \vdots\\ 0 & \cdots & 0 & X_{14}(N-1,l,m)\end{bmatrix},$$

$$\Upsilon_{l,m} = \begin{bmatrix} Y_1(1,l,m) & 0 & \cdots & 0 \\ Y_1(2,l,m) & \vdots & \vdots & \vdots \\ \vdots & \vdots & \vdots & \vdots \\ Y_1(N-1,l,m) & 0 & \vdots & \vdots \\ 0 & Y_2(1,l,m) & \vdots & \vdots \\ \vdots & Y_2(2,l,m) & \vdots & \vdots \\ \vdots & \vdots & \vdots & \vdots \\ \vdots & Y_2(N-1,l,m) & 0 & \vdots \\ \vdots & 0 & \ddots & 0 \\ \vdots & \vdots & 0 & Y_{14}(1,l,m) \\ \vdots & \vdots & \vdots & Y_{14}(2,l,m) \\ \vdots & \vdots & \vdots & \vdots \\ 0 & \cdots & 0 & Y_{14}(N-1,l,m) \end{bmatrix},$$

$$\Delta\Upsilon_{l,m} = \begin{bmatrix} \delta\Upsilon_{1,l,m} & 0 & \cdots & 0 \\ 0 & \delta\Upsilon_{2,l,m} & \ddots & \vdots \\ \vdots & \ddots & \ddots & 0 \\ 0 & \cdots & 0 & \delta\Upsilon_{14,l,m} \end{bmatrix}, \quad \Sigma\Upsilon_{i,m} = \begin{bmatrix} \sigma\Upsilon_{1,l,m} & 0 & \cdots & 0 \\ 0 & \sigma\Upsilon_{2,l,m} & \ddots & \vdots \\ \vdots & \ddots & \ddots & 0 \\ 0 & \cdots & 0 & \sigma\Upsilon_{14,l,m} \end{bmatrix},$$

$$\delta\Upsilon_{i,l,m} = \begin{bmatrix} \dfrac{Y_i(1,l-1,m)-2Y_i(1,l,m)+Y_i(1,l+1,m)}{h_x^2} + \dfrac{Y_i(1,l,m-1)-2Y_i(1,l,m)+Y_i(1,l,m+1)}{h_y^2} \\ \dfrac{Y_i(2,l-1,m)-2Y_i(2,l,m)+Y_i(2,l+1,m)}{h_x^2} + \dfrac{Y_i(2,l,m-1)-2Y_i(2,l,m)+Y_i(2,l,m+1)}{h_x^2} \\ \vdots \\ \dfrac{Y_i(N-1,l-1,m)-2Y_i(N-1,l,m)+Y_i(N-1,l+1,m)}{h_x^2} + \dfrac{Y_i(N-1,l,m-1)-2Y_i(N-1,l,m)+Y_i(N-1,l,m+1)}{h_x^2} \end{bmatrix},$$

$$\sigma\Upsilon_{i,l,m} = \begin{bmatrix} Y_1(1,l,m) & Y_2(1,l,m) & \cdots & Y_{14}(1,l,m) \\ Y_1(2,l,m) & Y_2(2,l,m) & \cdots & Y_{14}(2,l,m) \\ \vdots & \vdots & \cdots & \vdots \\ Y_1(N-1,l,m) & Y_2(N-1,l,m) & \cdots & Y_{14}(N-1,l,m) \end{bmatrix}, \quad D_{l,m} = \begin{bmatrix} d_{1,l,m} \\ d_{2,l,m} \\ \vdots \\ d_{14,l,m} \end{bmatrix},$$

$$\Omega_{l,m} = \begin{bmatrix} w_{1,l,m} \\ w_{2,l,m} \\ \vdots \\ w_{14,l,m} \end{bmatrix}, \quad M_{l,m} = \begin{bmatrix} a_{1,l,m} \\ a_{2,l,m} \\ \vdots \\ a_{14,l,m} \end{bmatrix}, \quad \mathrm{N}_{l,m} = \begin{bmatrix} c_{1,l,m} \\ c_{2,l,m} \\ \vdots \\ c_{14,l,m} \end{bmatrix}, \quad \Xi_{l,m} = \begin{bmatrix} c_{1,l,m} \\ c_{2,l,m} \\ \vdots \\ c_{14,l,m} \end{bmatrix}, \quad O_{l,m} = \begin{bmatrix} o_{1,l,m} \\ o_{2,l,m} \\ \vdots \\ o_{14,l,m} \end{bmatrix},$$

$$o_{i,l,m} = \begin{bmatrix} b_{i1,l,m} \\ b_{i2,l,m} \\ \vdots \\ b_{i14,l,m} \end{bmatrix}, \ \xi_{i,l,m} = \begin{bmatrix} \varepsilon_i(2,l,m) \\ \varepsilon_i(3,l,m) \\ \vdots \\ \varepsilon_i(N,l,m) \end{bmatrix}, \text{ and } \varsigma_{i,l,m} = \begin{bmatrix} \delta_i(2,l,m) \\ \delta_i(3,l,m) \\ \vdots \\ \delta_i(N-1,l,m) \end{bmatrix}.$$

Suppose the noise components $\varepsilon_i(k,l,m)$ and $\delta_j(k,l,m)$ are normally distributed, and the noise matrix $E_{l,m}$ has an unknown covariance matrix $\Sigma_{l,m}$ to be estimated. Then we use the ML method to solve the parameter estimation problem with the optimum solution $\hat{\Theta}_{l,m}$ and $\hat{\Sigma}_{l,m}$. The likelihood function of $Y_{l,m}$ is defined as follows [5]:

$$\begin{aligned} p_{l,m}\left(\mathrm{Y}_{l,m} \mid \phi_{l,m}, \Sigma_{l,m}\right) &= \frac{1}{\sqrt{2\pi}} \det\left(\Sigma_{l,m}\right)^{-M/2} \\ &\times \exp\left\{-\frac{1}{2}\left[\mathrm{Y}_{l,m} - \Phi_{l,m}\Theta_{l,m}\right]^T \Sigma_{l,m}^{-1} \left[\mathrm{Y}_{l,m} - \Phi_{l,m}\Theta_{l,m}\right]\right\} \end{aligned} \tag{4.11}$$

The log-likelihood function for the given M data points in $Y_{l,m}$ i.e., $M = 2\times14(N\text{-}1)$, can be defined as [5]

$$\begin{aligned} L_{l,m}(\Theta_{l,m}, \Sigma_{l,m}) &= \text{constant} - \frac{M}{2}\ln\left[\det \Sigma_{l,m}\right] \\ &- \frac{1}{2}\left[\mathrm{Y}_{l,m} - \Phi_{l,m}\Theta_{l,m}\right]^T \Sigma_{l,m}^{-1} \left[\mathrm{Y}_{l,m} - \Phi_{l,m}\Theta_{l,m}\right] \end{aligned} \tag{4.12}$$

We can estimate the unknown parameters $\Theta_{l,m}$ and the covariance matrices of noise $\Sigma_{l,m}$ by maximizing the log-likelihood function $L_{l,m}(\Theta_{l,m},\Sigma_{l,m})$, i.e., $\partial L_{l,m}\left(\Theta_{l,m}, \Sigma_{l,m}\right) / \partial \Theta_{l,m} = 0$ and $\partial L_{l,m}\left(\Theta_{l,m}, \Sigma_{l,m}\right) / \partial \Sigma_{l,m} = 0$, as follows:

$$\hat{\Sigma}_{l,m} = \frac{1}{M}\left[\mathrm{Y}_{l,m} - \Phi_{l,m}\Theta_{l,m}\right]^T \left[\mathrm{Y}_{l,m} - \Phi_{l,m}\Theta_{l,m}\right] \tag{4.13}$$

$$\hat{\Theta}_{l,m} = (\Phi_{l,m}^T \Phi_{l,m})^{-1} \Phi_{l,m}^T Y_{l,m} \tag{4.14}$$

In order to satisfy the following stability constraints $d_{i,l,m} \geq 0$, $w_{j,l,m} \geq 0$, $|1-\alpha_{j,l,m}| \leq 1$, $|c_{j,l,m} - 4\rho_{j,l,m}(h_x^{-2}+h_y^{-2})| \leq 1$, if $\rho_{j,l,m}<0$; $|c_{j,l,m}| \leq 1$, if $\rho_{j,l,m} \geq 0$ in the discrete 3-DEST model ((4.9)), a Matlab function, *lsqlin*, is used in the estimation procedure of the parameter identification of the stochastic 3-DEST model (see 4.7.3). For the stochastic 3-DEST model of gene/protein interaction network in each spatial region of embryo, the number of estimated regulatory parameters is 266. We have a total of 28 dynamic equations which will be solved simultaneously. To avoid over-fitting in parameter estimation and to find a more robust solution, we should interpolate these data points by the cubic spline method. Hence, we will test the robustness of system parameters on different numbers of interpolating data points from four times to six times the number of estimated parameters in the sequel.

According to the Akaike Information Criterion (AIC) method, we will let b=0 while the transcription regulation between the j-th transcription factor and the i-th target gene in the spatial region R is insignificant. We use the AIC to prune some insignificant regulatory parameters of TFs in (4.14). The AIC is defined to include both the residual variance in parameter estimation and the model complexity into one statistics for model order detection as [5]

$$AIC_{l,m}(p) = \log \frac{1}{M}(Y_{l,m} - \hat{Y}_{l,m})^T (Y_{l,m} - \hat{Y}_{l,m}) + \frac{2p}{M} \tag{4.15}$$

where p is the number of reserved parameters in the backward elimination method of the AIC. Regulatory parameters are pruned one by one as p is decreased until the smallest AIC in the smaller p is larger than the AIC value of the previous pruning step. While the minimum AIC is achieved, the most adequate transcription regulations for each target gene could be obtained from the most adequate model order point of view [90].

4.5.5. Data and Materials

In this chapter, we incorporate two spatial-temporal data, protein data (http://flyex.ams.sunysb.edu/flyex/) [52-55] mRNA data (http://flyex.ams.sunysb.edu/lab/ gaps.html) [43, 45] into the stochastic 3-DEST model to investigate how the transcription regulations and diffusion mechanisms cooperatively pattern eve stripes in the early embryogenesis of *Drosophila*. The spatial regions are first defined as shown in Figure 4.12 by Eve at cleavage cycle 14A and temporal class 8 (c14A8) in the embryo. Subsequently, the NN model combined with the method is trained by the available protein and mRNA data to simulate and mimic the unavailable mRNA data. The training of the NN model by the available data is achieved by minimizing the training error and maximizing the output correlations. Additionally, to avoid over-fitting in system identification, we must interpolate the data points to an adequate number. However, over-interpolated data will lose the low-frequency (or long-range) behavior of the development system. Moreover, using different numbers of interpolated data in system identification may also cause significant differences in parameter estimations, especially in the AIC method. Hence, the robustness of the stochastic 3-DEST model will also be tested by different numbers of interpolated data as an assessment to choose an adequate number of interpolation in the sequel, because the robustness principle has been employed to check if a model can work in the real cell and is employed to narrow down the range of models to the few in the modeling procedure of biological networks, i.e., robustness can help theorists identify the correct dynamic model [77]. From the ML parameter estimation method, the dynamic model in early *Drosophila* development is constructed. Then, we incorporate the AIC method into the identification process to prune the insignificant regulatory parameters and refine the model. This allows us to pick up the TFs, which are the most significant regulators for controlling the downstream genes in the early development of *Drosophila*.

The real biological systems are always robust. Therefore the model of a biological system should be robust and the robustness is a validation of dynamic models for biological systems [77]. To test the robust- ness of the 3-DEST model by different number of data points, we interpolate the time-course data from38 data points (i.e., four times the number of parameters) to 57 data points (i.e., six times the number of parameters), i.e., there are 20 test cases. However, among the 20 test cases only six test cases, which are respectively those with 38, 39, 40, 41, 42 and 44 interpolated data points, meet the model's stability constraints $d_{i,l,m}\geq 0$, $w_{j,l,m}\geq 0$, $|1-\alpha_{j,l,m}|\leq 1$, $|c_{j,l,m}-4\rho_{j,l,m}(h_x^{-2}+h_y^{-2})|\leq 1$, if $\rho_{j,l,m}<0$; $|c_{j,l,m}|\leq 1$, if $\rho_{j,l,m}\geq 0$ in (4.9) when the AIC values of the model are minimized. Therefore, only six kinds of data interpolations to meet robustness test can be used for the parameter identification of the 3-DEST model. Here, only the robustness test of the 3-DEST system in the spatial region $R_{2,22}$ is discussed for further choice of data interpolations. The robustness of the estimated basal levels and the regulatory abilities in the six test cases in the spatial region $R_{2,22}$ are shown in Table 4.1. As can be seen, there are a few changes such as basal levels ($\hat{K}_i(x_2, y_5)$) of *runt* at N =39 and *hairy* at N =41 (N represents number of interpolated data points), and there is no significant change in basal levels of protein ($\hat{\varpi}_j(x_2, y_5)$). In addition, the regulatory abilities ($\hat{\beta}_{eve,j}(x_2, y_5)$) of both *runt* and *slp* at N =38 and 39 are pruned by the AIC method when compared with the others. Therefore, except a few variations in N =38, 39 and 41 the other cases, i.e., N =40, 42 and 44, are robust for system identification. Here, mRNA and protein data with 44 interpolation time points are chosen for parameter estimation of the stochastic 3-DEST model of the whole embryo. When spatial-temporal data are ready and the number of interpolated data points is decided (N =44), system identification for parameter estimation in (4.13), (4.14), and (4.15) can be performed.

4.5.6. Results

After the parameters in (4.8) are estimated by ML and pruned by the AIC in (4.13), (4.14), and (4.15), the identified 3-DEST models for gene/protein interaction networks in the spatial regions of *Drosophila* embryo are given in the following.

$$
\begin{aligned}
\frac{\partial X_i(t,x_l,y_m)}{\partial t} &= \hat{K}_i(x_l,y_m) - \hat{\alpha}_i(x_l,y_m)X_i(t,x_l,y_m) + \sum_{j=1}^{14}\hat{\beta}_{ij} f\left(Y_j\left(t-\tau_j,x_l,y_m\right)\right) \\
&\quad +\phi_i(t,x_l,y_m) \\
\frac{\partial Y_j(t,x_l,y_m)}{\partial t} &= \hat{\varpi}_j(x_l,y_m) + \hat{\alpha}_j(x_l,y_m)X_j(t,x_l,y_m) - \hat{\lambda}_j(x_l,y_m)Y_j(t,x_l,y_m) \\
&\quad +\hat{\gamma}_j(x_l,y_m)\nabla^2 Y_j(t,x_l,y_m) + \varphi_j(t,x_l,y_m),
\end{aligned} \tag{4.16}
$$

where i,j =1, 2, ..., 14, l =1, 2, 3, m =1, 2, ..., 21. $\hat{K}_i$, $\hat{\varpi}_j$, $\hat{\alpha}_i$, $\hat{\beta}_{ij}$, $\hat{\lambda}_{ij}$, and $\hat{\gamma}_j$ are estimated by (4.14) and the covariance matrices of the stochastic noises ϕ_i and φ_j can be estimated in (4.13).

After system identification, the simulation results of the system model obtained using the ML estimation method and the AIC method are shown in Figure 4.13(b) (protein) and Figure 4.13(d) (mRNA) compared with the original data in Figure 4.13(a) (protein) and Figure 4.13(c) (mRNA), respectively. The 3-DEST gene/protein interaction networks in different spatial regions are constructed in Figure 4.14 through the diffusion coefficients $\hat{\gamma}_j$ and regulatory abilities $\hat{\beta}_{ij}$ of the identified 3-DEST dynamic model in (4.16). The changes in these diffusion coefficients and regulatory abilities in eve stripes will be simultaneously investigated to see whether there are some cooperative effects on them, which may give a clue of eve stripe formation.

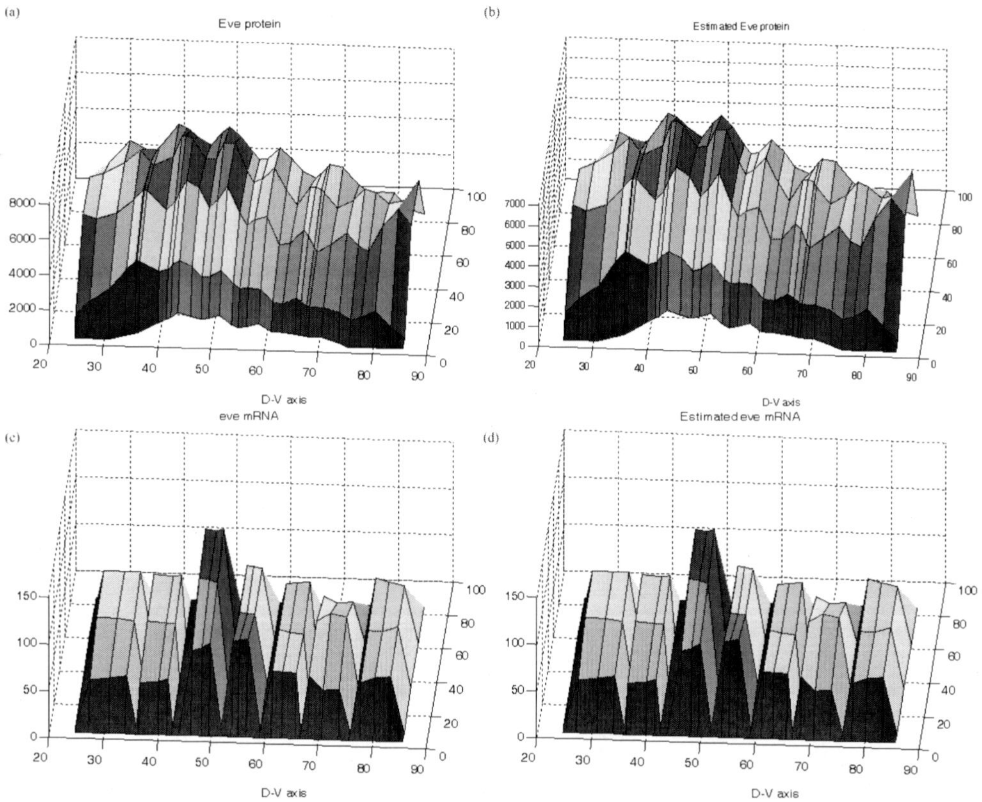

Figure 4.13. Original data and estimated results by our proposed dynamic model. The original eve mRNA and protein spatial data at c14A8 are shown in a) and c), respectively. After system identification, the estimated eve mRNA and protein spatial data generated by the dynamic model are shown in b) and d), respectively [85].

Table 4.1. Robustness tests of parameters [85], $\hat{K}_i(x_2, y_5)$, $\hat{\varpi}_j(x_2, y_5)$, and $\hat{\beta}_{eve,j}(x_2, y_5)$ (shown in (4.16)) of $R_{2,22}$ (i.e., l=2 and m=k+3×(*stripe*–1)=2+3×(2–1)=5) in the six test cases, i.e., the six test cases individually have 38, 39, 40, 41, 42 and 44 interpolated data points denoted by N=38, 39, 40, 41, 42, 44, respectively

parameters	$\hat{K}_i(x_2, y_5)$						$\hat{\varpi}_j(x_2, y_5)$						$\hat{\beta}_{eve,j}(x_2, y_5)$					
N	**38**	**39**	**40**	**41**	**42**	**44**	**38**	**39**	**40**	**41**	**42**	**44**	**38**	**39**	**40**	**41**	**42**	**44**
bicoid	8.18	8.26	8.34	8.41	8.47	8.60	29.54	29.15	28.75	28.39	28.05	27.37	584.11	580.97	714.09	794.65	721.02	719.16
caudal	30.73	30.69	30.59	85.11	30.39	30.22	58.60	58.12	57.78	22.43	57.25	56.67	0	0	0	0	0	0
eve	0	0	0	0	0	0	0	0	0	0	0	0	–1040.69	–1063.51	–1972.63	–1881.88	–1763.69	–1590.02
ftz	0	0	0	0	0	0	37.59	37.69	37.79	37.88	37.96	38.08	1340.85	1343.18	3338.48	3172.99	3112.51	2913.20
giant	0.72	0.72	0.72	1.65	2.07	0.72	19.82	19.64	19.43	13.85	15.33	16.72	–101.86	–106.81	–445.25	–414.64	–439.09	–429.87
hairy	18.58	17.65	60.80	0	65.51	66.21	58.03	58.66	80.99	80.91	80.84	80.69	0	0	–1178.98	–1039.98	–1101.73	–1033.91
hunchback	10.32	11.45	12.41	13.13	14.03	15.65	105.69	104.42	103.70	103.26	102.40	100.84	227.98	245.25	194.56	73.79	86.04	2.15
knirps	3.40	3.53	3.38	3.56	3.43	3.45	5.03	4.56	5.51	4.02	4.83	4.25	163.00	164.28	638.90	546.26	566.78	505.28
krüppel	0.86	0.86	0.86	0.86	0.86	0.86	23.05	22.54	22.05	21.60	21.16	20.33	–1010.23	–1003.42	–2153.69	–1946.24	–1980.59	–1827.30
odd	0	0	0	0	0	0	0	0	0	0	0	0	0	0	0	0	0	0
paired	0	0	0	0	0	0	0	0	0	0	0	0	0	0	557.86	497.67	555.37	549.33
runt	0	17.94	0	0	0	0	0	0	0	0	0	0	0	0	589.60	467.39	522.54	465.37
slp	0	0	0	0	0	0	0	0	0	0	0	0	0	0	0	0	0	0
tailless	0	0	0	0	0	0	20.74	20.50	20.29	20.09	19.89	19.53	0	0	0	0	0	0

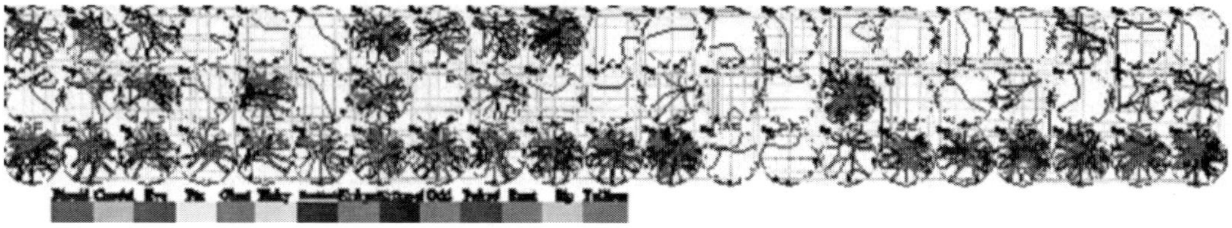

Figure 4.14. 3-DEST dynamic gene/protein interaction network for diffusion and transcriptional regulation mechanisms in different spatial regions in the whole embryo. The notations, $R_{1,11}$, …, $R_{7,33}$, are the 63 spatial regions of the whole embryo which is specified by Figure 4.12(a). in each spatial region $R_{stripe,ij}$, the colors of the outer ring in the color circle are specified by the 14 gene names, which are given by the color bar below the figure, respectively. Each color of the outer ring is specified by each gene. The solid lines that connect color circles stand for transcription regulation between genes in each spatial region based on regulatory abilities $\hat{\beta}_{ij}$ of the identified 3-DEST dynamic model in (4.16). Positive and negative regulations are denoted by arrows and bars at the end of solid lines, respectively. Additionally, the colors of the inner circle, i.e., the black and white circle, inside the color circle stand for the TFs' roles, i.e., donor or acceptor of the transcriptional regulation network, respectively. The bold color lines that connect the same genes in neighboring spatial regions with different roles stand for protein diffusions from donor (black inner circle) to acceptor (white inner circle) in neighboring spatial regions based on the diffusion coefficients $\hat{\gamma}_j$ of the identified 3-DEST dynamic model in (4.16). The specification of the colors in bold color lines is consistent with the colors in the outer ring of the color circle, which are specified by the color bar. For example (see also Figure 4.15a), Caudal in $R_{4,11}$ with green color in outer ring and black color in inner circle found regulates ftz (yellow) and runt (navy blue) and plays as a donor, which can diffuse to the neighboring regions. A clearer figure is available online at the website, http://www.ee.nthu.edu.tw/bschen/Drosophila_Fig4.pdf.

Previous research [84] on cis-regulatory module detection shows that the enhancer element of the second eve stripe contains the binding sequences of Krüppel, Giant, Bicoid and Hunchback, and the second eve stripe can be activated by Bicoid, Hunchback, and repressed by Giant and Krüppel (Table 4.1 and Figure 4.14).

In the analysis of eve stripe formation, the boundaries of eve stripe can be affected by diffusion from the neighboring regions where Eve serves as a donor to the egions where it plays the role of an acceptor. Figure 4.14 shows that Hunchback in $R_{2,22}$, $R_{3,11}$, $R_{3,33}$, $R_{4,31}$, $R_{6,11}$, $R_{6,13}$, and $R_{7,31}$ and Knirps in $R_{1,12}$, $R_{4,13}$, $R_{5,23}$, and $R_{7,13}$ positively and negatively regulate *eve*, respectively, and Eve in these regions simultaneously serves as a donor which diffuses through and affects the boundaries, i.e., stripe 1–2, stripe 3–4, stripe 4–5, stripe 5–6 and stripe 7-terminal Figure 4.14). Therefore, it shows that stripe boundaries are broken in the embryo with *hunchback* and *knirps* double mutant and the phenotype is similar to the embryo with a strong *eve* mutant [49]. In addition, *eve* in $R_{2,11}$ and $R_{2,23}$, which plays the role of donor and is repressed by Giant and Krüppel respectively, wouldlocally affect the anterior and posterior of the second eve stripe, respectively (Figure 4.14) [46, 81, 91]. Moreover, the effect of Giant and Krüppel respectively on the anterior and posterior of the second eve stripe should be diffusively reinforced by the same repressive transcription regulations in $R_{2,22}$. In the boundaries of the third eve stripe, *eve* in $R_{3,31}$ and $R_{4,31}$, which is negatively regulated by Hunchback and Knirps respectively, would diffuse to and effect on anterior and posterior boundaries of the third eve stripe, respectively (Figure 4.14) [46, 81, 91]. Moreover, we find that Giant and Hairy have no effect on the boundaries stripe 4–5 and stripe 5–6, respectively.

From the transcription regulations shown in Figure 4.14, we believe that most of them are new predicted except those discussed here because most of genetic studies in *Drosophila* are not easy to find direct transcription regulations without chromatin immune-precipitation microarray (ChIP-chip) experiments. In this chapter, we provide a direction for other biologists at the wet experiments of transcription regulations especially in ChIP-chip experiments. For example, according to the robustness tests in Table 4.1, we show that eve in $R_{2,22}$ is positively regulated by Ftz and Knirps and is negatively self-regulated. The robust regulations are the most probable suggestion in transcription regulations of eve stripe formation.

Moreover, in the large network, there exist a huge number of interaction patterns. Only a few types of interaction patterns called network motifs, which are embedded in the network and connected to each other, allow them to carry out their functions even in the presence of additional interactions. Mangan and Alon [92] have analyzed two feed-forward network motifs, i.e., coherent feed-forward loops (C-FFL) and incoherent feed-forward loops (I-FFL), and found that C-FFL acted as sign-sensitive delays, and I-FFL acted as sign-sensitive accelerators [92]. Moreover, Han et al. [93] propose that a signaling module composed of a C-FFL and an I-FFL causes an early transient response and a delayed prolonged response after a short stimulus [93]. The early transient responses and delayed prolonged responses plausibly depend on post-translation modification of existing proteins and new protein synthesis, respectively. The combinative signaling module is suggested and found in drug therapy. Therefore, we obtain C-FFL and I-FFL from the constructed network (Figure 4.15) according to the following rules. One relationship of the transcription regulations in Figure 4.15 serves as an edge of FFL, when the regulation relationship exists in at least four neighboring regions among its nine neighboring regions. For example (Figure 4.15a), a C-FFL C15 found in Figure 4.15 is composed of three transcription regulations (Caudal→Ftz in $R_{3,13}$, Runt→Ftz in $R_{3,13}$ and Caudal→Runt in $R_{4,11}$) and two diffusions (Caudal and Runt are both diffused from $R_{4,11}$ to $R_{3,13}$). In addition, these three regulatory relationships exist respectively in at least four neighboring regions, i.e., Caudal→Ftz in $R_{3,13}$, $R_{3,12}$, $R_{4,11}$ and $R_{4,31}$, Runt→Ftz in $R_{3,13}$, $R_{3,12}$, $R_{3,33}$, and $R_{4,31}$ and Caudal→Runt in $R_{4,11}$, $R_{3,13}$, $R_{4,12}$, and $R_{4,31}$. Therefore, C15 is one of the FFLs (Figure 4.15b) found in our network. By the same procedure, not only can we find 25 C-FFLs and 18 I-FFLs (Figure 4.15b) but also 13 possible combinative signaling modules among 25 C-FFLs and 18 I-FFLs, i.e., Odd in $R_{1,12}$, $R_{1,11}$ (C3 and I1 in Figure 4.15b), $R_{1,13}$ and $R_{1,23}$ (C1 and I1), and Slp in $R_{2,12}$, $R_{2,22}$ (C7 and I5), $R_{3,31}$ (I12 and C18), and $R_{3,12}$ (C19 and I4), Eve in $R_{3,12}$, $R_{3,13}$ (C14 and I9) and $R_{4,11}$ (C13 and I9), and Ftz in $R_{3,13}$ (C21 and I15) and $R_{6,13}$ (C25 and I17). Among these modules, we find that Hunchback acts as a source of FFLs to activate Ftz as an output expressed in eve stripes 3, 4, 6 and 7. From the embryo with *hb*$^-$mutants, eve stripes 2, 3, 4 and 7 are partially or completely deleted [49]. Although Ftz in $R_{6,12}$ and $R_{6,13}$ is activated respectively by I-FFL and combinative signaling module with Hunchback as an input source, Ftz in $R_{6,12}$ and $R_{6,13}$ is respectively negatively regulated and does not regulate *eve*. Therefore, we suggest that C-FFL, I-FFL and combinative signaling module are respectively important in activating speedy responses in $R_{4,11}$, $R_{4,12}$, and $R_{7,11}$, activating a delayed response in $R_{7,13}$ with the ability of noise filtering and activating a delayed prolonged response in $R_{3,13}$.

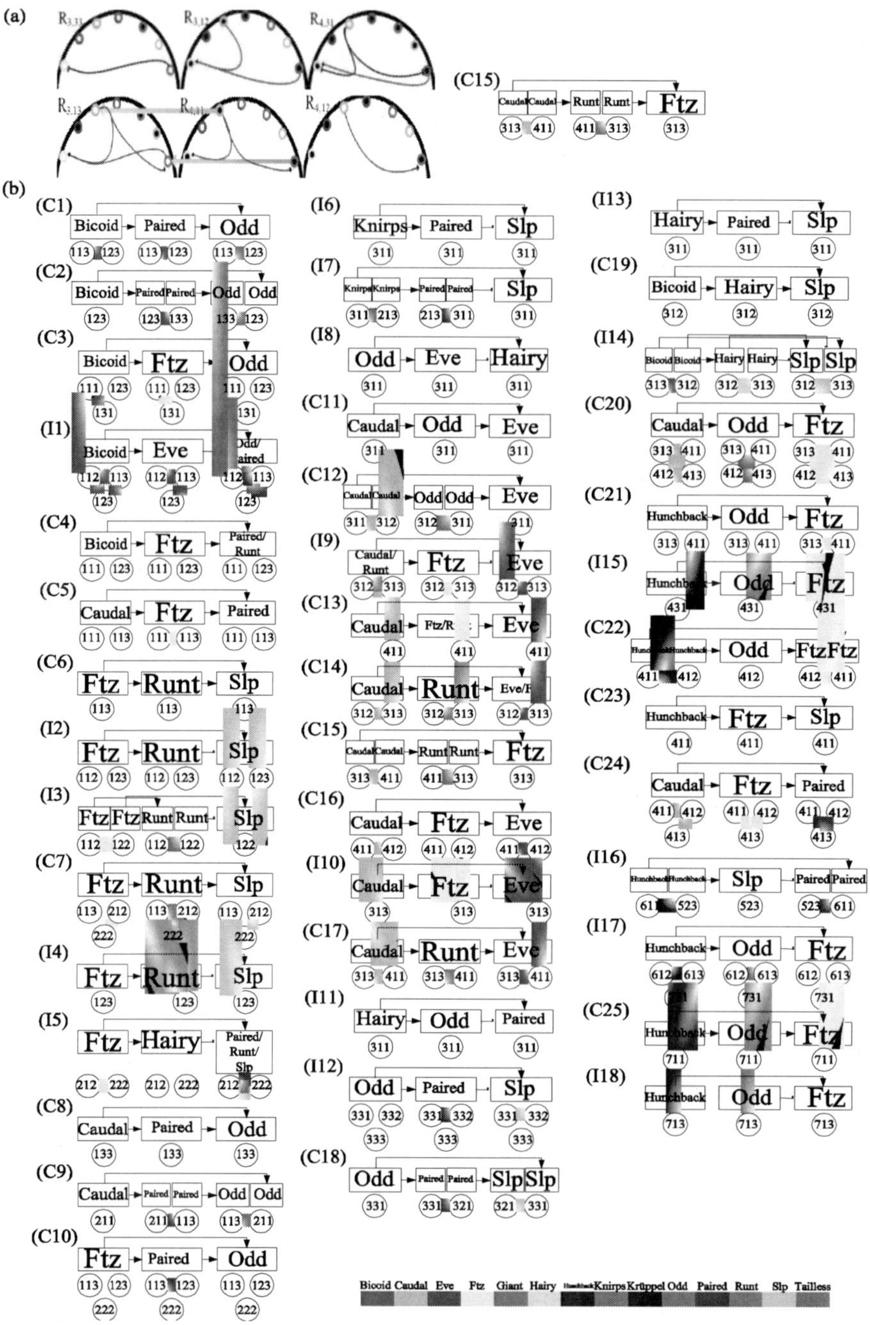

Figure 4.15. Coherence and incoherence feed-forward loops of 3-DEST dynamic gene/protein interaction network. (a) According to the rule that each of the regulation relationships of FFLs must exist in at least four neighboring spatial regions, parts of gene/protein interaction network (left) in $R_{3,2}$, $R_{3,3}$, $R_{4,1}$, and $R_{4,2}$ are examples of feed-forward loops, and can be redrawn as C15 (right). (b) From the above rule, we find the network motifs, i.e., 25 C-FFLs (C1~C25) and 18 I-FFLs (I1~I18), for the cooperation of transcription regulations with diffusions in early embryogenesis. The color bars denote diffusions, which are the same as those in Figure 4.14. A clearer figure is available online at the website, http://www.ee.nthu.edu.tw/bschen/Drosophila_Fig5.pdf.

4.5.7. Discussion

In this section, we are the first to combine mRNA dynamic equation with protein dynamic equation using spatial-temporal model to construct the gene/protein interaction network to investigate the gene/protein regulatory mechanisms of eve stripe formation in the early development of *Drosophila*. However, there are still three mechanisms of concern in *Drosophila* embryogenesis, i.e., protein-protein interactions, translation regulations and epigenetic regulations. In a recent study, protein-protein direct interactions are not found between the 14 early development-related TFs of *Drosophila* embryo [94], although there may exist some interactions which require a co-factor(s). For example, Bicoid has self-inhibitory property which requires a co-factor(s), and the binding site at the N-terminal region of Bicoid is evolutionarily conserved [95]. However, the understanding of protein–protein interactions via a co-factor(s) is limited. Moreover, cooperative bindings through sigmoid function have been implicitly concerned in previous models [76]. However, since the prior information of cooperative bindings in early embryogenesis is also limited, cooperative binding is not considered in our model. If the information of cooperative binding is most available, cooperative bindings can be considered easily as regulation candidates in the 3-DEST model, i.e., the cooperation regulation $\Sigma_{j,k}\beta_{ijk}f(Y_j(t,x,y))f(Y_k(t,x,y))$ could be considered in (4.8) In addition, there are two translation regulations of concern in early embryogenesis [96]. The first is Bicoid which binds to maternal caudal to repress its translation [97, 98], and the second is Nano which binds to the nanos response element (e.g. Pumilio) located within the 3' untranslated region of maternal hunchback and then results in maternal hunchback, which cannot be translated [99-101]. Since the understanding of translation regulations is limited so far, translation regulations are not yet included in the stochastic 3-DEST dynamic model yet. Finally, epigenetic regulations, such as DNA methylation, histone modification and RNAi, are able to play important roles in the regulation of gene expression, but they always interact to accomplish their responsibilities. Combinations of several epigenetic regulations conduct complex silencing such as chromosome inactivation and gene imprinting. For example, during *Drosophila* embryogenesis the proteins of the trithorax (trxG) and Polycomb groups (PcG) modify chromatin via interacting with chromosomal elements, Cellular Memory Modules (CMMs). A nearby gene can be continuously transcribed through mitotic cell division and meiosis by a switched activated state of CMMs during *Drosophila* embryogenesis. Thus, CMMs could affect the patterning of cells by the transcriptional control of genes involved in embryonic patterning. In conclusion, trxG and PcG confer epigenetic regulations for different binding affinities of transcription regulation, i.e., trans-effect, that result in embryonic patterning throughout *Drosophila* embryogenesis [102-104]. In the 3-DEST model, the space-variant parameters of regulatory abilities $\beta_{ij}(x,y)$ and basal levels of protein generation $\varpi_j(x,y)$ have implied the affection of epigenetic regulations on transcription regulations throughout eve stripe patterning of *Drosophila* embryogenesis. An example is shown in Table 4.1. As seen in N =40, 41, 42 and 44, epigenetic regulation of Hairy, which has been speculated by in the terminal system of the larvae [105], is probably identified that Hairy is encoded to transcriptionally regulate eve in $R_{2,22}$ in eve stripe formation.

In early embryogenesis, diffusion mechanism is needed not only for maternal genes but also for gap genes and pair-rule genes to regulate their target genes in the neighboring spatial regions, which can determine the roles of TFs in each region, i.e., donor/ acceptor. Without the dynamic space-time model, the dynamics of TFs' diffusions may not be easily observed from a system point of view, especially in 2-D space. The contributions of this chapter include the following. (1) Construction of a stochastic 3-DEST dynamic model for gene/protein interaction network which not only contains the concentration-dependent transcriptional abilities but also includes six stochastic processes to mimic the spatial-temporal dynamic interplay among the target genes and their regulatory TFs at the early embryonic stage (i.e., the following six processes (i) protein synthesis, (ii) protein decay, (iii) mRNA decay, (iv) protein diffusion, (v) transcription regulations, and (vi) auto-regulation are involved in our dynamic model). (2) Utilization of the AIC to refine the stochastic 3-DEST dynamic model for gene/protein interaction network via pruning the insignificant transcription regulations in each spatial region. (3) Findings of transcription regulations in the seven eve stripes in the stochastic 3-DEST gene/protein interaction network. (4) Validating of the identified gene/protein interaction network by literature reference with the wet experiments of gene mutations. (5) Inference of transcription regulations and diffusion mechanisms for playing a cooperative role in the creation of FFLs to build eve stripes by speedy responses, delayed responses with the ability of noise filtering and delayed and prolonged responses. For the possible experimental validation of the feed-forward loops (FFLs) in 3-DEST dynamic gene/protein interaction network, biologists can follow the similar experimental design in [106] and [48, 49, 79, 81, 82]. For example, if the two FFLs, X → Y → Z (X → Z) and X → Y —/ Z (X → Z) are considered, biologists can examine gene **Z**'s expression in the corresponding region found in Figure 4.15 of cellular blastoderm wild-type and $\mathbf{Y}^{-}$ embryos by filtered fluorescence imaging after immune-peroxidase staining with polyclonal antibodies specific for **Z**. By comparing gene **Z**'s expression in wild-type with $\mathbf{Y}^{-}$ embryos, the suggested FFLs in Figure 4.15 can be validated. In the future, the proposed spatial-temporal dynamic model and construction algorithm can be extended to gene/protein network construction of different biological phenotypes, which depend on compartments, especially in early embryonic development, e.g. post-natal stem/progenitor cell regulation and differentiation, differentiation of Hematopoietic stem cells (HSCs), the segmentally modulated Hox expression patterns and patterning of the wing in *Drosophila* development.

However, one of the weaknesses in system identification is the increase in computation burden due to the use of the AIC method. Because one of our main purposes is to extract the significant transcription/translation regulations via pruning the insignificant transcription/translation regulations by using the AIC method, we use an explicit scheme with some stability constraints on the parameters to construct and then refine the gene/protein interaction network. Additionally, computation complexity will be increased, when the spatial regions are precisely specified. Moreover, a plenty of spatial-temporal data are needed in parameter estimation of the 3-DEST model. Although we know that eve stripes of the *Drosophila* embryo is probably not just built by the 14 early development-related genes, it is not a problem to estimate a more complicated dynamic regulatory network by the proposed method if much more mRNA and protein data are available in the future.

4.6. Conclusion

By constructing GR and PPI networks, integrated GR and PPI networks, interspecies host-pathogen PPI networks, and gene/protein interaction networks in spatial-temporal *Drosophila* development, cellular functions and molecular mechanisms of any specific biological condition can be investigated using systems biology methods such as dynamic modeling, microarray data, and large database mining.

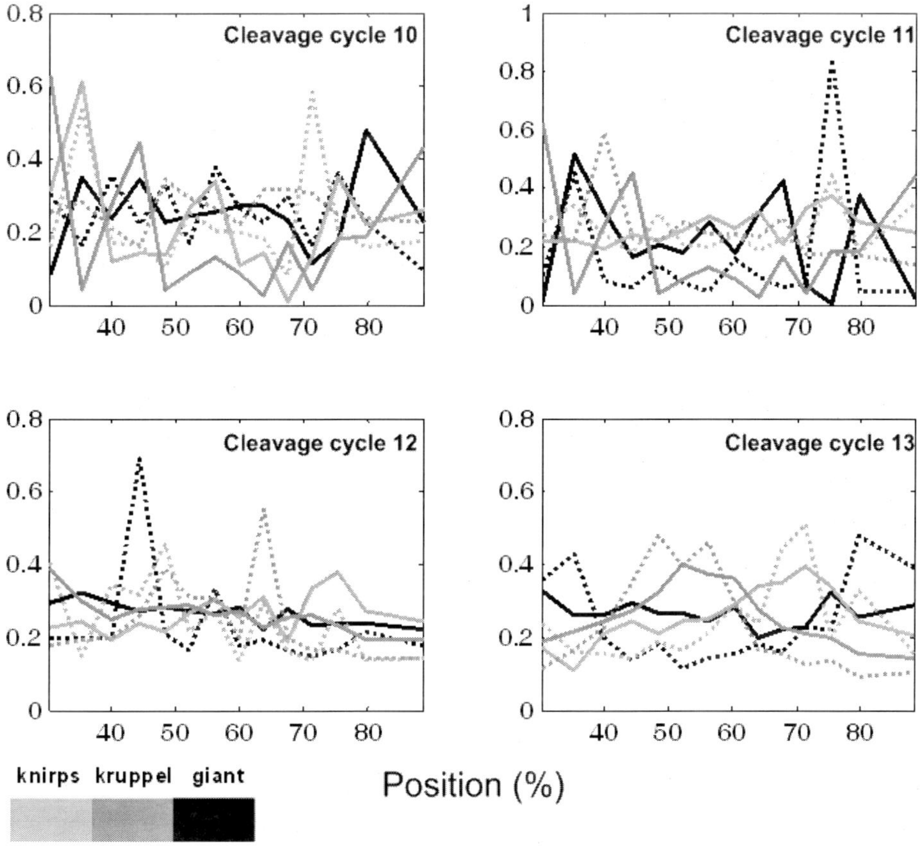

Figure 4.16. Normalized mRnA and protein expressions. Solid line and dashed line denote protein and mRnA expressions, respectively. The expressions of *knirps* (cyan line), *krüppel* (green line) and *giant* (black line) are plotted in time profiles.

4.7. Appendix

4.7.1. Reconstruction of Unavailable Data by Neural-Network Learning

Because there are some deficiencies in the mRNA data, recovery of these missing data is needed when the data are used for system identification. For the deficiencies of mRNA data, a back-propagation NN training method is employed to reconstruct them. Three classes of genes, i.e., maternal genes, gap genes, and pair-rule genes, into which the 14 early development-related genes are divided, are utilized for the reconstruction of these unmeasured data in their classes of genes, individually. For each data reconstruction process, we individually train and reconstruct these data in each class of genes along the A-P axis, since the protein and mRNA data of each class of genes along the A-P axis are roughly similar (Figure 4.16) [42]. Note that the downstream class of genes with missing mRNA data is reconstructed by the upstream class of genes via the back-propagation NN training method. The training methods, i.e., Broyden, Fletcher, Goldfard and Shanno (BFGS), Levenberg-Marquardt, Powell-Beale Restarte, Polak-Ribiere, Fletcher-Reeves, and Rprop, have been employed to test the performance of data reconstruction. In this chapter, NN combined with the BFGS method is used for training and simulating the unmeasured mRNA data [107], because the NN plus BFGS method has the best performance in our tests (data not shown). In order to obtain an optimal NN training results, we maximize the output correlations and minimize the training errors in the training processes. A few of the unmeasured protein data points are also reconstructed by the same learning and simulating processes. For example, if the mRNA data of gene **A** is unknown, a back-propagation NN is trained by the protein data of the upstream class of gene **A** as input and the protein data of gene **A** as output. Then the mRNA data of gene **A** can be simulated by the mRNA data of the upstream class of gene **A** through the well-trained back-propagation NN. After these missing data are simulated, the parameter estimation for the system identification of stochastic 3-DEST gene/protein interaction network model in (4.8) or (4.9) is introduced in the following section.

4.7.2. Stability of Discrete 3-DEST Model

Lemma 1 [108]

Let $X(l,k,m)=g_1^k e^{il\Phi} e^{im\Phi}$ and $Y(k,l,m)=g_2^k e^{il\Theta} e^{im\Theta}$ where $X(k,l,m)\in\Omega_x:=\{X_i(k,l,m)\}$ and $Y(k,l,m)\in\Omega_y:=\{Y_j(k,l,m)\}$ for $i,j=1,\ldots,14$ in (4.9). An one-step finite difference scheme (with constant coefficient) is stable if and only if $|g_1(\Phi)|\leq 1$, $|g_2(\Theta)|\leq 1$ and $g_1(\Phi)$ and $g_2(\Theta)$ both are independent of l and m.

Substituting $X(l,k,m)=g_1^k e^{il\Phi} e^{im\Phi}$ and $Y(k,l,m)=g_2^k e^{il\Theta} e^{im\Theta}$ in (4.9), we obtain $g_1^{k+1} e^{il\Phi} e^{im\Phi}=(1-a_{l,m})g_1^k e^{il\Phi} e^{im\Phi}$ and

$$g_2^{k+1} e^{il\Theta} e^{im\Theta} = c_{l,m} g_2^k e^{il\Theta} e^{im\Theta} + \rho_{l,m}\left[\frac{g_2^k e^{i(l-1)\Theta} e^{im\Theta} - 2g_2^k e^{il\Theta} e^{im\Theta} + g_2^k e^{i(l+1)\Theta} e^{im\Theta}}{h_x^2} + \frac{g_2^k e^{il\Theta} e^{i(m-1)\Theta} - 2g_2^k e^{il\Theta} e^{im\Theta} + g_2^k e^{il\Theta} e^{i(m+1)\Theta}}{h_y^2}\right]$$

We can obtain $g_1=1-a_{l,m}$ and $g_2=c_{l,m}-4\rho_{l,m}(h_x^{-2}+h_y^{-2})\sin^2(\Theta/2)$.

Thus, according to Lemma 1, the scheme is stable if and only if $|1-a_{l,m}|\leq 1$ and

$$\begin{cases} \left|c_{l,m} - 4\rho_{l,m}\left(h_x^{-2} + h_y^{-2}\right)\right| \leq 1 & \text{if } \rho_{l,m} < 0 \\ \left|c_{l,m}\right| \leq 1 & \text{if } \rho_{l,m} \geq 0 \end{cases}.$$

4.7.3. Procedure of Stability Constrained Estimation

The stability constrained estimation of parameters can be performed by a Matlab function, *lsqlin*.

A procedure for the parameter estimation is given as follows:

1. First, assume $\rho_{j,l,m}\geq 0$ with the constraint $|c_{j,l,m}|<1$ for j=1,...,14 in (4.9) and estimate the parameters, { $\hat{d}_{i,l,m}$, $\hat{w}_{j,l,m}$, $\hat{a}_{i,l,m}$, $\hat{c}_{i,l,m}$, $\hat{\rho}_{j,l,m}$, $\hat{b}_{ij,l,m}$ }, under the constraints, $d_{i,l,m}\geq 0$, $w_{j,l,m}\geq 0$, $|c_{j,l,m}|\leq 1$ and $|1-a_{j,l,m}|\leq 1$.
2. Change the assumption of $\rho_{j,l,m}$ according to the estimated $\hat{\rho}_{j,l,m}$ of the previous step, and estimate the parameters under the constraints, $d_{i,l,m}\geq 0$, $w_{j,l,m}\geq 0$, $|c_{j,l,m}|\leq 1$ and $|1-a_{j,l,m}|\leq 1$ if the diffusion coefficient estimated by the previous step for some j (i.e., some TFs) is non-negative, $\hat{\rho}_{j,l,m}\geq 0$. But otherwise the constraints of the other TFs with negative diffusion coefficients, $\hat{\rho}_{j,l,m}<0$, estimated by the previous step are changed to be $d_{i,l,m}\geq 0$, $w_{j,l,m}\geq 0$, $|c_{j,l,m}-4\rho_{j,l,m}(h_x^{-2}+h_y^{-2})|\leq 1$ and $|1-a_{j,l,m}|\leq 1$.
3. Stop and obtain the estimated parameters if the sign of the estimated $\hat{\rho}_{j,l,m}$ for each j is not changed anymore in the last two estimation steps, otherwise go back to step 2.

REFERENCES

[1] Harbison, CT; Gordon, DB; Lee, TI; Rinaldi, NJ; Macisaac, KD; Danford, TW; et al., "Transcriptional regulatory code of a eukaryotic genome," *Nature*, vol. 431, 99-104, Sep 2 2004.

[2] Hughes, TR; Marton, MJ; Jones, AR; Roberts, CJ; Stoughton, R; Armour, CD; et al., "Functional discovery via a compendium of expression profiles," *Cell*, vol. 102, 109-126, Jul 7 2000.

[3] Lin, LH; Lee, HC; Li, WH; Chen, BS. "Dynamic modeling of cis-regulatory circuits and gene expression prediction via cross-gene identification," *BMC Bioinformatics*, vol. 6, Oct 18 2005.

[4] Chang, YH; Wang, YC; Chen, BS. "Identification of transcription factor cooperativity via stochastic system model," *Bioinformatics*, vol. 22, 2276-2282, Sep 15 2006.

[5] Johansson, R. *System modeling and identification*. Englewood Cliffs, NJ: Prentice Hall, 1993.

[6] Chen, HC; Lee, HC; Lin, TY; Li, WH; Chen, BS. "Quantitative characterization of the transcriptional regulatory network in the yeast cell cycle," *Bioinformatics*, vol. 20, 1914-27, Aug 12 2004.

[7] Wu, WS; Li, WH; Chen, BS. "Computational reconstruction of transcriptional regulatory modules of the yeast cell cycle," *BMC Bioinformatics*, vol. 7, Sep 29 2006.

[8] Wu, WS; Li, WH; Chen, BS. "Identifying regulatory targets of cell cycle transcription factors using gene expression and ChIP-chip data," *BMC Bioinformatics*, vol. 8, Jun 8 2007.

[9] Li, CW; Chu, YH; Chen, BS. "Construction and clarification of dynamic gene regulatory network of cancer cell cycle via microarray data," *Cancer Inform*, vol. 2, 223-41, 2006.

[10] Li, CW; Chen, BS. "Identifying Functional Mechanisms of Gene and Protein Regulatory Networks in Response to a Broader Range of Environmental Stresses," *Comparative and Functional Genomics*, 2010.

[11] Chen, BS; Yang, SK; Lan, CY; Chuang, YJ. "A systems biology approach to construct the gene regulatory network of systemic inflammation via microarray and databases mining," *BMC Medical Genomics*, vol. 1, Sep 30 2008.

[12] Stark, C; Breitkreutz, BJ; Reguly, T; Boucher, L; Breitkreutz, A; Tyers, M. "BioGRID: a general repository for interaction datasets," *Nucleic Acids Research*, vol. 34, D535-9, Jan 1 2006.

[13] Ashburner, M; Ball, CA; Blake, JA; Botstein, D; Butler, H; Cherry, JM; et al., "Gene ontology: tool for the unification of biology. The Gene Ontology Consortium," *Nat Genet*, vol. 25, 25-9, May 2000.

[14] Chen, BS; Wang, YC; Wu, WS; Li, WH. "A new measure of the robustness of biochemical networks," *Bioinformatics*, vol. 21, 2698-2705, Jun 1 2005.

[15] Chang, WC; Li, CW; Chen, BS. "Quantitative inference of dynamic regulatory pathways via microarray data," *BMC Bioinformatics*, vol. 6, Mar 7 2005.

[16] Voit, EO. *Computational analysis of biochemical systems : a practical guide for biochemists and molecular biologists*. New York: Cambridge University Press, 2000.

[17] Chu, LH; Chen, BS. "Construction of a cancer-perturbed protein-protein interaction network for discovery of apoptosis drug targets," *BMC Syst Biol*, vol. 2, p. 56, 2008.

[18] Kuo, ZY; Chuang, YJ; Chao, CC; Liu, FC; Lan, CY; Chen, BS. "Identification of Infection- and Defense-Related Genes via a Dynamic Host-Pathogen Interaction Network Using a Candida Albicans -Zebrafish Infection Model," *J Innate Immun*, vol. 5, 137-52, 2013.

[19] Wang, YC; Lin, C; Chuang, MT; Hsieh, WP; Lan, CY; Chuang, YJ. et al., "Interspecies protein-protein interaction network construction for characterization of host-pathogen interactions: a Candida albicans-zebrafish interaction study," *BMC Syst Biol*, vol. 7, p. 79, 2013.

[20] Wang, YC; Chen, BS. "A network-based biomarker approach for molecular investigation and diagnosis of lung cancer," *BMC Medical Genomics*, vol. 4, Jan 6 2011.

[21] Wang, YC; Chen, BS. "Integrated cellular network of transcription regulations and protein-protein interactions," *BMC Systems Biology*, vol. 4, Mar 8 2010.

[22] Dyer, SA; Dyer, JS. "Cubic-spline interpolation: Part 1," *IEEE Instrumentation & Measurement Magazine*, vol. 4, 44-46, Mar 2001.

[23] Akaike, H. "New Look at Statistical-Model Identification," *IEEE Transactions on Automatic Control*, vol. Ac19, 716-723, 1974.

[24] Esser, K. *The mycota : a comprehensive treatise on fungi as experimental systems for basic and applied research*, 2nd ed. Berlin ; New York: Springer, 2008.

[25] Uhl, MA; Biery, M; Craig, N; Johnson, AD. "Haploinsufficiency-based large-scale forward genetic analysis of filamentous growth in the diploid human fungal pathogen C.albicans," *Embo Journal*, vol. 22, 2668-2678, Jun 2 2003.

[26] Yaar, L; Mevarech, M; Koltin, Y. "A Candida albicans RAS-related gene (CaRSR1) is involved in budding, cell morphogenesis and hypha development," *Microbiology-Uk*, vol. 143, 3033-3044, Sep 1997.
[27] Woo, M; Lee, K; Song, K. "MYO2 is not essential for viability, but is required for polarized growth and dimorphic switches in Candida albicans," *Fems Microbiology Letters*, vol. 218, 195-202, Jan 21 2003.
[28] Dunkler A; Wendland, J. "Candida albicans RHO-type GTPase-encoding genes required for polarized cell growth and cell separation," *Eukaryotic Cell*, vol. 6, 844-854, May 2007.
[29] Spooner R; Yilmaz, O. "The Role of Reactive-Oxygen-Species in Microbial Persistence and Inflammation," *International Journal of Molecular Sciences*, vol. 12, 334-352, Jan 2011.
[30] Schoonbroodt S; Piette, J. "Oxidative stress interference with the nuclear factor-kappa B activation pathways," *Biochemical Pharmacology*, vol. 60, 1075-1083, Oct 15 2000.
[31] Allen, LAH; Beecher, BR; Lynch, JT; Rohner, OV; Wittine, LM. "Helicobacter pylori disrupts NADPH oxidase targeting in human neutrophils to induce extracellular superoxide release," *Journal of Immunology*, vol. 174, 3658-3667, Mar 15 2005.
[32] Harada, T; Miyake, M; Imai, Y. "Evasion of Legionella pneumophila from the bactericidal system by reactive oxygen species (ROS) in macrophages," *Microbiology and Immunology*, vol. 51, 1161-1170, 2007.
[33] Tsunawaki, S; Yoshida, LS; Nishida, S; Kobayashi, T; Shimoyama, T. "Fungal metabolite gliotoxin inhibits assembly of the human respiratory burst NADPH oxidase," *Infection and Immunity*, vol. 72, 3373-3382, Jun 2004.
[34] Wellington, M; Dolan, K; Krysan, DJ. "Live Candida albicans Suppresses Production of Reactive Oxygen Species in Phagocytes," *Infection and Immunity*, vol. 77, 405-413, Jan 2009.
[35] Nakagawa, Y; Kanbe, T; Mizuguchi, I. "Disruption of the human pathogenic yeast Candida albicans catalase gene decreases survival in mouse-model infection and elevates susceptibility to higher temperature and to detergents," *Microbiology and Immunology*, vol. 47, 395-403, 2003.
[36] Frohner, IE; Bourgeois, C; Yatsyk, K; Majer, O; Kuchler, K. "Candida albicans cell surface superoxide dismutases degrade host-derived reactive oxygen species to escape innate immune surveillance," *Molecular Microbiology*, vol. 71, 240-252, Jan 2009.
[37] Sim, S; Yong, TS; Park, SJ; Im, KI; Kong, Y; Ryu, JS; et al. "NADPH oxidase-derived reactive oxygen species-mediated activation of ERK1/2 is required for apoptosis of human neutrophils induced by Entamoeba histolytica," *Journal of Immunology*, vol. 174, 4279-4288, Apr 1 2005.
[38] Yang, TC; Lai, CC; Shiu, SL; Chuang, PH; Tzou, BC; Lin, YY; et al. "Japanese encephalitis virus down-regulates thioredoxin and induces ROS-mediated ASK1-ERK/p38 MAPK activation in human promonocyte cells," *Microbes and Infection*, vol. 12, 643-651, Aug 2010.
[39] Williams, B; Kabbage, M; Kim, HJ; Britt, R; Dickman, MB. "Tipping the Balance: Sclerotinia sclerotiorum Secreted Oxalic Acid Suppresses Host Defenses by Manipulating the Host Redox Environment," *Plos Pathogens*, vol. 7, Jun 2011.
[40] Driever, W; Nussleinvolhard, C. "A Gradient of Bicoid Protein in Drosophila Embryos," *Cell*, vol. 54, 83-93, Jul 1 1988.

[41] Eldar, A; Dorfman, R; Weiss, D; Ashe, H; Shilo, BZ; Barkai, N. "Robustness of the BMP morphogen gradient in Drosophila embryonic patterning," *Nature*, vol. 419, 304-308, Sep 19 2002.

[42] Jaeger, J; Blagov, M; Kosman, D; Kozlov, KN; Manu, Myasnikova, E; et al., "Dynamical analysis of regulatory interactions in the gap gene system of Drosophila melanogaster," *Genetics*, vol. 167, 1721-1737, Aug 2004.

[43] Jaeger, J; Sharp, DH; Reinitz, J. "Known maternal gradients are not sufficient for the establishment of gap domains in Drosophila melanogaster," *Mechanisms of Development*, vol. 124, 108-128, Feb 2007.

[44] Gilbert, SF. *Developmental biology*, 8th ed. Sunderland, Mass.: Sinauer Associates, Inc. Publishers, 2006.

[45] Jaeger, J; Surkova, S; Blagov, M; Janssens, H; Kosman, D; Kozlov, KN; et al. "Dynamic control of positional information in the early Drosophila embryo," *Nature*, vol. 430, 368-371, Jul 15 2004.

[46] Wu, XL; Vakani, R; Small, S. "Two distinct mechanisms for differential positioning of gene expression borders involving the Drosophila gap protein giant," *Development*, vol. 125, 3765-3774, Oct 1998.

[47] Gaul U; Jackle, H. "Role of gap genes in early Drosophila development," *Adv Genet*, vol. 27, 239-75, 1990.

[48] Kraut, R; Levine, M. "Spatial Regulation of the Gap Gene Giant during Drosophila Development," *Development*, vol. 111, 601-609, Feb 1991.

[49] Frasch, M; Levine, M. "Complementary Patterns of Even-Skipped and Fushi-Tarazu Expression Involve Their Differential Regulation by a Common Set of Segmentation Genes in Drosophila," *Genes & Development*, vol. 1, 981-995, Nov 1987.

[50] Ingham, PW; Baker, NE; Martinezarias, A. "Regulation of Segment Polarity Genes in the Drosophila Blastoderm by Fushi-Tarazu and Even-Skipped," *Nature*, vol. 331, 73-75, Jan 7 1988.

[51] Jiang, J; Levine, M. "Binding Affinities and Cooperative Interactions with Bhlh Activators Delimit Threshold Responses to the Dorsal Gradient Morphogen," *Cell*, vol. 72, 741-752, Mar 12 1993.

[52] Kozlov, K; Myasnikova, E; Samsonova, M; Reinitz, J; Kosman, D. "Fast redundant dyadic wavelet transform in application to spatial registration of the expression patterns of Drosophila segmentation genes," *15th International Conference on Pattern Recognition, Vol 3, Proceedings*, 459-462, 2000.

[53] Myasnikova, E; Samsonova, A; Kozlov, K; Samsonova, M; Reinitz, J. "Registration of the expression patterns of Drosophila segmentation genes by two independent methods," *Bioinformatics*, vol. 17, 3-12, Jan 2001.

[54] Myasnikova, EM; Kosman, D; Reinitz, J; Samsonova, MG. "Spatio-temporal registration of the expression patterns of Drosophila segmentation genes," *Proc Int Conf Intell Syst Mol Biol*, 195-201, 1999.

[55] Surkova, S; Kosman, D; Kozlov, K; Manu, Myasnikova, E; Samsonova, AA; et al., "Characterization of the Drosophila segment determination morphome," *Developmental Biology*, vol. 313, 844-862, Jan 15 2008.

[56] Arbeitman, MN; Furlong, EEM; Imam, F; Johnson, E; Null, BH. Baker, BS; et al., "Gene expression during the life cycle of Drosophila melanogaster," *Science*, vol. 297, 2270-2275, Sep 27 2002.

[57] Hart, CE; Mjolsness, E; Wold, BJ. "Connectivity in the yeast cell cycle transcription network: inferences from neural networks," *Plos Computational Biology*, vol. 2, 1592-1607, Dec 2006.

[58] Cheron, G; Draye, JP; Bourgeios, M; Libert, G. "A dynamic neural network identification of electromyography and arm trajectory relationship during complex movements," *IEEE Transactions on Biomedical Engineering*, vol. 43, 552-558, May 1996.

[59] Koike, Y; Kawato, M. "Estimation of Dynamic Joint Torques and Trajectory Formation from Surface Electromyography Signals Using a Neural-Network Model," *Biological Cybernetics*, vol. 73, 291-300, Sep 1995.

[60] Linko, P; Zhu, YH. "Neural Network Programming in Bioprocess Variable Estimation and State Prediction," *Journal of Biotechnology*, vol. 21, 253-269, Dec 1991.

[61] Liu, MM; Herzog, W; Savelberg, HHCM. "Dynamic muscle force predictions from EMG: an artificial neural network approach," *Journal of Electromyography and Kinesiology*, vol. 9, 391-400, Dec 1999.

[62] Ressom, H; Reynolds, R; Varghese, RS. "Increasing the efficiency of fuzzy logic-based gene expression data analysis," *Physiological Genomics*, vol. 13, 107-117, Apr 16 2003.

[63] Woolf, PJ; Wang, YX. "A fuzzy logic approach to analyzing gene expression data," *Physiological Genomics*, vol. 3, 9-15, Jun 29 2000.

[64] Chiang, JH; Chao, SY. "Modeling human cancer-related regulatory modules by GA-RNN hybrid algorithms," *BMC Bioinformatics*, vol. 8, Mar 14 2007.

[65] Maraziotis, I; Dragomir, A; Bezerianos, A. "Recurrent neuro-fuzzy network models for reverse engineering gene regulatory interactions," *Computational Life Sciences, Proceedings*, vol. 3695, 24-34, 2005.

[66] Vohradsky, J. "Neural model of the genetic network," *Journal of Biological Chemistry*, vol. 276, 36168-36173, Sep 28 2001.

[67] Armananzas, R; Inza, I; Larranaga, P. "Detecting reliable gene interactions by a hierarchy of Bayesian network classifiers," *Computer Methods and Programs in Biomedicine*, vol. 91, 110-121, Aug 2008.

[68] Djebbari A; Quackenbush, J. "Seeded Bayesian Networks: Constructing genetic networks from microarray data," *BMC Systems Biology*, vol. 2, Jul 4 2008.

[69] Sahoo, D; Dill, DL; Gentles, AJ; Tibshirani, R; Plevritis, SK. "Boolean implication networks derived from large scale, whole genome microarray datasets," *Genome Biology*, vol. 9, 2008.

[70] Shmulevich, I; Dougherty, ER; Kim, S; Zhang, W. "Probabilistic Boolean networks: a rule-based uncertainty model for gene regulatory networks," *Bioinformatics*, vol. 18, 261-274, Feb 2002.

[71] Brown, PO; Botstein, D. "Exploring the new world of the genome with DNA microarrays," *Nature Genetics*, vol. 21, 33-37, Jan 1999.

[72] Mestl, T; Plahte, E; Omholt, SW. "A Mathematical Framework for Describing and Analyzing Gene Regulatory Networks," *Journal of Theoretical Biology*, vol. 176, 291-300, Sep 21 1995.

[73] Gursky, VV; Reinitz, J; Samsonov, AM. "How gap genes make their domains: An analytical study based on data driven approximations," *Chaos*, vol. 11, 132-141, Mar 2001.

[74] Perkins, TJ; Jaeger, J; Reinitz, J; Glass, L. "Reverse engineering the gap gene network of Drosophila melanogaster," *Plos Computational Biology*, vol. 2, 417-428, May 2006.

[75] Reinitz, J; Kosman, D; Vanario-Alonso, CE; Sharp, DH. "Stripe forming architecture of the gap gene system," *Developmental Genetics*, vol. 23, 11-27, 1998.

[76] Reinitz, J; Sharp, DH. "Mechanism of Eve Stripe Formation," *Mechanisms of Development*, vol. 49, 133-158, Jan 1995.

[77] Alon, U. *An introduction to systems biology : design principles of biological circuits.* Boca Raton, FL: Chapman & Hall/CRC, 2007.

[78] Cadigan, KM; Grossniklaus, U; Gehring, WJ. "Localized Expression of Sloppy Paired Protein Maintains the Polarity of Drosophila Parasegments," *Genes & Development*, vol. 8, 899-913, Apr 15 1994.

[79] Carroll SB; Vavra, SH. "The Zygotic Control of Drosophila Pair-Rule Gene-Expression .2. Spatial Repression by Gap and Pair-Rule Gene-Products," *Development*, vol. 107, 673-683, Nov 1989.

[80] Schulz, C; Tautz, D. "Zygotic Caudal Regulation by Hunchback and Its Role in Abdominal Segment Formation of the Drosophila Embryo," *Development*, vol. 121, 1023-1028, Apr 1995.

[81] Small, S; Blair, A; Levine, M. "Regulation of two pair-rule stripes by a single enhancer in the Drosophila embryo," *Developmental Biology*, vol. 175, 314-324, May 1 1996.

[82] Yu, Y; Pick, L. "Non-periodic cues generate seven ftz stripes in the Drosophila embryo," *Mech Dev*, vol. 50, 163-75, Apr 1995.

[83] Schroeder, MD; Pearce, M; Fak, J; Fan, HQ; Unnerstall, U; Emberly, E; et al., "Transcriptional control in the segmentation gene network of Drosophila," *Plos Biology*, vol. 2, 1396-1410, Sep 2004.

[84] Rajewsky, N; Vergassola, M; Gaul, U; Siggia,v "Computational detection of genomic cis-regulatory modules applied to body patterning in the early Drosophila embryo," *BMC Bioinformatics*, vol. 3, 2002.

[85] Li CW; Chen, BS. "Stochastic spatio-temporal dynamic model for gene/protein interaction network in early Drosophila development," *Gene Regul Syst Bio*, vol. 3, 191-210, 2009.

[86] Brandman; O; Meyer, T. "Feedback loops shape cellular signals in space and time," *Science*, vol. 322, 390-395, Oct 17 2008.

[87] Klipp, E. *Systems biology in practice : concepts, implementation and application.* Weinheim: Wiley-VCH, 2005.

[88] Keener JP; Sneyd, J. *Mathematical physiology*. New York: Springer, 1998.

[89] Mitchell, AR. Griffiths, DF. *The finite difference method in partial differential equations*. Chichester Eng. ; New York: Wiley, 1980.

[90] Miller, AJ. *Subset selection in regression*, 2nd ed. Boca Raton: Chapman & Hall/CRC, 2002.

[91] Small, S; Blair, A; Levine, M. "Regulation of Even-Skipped Stripe-2 in the Drosophila Embryo," *Embo Journal*, vol. 11, 4047-4057, Nov 1992.

[92] Mangan, S; Alon, U. "Structure and function of the feed-forward loop network motif," *Proceedings of the National Academy of Sciences of the United States of America*, vol. 100, 11980-11985, Oct 14 2003.

[93] Han, ZG; Vondriska, TM; Yang, L; MacLellan, WR; Weiss, JN; Qu, ZL."Signal transduction network motifs and biological memory," *Journal of Theoretical Biology*, vol. 246, 755-761, Jun 21 2007.

[94] Lin, CY; Chen, SH; Cho, CS; Chen, CL; Lin, FK; Lin, CH; et al. "Fly-DPI: database of protein interactomes for D-melanogaster in the approach of systems biology," *BMC Bioinformatics*, vol. 7, 2006.

[95] Zhao, C; York, A; Yang, F; Forsthoefel, DJ; Dave, V; Fu, DC; et al., "The activity of the Drosophila morphogenetic protein Bicoid is inhibited by a domain located outside its homeodomain," *Development*, vol. 129, 1669-1680, Apr 2002.

[96] Niessing, D; RiveraPomar, R; LaRosee, A; Hader, T; Schock, F; Purnell, BA; et al. "A cascade of transcriptional control leading to axis determination in Drosophila," *Journal of Cellular Physiology*, vol. 173, 162-167, Nov 1997.

[97] Dubnau, J; Struhl, G. "RNA recognition and translational regulation by a homeodomain protein," *Nature*, vol. 379, 694-699, Feb 22 1996.

[98] RiveraPomar, R; Niessing, D; SchmidtOtt, U; Gehring, WJ; Jackle, H. "RNA binding and translational suppression by bicoid," *Nature*, vol. 379, 746-749, Feb 22 1996.

[99] Murata, Y; Wharton, RP. "Binding of Pumilio to Maternal Hunchback Messenger-Rna Is Required for Posterior Patterning in Drosophila Embryos," *Cell*, vol. 80, 747-756, Mar 10 1995.

[100] Stjohnston D; Nussleinvolhard, C. "The Origin of Pattern and Polarity in the Drosophila Embryo," *Cell*, vol. 68, 201-219, Jan 24 1992.

[101] Tautz, D. "Regulation of the Drosophila Segmentation Gene Hunchback by 2 Maternal Morphogenetic Centers," *Nature*, vol. 332, 281-284, Mar 17 1988.

[102] Dejardin J; Cavalli, G. "Chromatin inheritance upon Zeste-mediated Brahma recruitment at a minimal cellular memory module," *Embo Journal*, vol. 23, 857-868, Feb 25 2004.

[103] Maurange C; Paro, R. "A cellular memory module conveys epigenetic inheritance of hedgehog expression during Drosophila wing imaginal disc development," *Genes & Development*, vol. 16, 2672-2683, Oct 15 2002.

[104] Rank, G; Prestel, M; Paro, R. "Transcription through intergenic chromosomal memory elements of the Drosophila bithorax complex correlates with an epigenetic switch," *Molecular and Cellular Biology*, vol. 22, 8026-8034, Nov 2002.

[105] Kim, J; Kerr, JQ. Min, GS. "Molecular heterochrony in the early development of Drosophila," *Proceedings of the National Academy of Sciences of the United States of America*, vol. 97, 212-216, Jan 4 2000.

[106] Mangan, S; Itzkovitz, S; Zaslaver, A; Alon, U. "The incoherent feed-forward loop accelerates the response-time of the gal system of Escherichia coli," *Journal of Molecular Biology*, vol. 356, 1073-1081, Mar 10 2006.

[107] Gilbert, JC, Lemarechal, C. "Some Numerical Experiments with Variable Storage Quasi-Newton Algorithms," *Mathematical Programming*, vol. 45, 407-435, Dec 1989.

[108] Strikwerda, JC. *Finite difference schemes and partial differential equations*. Pacific Grove, Calif.: Wadsworth & Brooks/Cole Advanced Books & Software, 1989.

Chapter 5

MEASUREMENT OF SYSTEM CHARACTERS OF BIOLOGICAL NETWORKS VIA MICROARRAY DATA

ABSTRACT

Following the construction of GRNs, PPI networks or their integrated cellular networks as described in the previous chapter, some characteristics of these biological networks need to be estimated to gain further insight into their systematic mechanisms. Robustness, the network's stability in maintaining systematic performance under intrinsic perturbations, and response ability, the network's sensitivity of response to external stimuli or of transduction to downstream regulators, are two important complementary or antagonistic system characteristics to consider when discussing biological network performance. These features can however not be measured directly for all networks in the experimental procedure. Even biological processes such as development, differentiation, tumorigenesis, and aging are increasingly being described in terms of temporal ordering of highly orchestrated transcriptional programs [1, 2]. The term "robustness" is present in a diversity of scientific fields, from engineering and control theory [3-5] to dynamic systems [6] and biology [7]. It should be noted that robustness describes a relative property, not an absolute one, as no system can maintain stability in all aspects when it is perturbed. In other words, robustness is not immutable. Since network sensitivity (response ability) is antagonistic to network robustness, a robust biological network is less sensitive, and vice versa. Measuring network sensitivity can therefore indicate a network's robustness. Finally, a network-based biomarker measurement for cancer is also introduced by comparing PPI networks of lung cancer cells and normal cells using a systems biology method through corresponding sample microarray data.

INTRODUCTION

In this chapter, we will investigate system characteristics of biological networks such as network robustness, network response ability (sensitivity), and network-based biomarkers, using microarray data and systems biology methods. First, we construct the dynamic model of a GRN associated with aging-related pathophysiological phenotypes. Based on the GRNs identified at young and aged stages, we then measure network robustness by the GRN's ability to tolerate network perturbation. It is found that the GRN at the aged stage is more robust than the GRN at young stage, while response ability (sensitivity) shows a reverse trend. Thus, aging seems to increase robustness while decreasing response ability.

Finally, a network-based biomarker identification method is introduced by comparing the PPI networks of normal and cancer cells, which are identified by their corresponding sample microarray data. Such network biomarkers open up possibilities for identifying multiple drug targets for cancer therapy.

5.1. MEASUREMENTS OF NETWORK ROBUSTNESS AND SENSITIVITY ESTIMATION VIA MICROARRAY DATA USING A SYSTEMS BIOLOGY APPROACH

Let the GRN of interest being consisted of N genes. After parameter identification via microarray data, the GRN in (4.1) can be represented by the following linear discrete-time dynamic system [1, 2]

$$x(t+1) = \hat{A}x(t) + \hat{K} + n(t)$$
$$\hat{A} = \begin{bmatrix} \hat{a}_{11} & \cdots & \hat{a}_{1N} \\ \vdots & \ddots & \\ \hat{a}_{N1} & & \hat{a}_{NN} \end{bmatrix}, \quad \hat{K} = \begin{bmatrix} \hat{k}_1 \\ \vdots \\ \hat{k}_N \end{bmatrix} \tag{5.1}$$

where the state vector $x(t) = [x_1(t) \cdots x_N(t)]^T$ stands for the discrete-time mRNA expression levels of total N genes at times $t = 1, 2, ..., K$. The system matrix $A^{\wedge}$denotes gene interactions in the gene network estimated by microarray data via the least squares or maximum likelihood parameter estimation method in Chapter 3, that is, $\hat{a_{ij}}$ denotes the estimated interaction of gene j with gene i. If $i \neq j$, then $\hat{k_i}$ denotes the estimated basal level of the i-th gene. n(t) denotes the model residual and measurement noise. The steady state x_s of $x(t)$ is obtained as $t \rightarrow \infty$:

$$x_s = \hat{A}x_s + \hat{K} \quad \text{or} \quad x_s = \left(1 - \hat{A}\right)^{-1} \hat{K} \tag{5.2}$$

To simplify analysis of "robustness" of the steady state (phenotype), the origin of the dynamic system is shifted to the steady state x_s, that is, $x(t) = \tilde{x}(t) + x_s$. This shift allows the following shifted dynamic system to be achieved by subtracting equation (5.1) from equation (5.2) [2]:

$$\tilde{x}(t+1) = \hat{A}\tilde{x}(t) \tag{5.3}$$

Therefore, the robustness of the phenotype (i.e., the steady state) of the gene network becomes the robustness of the shifted gene network in equation (5.3) at the origin $\tilde{x}(t) \equiv 0$. In the following, its network robustness (tolerance ability to intrinsic network perturbation) is tested. Suppose the linear GRN suffers from intrinsic molecular perturbations mainly due to process noise, thermal fluctuation, or genetic mutations. The interactive matrix $\hat{A}$ is then perturbed as $\hat{A}(1 + \eta)$, where η denotes the ratio of intrinsic perturbation. That is, the corresponding additional system perturbation is $\Delta\hat{A}=\eta\hat{A}$ [6]. A GRN with intrinsic perturbation can then be represented by

$$\tilde{x}(t+1) = \hat{A}(1+\eta)\tilde{x}(t) = \left(\hat{A}+\Delta\hat{A}\right)\tilde{x}(t) \tag{5.4}$$

Because quadratic stability with the Lyapunov energy-like function $V(\tilde{x}) = \tilde{x}^T(t)P\tilde{x}(t)>0$, with $V(0) = 0$ for a positive symmetric matrix, $P = P^T > 0$. The perturbative GRN in equation (5.4) is robustly stable if $\Delta V(\tilde{x}) = V(\tilde{x}(t+1)) - V(\tilde{x}(t)) \leq 0$, i.e., the energy of the gene network is not increased by intrinsic perturbations [6].

Proposition 5.1 [2]

Based on this idea of robust stability, if the following inequality has a positive definite solution $P = P^T > 0$ [2],

$$\left[\hat{A}(1+\eta)\right]^T P\left[\hat{A}(1+\eta)\right] \leq P \tag{5.5}$$

then the perturbative gene network is robustly stable or the phenotype of the gene network is maintained under parametric perturbation ratio η.

proof: see 5.6.1.

The network robustness η^0 of the perturbative gene network is equal to the tolerance of the largest perturbation that does not violate network stability or still maintains the phenotype of the perturbative gene network:

$$\begin{aligned}&\eta^0 = \max \eta \\ &\text{subject to } P > 0,\ (5.5)\end{aligned} \tag{5.6}$$

That is, the network robustness is the maximum perturbation ratio η^0 tolerated by the GRN such that network stability (or phenotype) is still maintained. The constrained optimization problem in (5.6) can be solved by increasing η until no positive solution P exists in equation (5.5) up to the highest value possible without violating the robust stability in equation (5.5). A positive definite solution $P > 0$ in equation (5.5) can be easily obtained by using the linear matrix inequality (LMI) Toolbox of Matlab. This network robustness method has been used to measure the relative network robustness of multiple loops of a gene regulatory network associated with aging-related pathophysiological phenotypes by using previously reported microarray data. It profiles the effects of aging on gene expression in the thymus, spinal cord, and eye tissues in mice [2]. This aging-related GRN includes 16 genes (Figure 5.1). The relative robustness η^0 toward normalized perturbation for GRNs of young, aged, and calorie-restriction (CR) groups is shown in Table 5.1, where young mice are 3, 6,

25, 28 weeks old animals; aged mice are 68, 71, 102, 105 weeks old animal; calorie-restriction (CR) are 68, 71, 96 weeks old animals, fed with 60% CR diet beginning at 24 weeks of age. Obviously, GRNs for the aged group are more robust than for young and CR groups in different tissues.

After the network robustness of aging-related GRNs is measured, the response abilities of genes to external stimuli is examined. Assuming that the GRN responds to external stimuli $U(t)$, including upstream regulatory signals and external signals outside the network (e.g. hormones, carcinogens, oxidative stress, or ambient biomedical molecules) that could produce an effect on the gene network, the dynamic equation (5.3) could be modified as follows:

$$\begin{aligned}\tilde{x}(t+1) &= \hat{A}\tilde{x}(t) + U(t) \\ Y(t) &= C\tilde{x}(t)\end{aligned} \tag{5.7}$$

where $U(t) = [u_1(t) \cdots u_N(t)]^T$ represents external stimuli and $Y(t)$ denotes the output signal response of specific genes of interest. For example, if the output signal response of the gene i to external stimuli $U(t)$ is to be analyzed, then C = diag(0, 0, ..., 0, 1, 0, 0, ..., 0), i.e., all elements of C are zero except for a single element at the i-th diagonal component. For example, C = diag(0, 0, 1, 0, ..., 0) describes the gene response of the *P53* gene. If the network response ability of the entire GRN to external stimuli is analyzed, then $C = I$ (identity matrix).

The effect of input signals $U(t)$ on output signal $Y(t)$ is less than or equal to a positive value δ, if the following inequality holds [6]

$$\frac{\sum_{t=0}^{\infty} Y^T(t)Y(t)}{\sum_{t=0}^{\infty} U^T(t)U(t)} \le \delta^2 \tag{5.8}$$

δ denotes the upper bound of the effect of $U(t)$ on $Y(t)$ for all bounded input signals $U(t)$. δ^0, the smallest upper bound of δ in (5.8), is called the "gene response ability (or sensitivity)" of the GRN. δ^0 permits us to obtain a more systematic insight into the ability of gene response to external stimuli for individual genes or the entire GRN.

Proposition 5.2 [2]

From the system gain result in [6], the network response of the dynamic GRN has an upper bound δ in (5.8), if there exists a positive definite $P = P^T > 0$ solution to the following LMI [2]:

$$\begin{bmatrix} \hat{A}^T P\hat{A} + C^T C - P & \hat{A}^T P \\ P\hat{A}^T & P - \delta^2 I \end{bmatrix} \le 0 \tag{5.9}$$

That is, if the above LMI holds for some $P > 0$, then the effect of $U(t)$ on $Y(t)$ must be less than or equal to δ, and equation (5.8) holds. The response ability (minimal δ) of the GRN to external stimuli can be obtained by solving the following constrained optimization problem:

$$\delta^0 = \min_{P>0} \delta \quad \text{subject to (5.9)} \tag{5.10}$$

proof: see 5.6.2.

This can be solved by decreasing the upper bound δ in equation (5.8) until no $P > 0$ exists in equation (5.9) by using Matlab LMI Toolbox. By following equation (5.10), the network response abilities of aging-related GRNs in Figure 5.1 can be measured and compared at different life stages. Results are shown in Table 5.1. Obviously, the network response ability of aged group is less than young and CR groups.

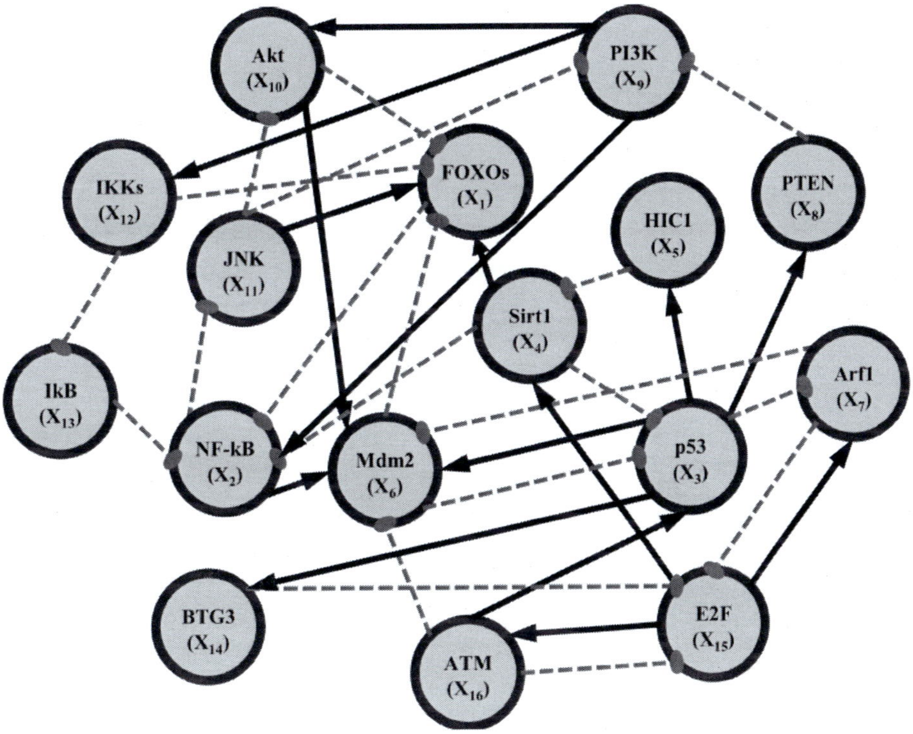

Figure 5.1. Multiple loops of a gene regulatory network associated with aging-related pathophyiological phenotypes. This aging-related gene regulatory network includes 16 genes: *FOXO$_s$*, *NF-kB*, *P53*, *SIRT1*, *HIC1*, *Mdm2*, *Arf1*, *PTEN*, *P13K*, *Akt*, *JNK*, *IKK*, *IkB*, *BTG3*, *E3F1*, and *ATM*. Dashed red lines and black arrows indicate negative and positive parameters of regulated interaction, respectively.

Table 5.1. The network robustness (η^0) and network response ability (δ^0) of a gene regulatory network with 16 aging-related genes (Figure 5.1) across different tissues at young, aged, and CR stages. Since the microarray data of CR for eye tissue are unavailable, we have not estimated the network robustness and response ability of eye tissue in CR group [2]

Tissue	Young	Aged	CR
Thymas			
η^0	0.2310	0.3750	0.2050
δ^0	1.1653	0.9463	1.2279
Spinal Cord			
η^0	0.2410	0.6270	0.1400
δ^0	1.1630	0.9367	1.2645
Eye			
η^0	0.1600	0.2910	
δ^0	1.2376	0.8798	

5.2. Tradeoff between Network Robustness and Network Response Ability during Aging Process

It has previously been shown that gene networks, associated with aging-related pathophysiological phenotypes suffer intrinsic perturbations mainly due to process noise, molecular fluctuations or genetic mutations during aging. The network robustness of GRN is defined as its ability to tolerate intrinsic perturbation, and it plays a principal role in the fail-safe mechanism of the evolutionary process. Therefore, comparison of network robustness η^0 for different gene networks pertaining to the young, aged, and CR stages shown in Table 5.1 may clearly depict their relative ability to tolerate intrinsic perturbations that the system can tolerate. Table 5.1 shows that the "aged" group has higher network robustness but lower network response ability. In order to tolerate the intrinsic parameter fluctuations accumulated due to genetic mutation, the elderly GRN becomes more robust than the young GRN. Robust GRNs are less responsive to external stimuli; consequently, the network protective ability against external stimuli decreases in the older gene networks. However, a more robust GRN may harbor more accumulated genetic mutations, which through random drift might provide more evolutionary paths to other phenotypes of gene network and thus lead to some aging-related chronic diseases like cancers, metabolic disorders, and arthritis. Table 5.1 also shows that the "young" network groups are less robust, with greater response to external stimuli. These GRNs are therefore also less robust to intrinsic perturbations and elicit a strong response toward external stimuli. This might imply that some gene expressions of such networks could be easily reprogrammed to mediate downstream genes or regulators for further reactions. It also allows for modulation of gene expression in response to external stimuli, such as exposure to oxidative stress, carcinogens, and pro-inflammation molecules. These observations suggest that the gene regulatory network at the earlier stage of life or under conditions of CR may opportunely have protective adaptations to maintain intact regulatory structures and homeostasis of cellular functions. The purpose of the above analysis is to gain a greater understanding of the systematic protective and/or defensive mechanisms inherent to aging-related GRNs. The proposed robustness measurement methods may be used for future studies of GRNs involved in various biological processes and may have potential applications in gene therapy and drug target selection.

Based on the tradeoff between network robustness and network response ability of gene networks in Table 5.1 in the aging process, we can make a network robustness-based hypothesis on network aging: In the aging process, the network robustness of the GRN is gradually increased to tolerate the accumulated genetic variations, and simultaneously, the antagonistic network response ability to external stimuli is gradually decreased. Because the power of natural selection fades with age, it is possible for hazardous late-acting genes to exist due to the accumulation of genetic variations. Observations to support these hypotheses have been found in late-life mortality plateaus [8] and in *Drosophila melanogaster* testing [9]. To buffer the effects of these accumulated genetic variations, gene networks need more robustness to retain their functions. However, these robust gene networks with large amounts of harbored genetic variations at the late stage of life might provide the gene networks with more paths for other phenotypes by random drift. This is the cause of aging-related diseases like atherosclerosis, prostate hypertrophy, Alzheimer's disease, and cancers. Furthermore, a large response ability to external stimuli, such as to growth factors at a young stage, may be useful for growth and reproduction. With increasing age, the decrease in the network response ability to external stimuli, such as free radicals, inflammation, infection, and pathogenic invasion, results in failure to trigger protective mechanisms and might be the cause of aging.

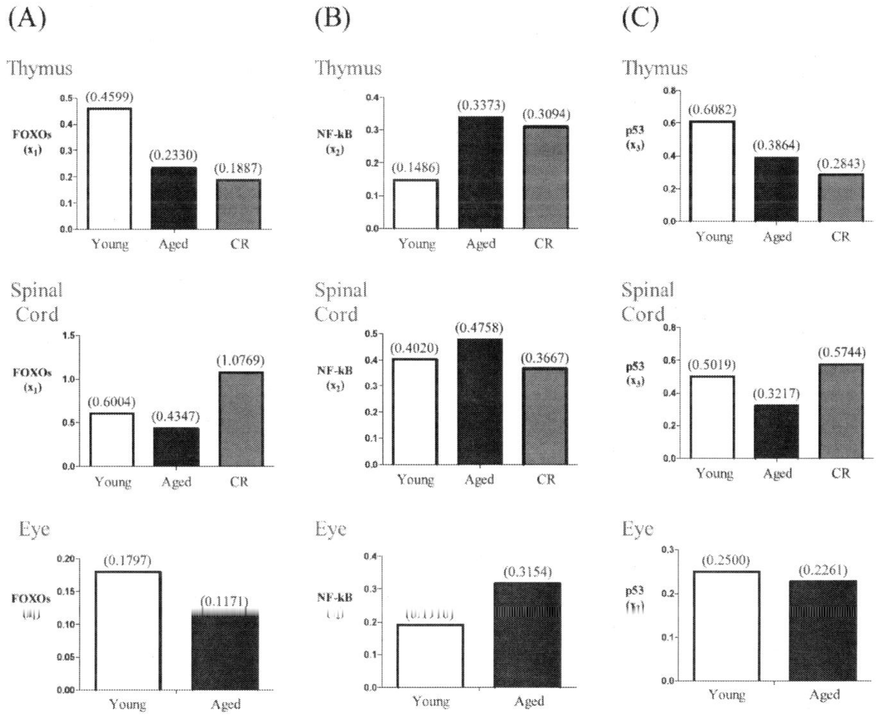

Figure 5.2. Comparison of individual gene response abilities (δ^0) of all genes in the aging-related GRN in Figure 5.1 across different tissues at the young, aged, and CR stages. The most conspicuous difference and parallel analytic results among different tissues are the genes FOXOs (A), NF-κB (B), and p53 (C), which is discussed further in Figure 5.3.

5.3. Gene Response Ability of Individual Genes in the Aging-Related Regulatory Network in the Young, Aged, and CR Groups

After analyzing network robustness and network response ability of the entire GRN associated with aging, we focus on the response abilities of individual genes with $C = \text{diag}(0, \ldots, 0, 1, 0, \ldots, 0)$ in (5.7) for the i-th gene to be analyzed, i.e., all are zero except the i-th diagonal term. The purpose of this analysis is to gain a greater understanding of the protective and/or defensive genetic mechanisms inherent to the aging-related GRN. If a gene has greater response ability to external stimuli, the gene should be more efficient in the transduction of active signals to downstream genes for further reaction. For this reason, we measure downstream transduction ability of each individual gene (called individual gene response ability) for a number of genes in the aging-related gene network (shown in Figure 5.1) in the "young","aged", and "CR" groups. Response abilities of these individual genes across different tissues are summarized in Figure 5.2. Each subgraph represents response abilities of the individual genes with respect to the young, aged, and CR groups in different tissues. We find that these genes examined have different gene response abilities at different stages of life, with the most conspicuous difference and parallel analytic results among different tissues for 1) the gene FOXOs (Figure 5.2A) in the thymus ($\delta^o_{\text{FOXOs\_young}} = 0.4599$, $\delta^o_{\text{FOXOs\_aged}} = 0.2330$, $\delta^o_{\text{FOXOs\_CR}} = 0.1887$), spinal cord ($\delta^o_{\text{FOXOs\_young}} = 0.6004$, $\delta^o_{\text{FOXOs\_aged}} = 0.4347$, $\delta^o_{\text{FOXOs\_CR}} = 1{:}0769$), and eye ($\delta^o_{\text{FOXOs\_young}} = 0{:}1797$, $\delta^o_{\text{FOXOs\_aged}} = 0.1171$); 2) the gene NF-κB (Figure 5.2B) in the thymus ($\delta^o_{\text{NF-}\kappa\text{B\_young}} = 0.1486$, $\delta^o_{\text{NF-}\kappa\text{B\_aged}} = 0.3373$, $\delta^o_{\text{NF-}\kappa\text{B\_CR}} = 0.3094$), spinal cord ($\delta^o_{\text{NF-}\kappa\text{B\_young}} = 0.4020$, $\delta^o_{\text{NF-}\kappa\text{B\_aged}} = 0.4758$, $\delta^o_{\text{NF-}\kappa\text{B\_CR}} = 0.3667$), and eye ($\delta^o_{\text{NF-}\kappa\text{B\_young}} = 0.1918$, $\delta^o_{\text{NF-}\kappa\text{B\_aged}} = 0.3154$); 3) the gene p53 (Figure 5.2C) in the thymus ($\delta^o_{\text{p53\_young}} = 0.6082$, $\delta^o_{\text{p53\_aged}} = 0.3864$, $\delta^o_{\text{p53\_CR}} = 0.2843$) , spinal cord ($\delta^o_{\text{p53\_young}} = 0.5019$, $\delta^o_{\text{p53\_aged}} = 0.3217$, $\delta^o_{\text{p53\_CR}} = 0.5744$), and eye ($\delta^o_{\text{p53\_young}} = 0.2500$, $\delta^o_{\text{p53\_aged}} = 0.2261$). Further comparisons and discussion are given in the following sections.

5.3.1. Comparison of Individual Gene Response Abilities for FOXOs among the Young, Aged, and CR Groups

The free radical theory of aging postulates a direct causal relationship between aging and accumulation of damage due to ROS [10]. The role of FOXOs in controlling cellular oxidative stress and lifespan regulation notably substantiates this theory. According to our measurement of individual gene response ability for FOXOs (Figure 5.2A), we have obtained a conspicuously lower gene response ability for the "aged" group ($\delta^o_{\text{FOXOs\_aged}} = 0.2330$ in the thymus, $\delta^o_{\text{FOXOs\_aged}} = 0.4347$ in the spinal cord, $\delta^o_{\text{FOXOs\_aged}} = 0.1171$ in the eye) versus the "young" group ($\delta^o_{\text{FOXOs\_young}} = 0.4599$ in the thymus, $\delta^o_{\text{FOXOs\_young}} = 0.6004$ in the spinal cord,

$\delta^{o}_{FOXOs\_young} = 0.1797$ in the eye). This may imply that FOXOs has less efficient transduction for orchestrating the programs of gene expression that regulate apoptosis, cell cycle arrest, and stress resistance during aging. Changes in several key FOXOs target genes mediate the ability of FOXOs to induce cell death, including: upregulation of BIM, which is a proapoptotic BCL-2 family member; downregulation of BCL-XL, which is an antiapoptotic BCL- 2 family member [11]; and upregulation of genes that encode death cytokines, such as Fas ligand and TRAIL [12] (see Figure 5.3A). In addition to upregulating cell apoptosis, activated FOXOs up-regulates its downstream genes that encode the Cdk inhibitor p27KIP1, p21, and the Rb family member p130 to mediate G1 arrest as well as to promote cell cycle arrest by repressing the expression of two positive cell cycle regulators, cyclin D1 and D2 [13]. Several FOXOs target genes, such as Growth Arrest and DNA Damage 45 (GADD45) and cyclin G2, have been shown to mediate the effect of FOXOs on the G2/M boundary to stimulate DNA repair or apoptosis and to inhibit cell proliferation [14]. Hence, the lower gene response ability of FOXOs that we estimate during the aging process may imply that poor efficiency for induction of cell cycle arrest and apoptosis not only results in a decreased ability of cells to arrest at key checkpoints, leading to tumor development, but also leads to the survival of abnormal or mutated cells that would normally die, resulting in tumorigenesis. Moreover, a lower gene response ability of FOXOs in response to cumulative oxidative stress does not allow time for repair of damaged DNA and detoxification of cells from ROS. Adequate induction of cell apoptosis for eliminating damaged or abnormal cells may also not exist in the aged GRNs due to the reduced response ability of FOXOs, which may lead to a reduced transduction of cell damage information to the downstream corresponding regulators of cell death. Therefore, many damaged or abnormal cells accumulate in the body, resulting in aging.

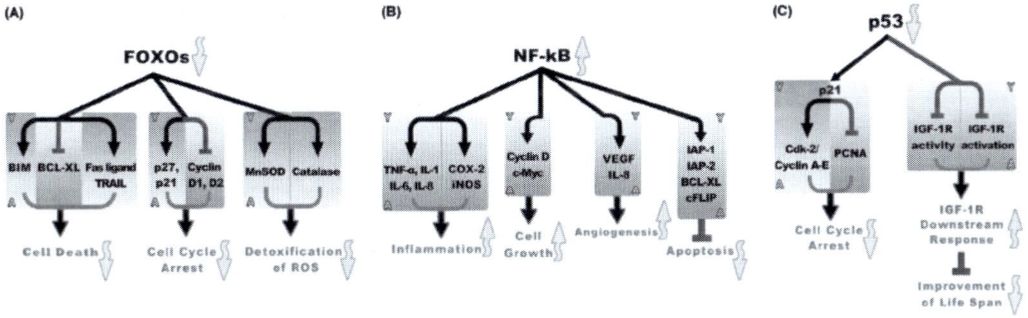

Figure 5.3. The effect of changing gene response ability of an individual gene on its downstream regulators and cellular functions during aging processes. Some major downstream regulators of FOXOs (A), NF-κB (B), and p53 (C) are shown to mediate distinct downstream cellular functions. Arrows in black indicate activation; blunt arrows in red indicate suppression. Wavy arrows in yellow represent the changes to gene response ability and the downstream cellular functions such as cell death, cell cycle arrest, detoxification of ROS, and so on. The upward arrows indicate an increase and the downward arrows indicate a decrease. The gradual changes between light blue and dark blue in rectangle indicate the dynamic changing process of the gene response ability on its downstream regulators from the young stage to aged stage. All detailed regulatory changes occurring due to the change of gene response ability are described in the text. [Y: young stage; A: aged stage].

In addition, active FOXOs permit detoxification of ROS, thus increasing cellular resistance by upregulating downstream genes that encode free radical scavenging enzymes, such as Mn superoxide dismutase (MnSOD), which converts superoxide byproducts to hydrogen peroxide and oxygen, and catalase, which transforms hydrogen peroxide into water and oxygen [13, 15]. Thus, FOXOs control two aspects of cellular resistance to stress: repair of damage resulting from ROS and detoxification of ROS (see Figure 5.3A). Stress resistance is highly coupled with organismal longevity, raising the possibility of a stress resistance mechanism by which FOXOs extends lifespan. Accordingly, a poor response ability of FOXOs implies deterioration of longevity and tumor suppression.

5.3.2. Individual Gene Response Ability of NF-κB in the Young, Aged, and CR Groups

Several studies have shown oxidative stress to be a major factor in the aging process. NF-κB, a pleiotropic and ubiquitous regulator, is well known to be sensitive to various external stimuli, particularly oxidative stress. Activation of NF-κB mediates expression of many genes that are central to the production of inflammatory cytokines and tumorigenesis; these include interleukins (IL-1β, IL-2, and IL-6), tumor necrosis factors (TNF-α), adhesion molecules (ICAM-1, V-CAM, and E-selectin), enzymes (iNOS and COX-2), and gene products that negatively regulate apoptosis (IAP-1, IAP-2, Bcl-xL, and cFLIP). Based on our estimation of individual gene response ability for NF-κB (illustrated in Figure 5.2B), its gene response abilities in the "aged" group ($\delta^{o}_{NF\text{-}\kappa B\_aged}$ = 0.3373 in the thymus, $\delta^{o}_{NF\text{-}\kappa B\_aged}$ = 0.4758 in the spinal cord, δ = 0.3154 in the eye) are much higher than the corresponding values for the "young" group ($\delta^{o}_{NF\text{-}\kappa B\_young}$ = 0.1486 in the thymus, $\delta^{o}_{NF\text{-}\kappa B\_young}$ = 0.4020 in the spinal cord, $\delta^{o}_{NF\text{-}\kappa B\_young}$ = 0.1918 in the eye). Furthermore, this gene response ability is reversible by CR ($\delta^{o}_{NF\text{-}\kappa B\_CR}$ = 0.3094 in the thymus, $\delta^{o}_{NF\text{-}\kappa B\_CR}$ = 0.3667 in the spinal cord) when compared with the response values for the "aged" group. Lowered gene response ability in the young group indicates less transduction to the downstream regulators of NF-κB signaling. This observation may conform with evidence corroborating the idea that NF-κB activation from external stimuli is an inducible but transient event, where the IκB gene is first activated and then NF-κB is transported from the nucleus to the cytoplasm [16]. Additionally, in multiple tissues, CR could directly or indirectly increase gene expression of Nfkbia, an inhibitor of NF-κB, thus eliminating the downstream responses originating from NF-κB [17].

In comparison with the young group, however, the higher gene response ability of NF-κB in the aged group is consistent with previous studies. Much evidence in the literature has shown that NF-κB can be activated during the aging process and be considered as a signaling pathway to regulate several downstream gene expressions. Moreover, the binding activity of NF-κB is highly increased during the aging process, a fact that has been also observed in multiple tissues, such as the brain, liver, and kidney. Accordingly, the elevated gene response ability of NF-κB may imply a general mechanism with increasing age [18-21]. Thus, NF-κB may lose its transient activation nature and become constitutively activated during the aging process [22]. Higher gene response ability also indicates a better efficiency of signal transduction to its downstream regulators, which may be associated with the increase in inflammatory cytokines that contribute to the progression of many degenerative diseases and tumorigenesis. Higher gene response ability of NF-κB may efficiently induce expression of certain genes that encode several inflammatory cytokines, including IL-1, IL-6, IL-8, and TNF-α, as well as the inflammatory enzymes cyclooxygenase-2 (COX-2) and inducible nitric oxide synthase (iNOS) [23, 24]. It is important to note that these inflammatory proteins contain redox-sensitive NF-κB-specific DNA binding sites in the promoter regions and that their production is greatly influenced by the oxidative stress that persists during aging [25] (see Figure 5.3B). On the other hand, the activated NF-κB may induce several genes, so it not only acts as a potent NF-κB activator to form an autoactivating loop that sustains its downstream regulation, but also promotes cell proliferation via growth factors such as TNF-α, IL-1, and IL-6. In addition to the growth factors, certain cell-cycle regulatory proteins, such as cyclin D and cMyc, two NF-κB downstream regulators, have an important role in cell growth and proliferation [26], and key angiogenesis factors, such as vascular endothelial growth factor (VEGF) and interleukin-8 [27], are directly or indirectly enhanced by NF-κB activation. Several gene products, including IAP-1, IAP-2, BCL-XL, and cFLIP, that negatively regulate apoptosis are mediated by NF-κB activation as well [28]. Therefore, the elevated NF-κB gene response ability we estimate for the aged group seems to mimic the gain of function in NF-κB, which not only brings about dysregulation of proliferation and induction of antiapoptosis, but also facilitates cancer maintenance and invasion, and may be involved in the increased incidence of tumorigenesis in old age.

5.3.3. Comparison of Individual Gene Response Abilities of the p53 Gene among the Young, Aged, and CR Groups

Throughout life, all organisms incessantly suffer from various stresses, which induce counteracting cellular responses to prevent further deterioration. p53 plays a central role in integrating environmental stresses and cellular responses [29]. Significant stress and DNA damage may cause either cell cycle arrest so that DNA repair may be performed or the induction of apoptosis and promotion of cellular senescence [30, 31]. According to GRAM measurements, the response ability of the p53 gene (shown in Figure 5.2C) is conspicuously lower in the aged group ($\delta^{o}_{p53\_aged} = 0.3864$ in the thymus, $\delta^{o}_{p53\_aged} = 0.3217$ in the spinal cord, $\delta^{o}_{p53\_aged} = 0.2261$ in the eye) than in the young group ($\delta^{o}_{p53\_young} = 0.6082$ in the thymus, $\delta^{o}_{p53\_young} = 0.5019$ in the spinal cord, $\delta^{o}_{p53\_young} = 0.2500$ in the eye), manifesting that p53 in older organisms has a poor ability to efficiently transduce active regulatory signals to downstream regulators in response to external stimuli.

The function of p53, particularly in tumor suppression, involves transactivation or transrepression of a large number of genes involved in cell cycle regulation and apoptosis, such as the cyclin-dependent kinase (Cdk) inhibitor p21 and the insulin-like growth factor receptor IGF-1R. p53 appears to be essential for upregulation of p21 following DNA damage [32], and p21 subsequently encodes a checkpoint protein that interacts with cell cycle proteins such as Cdks and blocks activity of cyclin/Cdk complexes. p21 not only binds to Cdk2/cyclin A-E complexes and prevents exit from G1 into S [33], but can also pause the cell cycle during

S-phase by inhibiting proliferating cell nuclear antigen (PCNA), which is required for replication of DNA and DNA repair [84] (see Figure 5.3C). The response ability of the p53 gene that we estimate for the aged group may imply less expression of p21 and, thus, abnormal regulation of the cell cycle, which may affect the efficiency of cells to repair damaged DNA. Loss of p53 function suggests that not only tumor risk may increase, but also that tissue homeostasis may be impaired during the aging process.

The level of IGF-1R is controlled by several tumor suppressors, including WT-1, BRCA1, and p53 [34]. p53 suppresses IGF-1R activity by reducing endogenous levels of IGF-1R mRNA [35] and inhibiting ligand-induced activation of IGF-1R by stimulating transcription of IGFbinding protein-3 (IGFBP-3). Loss of p53 function due to DNA mutations has been associated with overexpression of IGF-1R and an increased risk of tumorigenesis. High serum levels of the IGF-1 ligand have been correlated to increased risk for prostate and breast cancers in human population studies [87]. In addition, previous evidence suggests that DNA mutations that cause a decline in IGF-1R levels or block signal transduction by this receptor also significantly improve lifespan in the nematode *Caenorhabditis elegans* and the fruit fly *Drosophila melanogaster*. A similar result has been obtained from a cohort study of elderly Dutch citizens, revealing that genetic variations causing reduced insulin/IGF-1 signaling are beneficial for survival in old age [36]. Therefore, the attenuated response ability of p53 that we estimate from the aged group may be the main cause of loss of p53 function, resulting in increased responsiveness of IGF-1R. This may also explain the increased risk of tumorigenesis and other associated deleterious effects on longevity during the aging process.

5.4. Network-Based Biomarker Determination via Sample Microarray Data Using a Systems Biology Approach

Biomarkers are used either in diagnostic evaluation to determine the health of a patient with or without a disease, or as a prognostic indicator to determine a patient's prognosis. Present biomarker identification methods, which strictly use gene expression profiles, cannot show how the different genes within the biomarker gene set are related to each other; that is, biomarkers are not identified from a systems biology perspective. Furthermore, the gene lists obtained for similarly diagnosed patients by different research groups differ widely and share few common genes. Here, a systems biology approach is introduced for the integration of microarray data and PPI information to develop a network-based biomarker for systematic investigation into the network mechanism of lung carcinogenesis and the diagnosis of lung cancer [37]. The network-based biomarker consists of two protein association networks constructed from cancer and noncancer samples. The proposed method can be widely applied to determining network-based biomarkers for other diseases. The overall flowchart of the proposed network-based biomarker approach is shown in Figure 5.4 [37]. First, PPI information and microarray sample data for smokers with and without cancer are used to construct candidate PPI networks for cancer and noncancer samples. Since data for both samples are limited, the number of proteins used in candidate PPI network construction is also restricted: to avoid overfitting in network construction, the maximum degree of proteins in the candidate network are less than the number of samples, imposing a limit on the size of the candidate network.

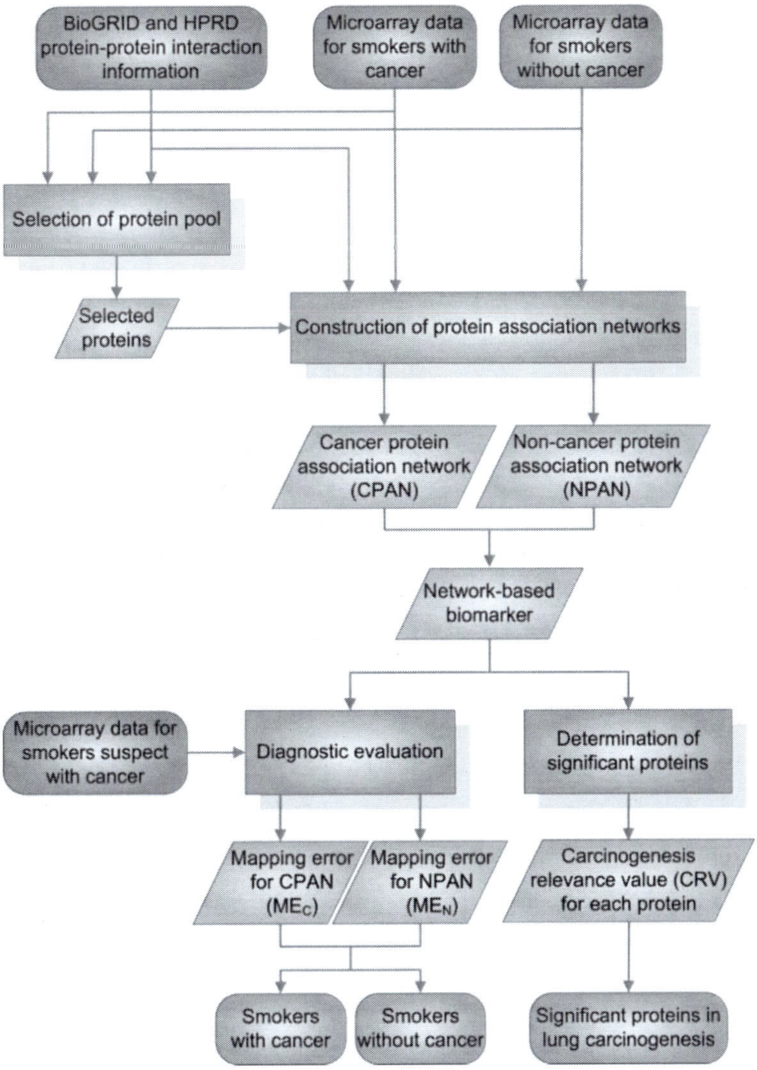

Figure 5.4. Flowchart for constructing a network-based biomarker for lung cancer investigation and diagnosis. By using microarray data for smokers with and without cancer, PPI networks with and without cancer are identified (Figure 5.5) through equation (5.11). The network-based biomarker is obtained from the difference in network structure between the two networks (Figure 5.6). Significant proteins are determined from their carcinogenesis relevance value (CRV) through equations (5.14) and (5.15).

5.4.1. Construction of Network-Based Biomarker by Microarray Data

Using a simple regression model, the candidate PPI network is then further validated by the sample microarray data to highlight independent protein associations of both samples relative to their respective data sets. For a target protein i in the candidate PPI network, the protein is described using the following protein association model [38]:

$$y_i(n) = \sum_{k=1}^{N_i} \alpha_{ik} y_k(n) + \varepsilon_i(n) \tag{5.11}$$

where $y_i(n)$ represents the gene expression level of the target protein i for the sample n, and α_{ik} denotes the association ability between the target protein $y_i(n)$ and protein $y_k(n)$ for sample n. N_i represents the number of proteins interacting with the target protein i; it can be obtained from the rough PPI network. $\varepsilon_i(n)$ denotes stochastic noise associated with other factors or model uncertainty. Equation (5.11) states that, biologically, the expression level of the target protein i is associated with the expression levels of interacting proteins.

Based on the reverse engineering methods in chapter 3, the associated interaction parameter α_{ik} in equation (5.11) is identified through maximum likelihood estimation [26] on microarray data. AIC and Student's t-test were employed for model order selection and for tests on the statistical significance of protein associations. Based on α_{ik}, two matrices are established to represent the cancer protein association network (CPAN) and the non-cancer protein association network (NPAN) as follows [37]

$$C = \begin{bmatrix} \alpha_{11,\mathrm{C}} & \alpha_{12,\mathrm{C}} & \cdots & \alpha_{1K,\mathrm{C}} \\ \alpha_{21,\mathrm{C}} & \alpha_{22,\mathrm{C}} & \cdots & \alpha_{2K,\mathrm{C}} \\ \vdots & \vdots & \ddots & \vdots \\ \alpha_{K1,\mathrm{C}} & \alpha_{K2,\mathrm{C}} & \cdots & \alpha_{KK,\mathrm{C}} \end{bmatrix}, \quad N = \begin{bmatrix} \alpha_{11,\mathrm{N}} & \alpha_{12,\mathrm{N}} & \cdots & \alpha_{1K,\mathrm{N}} \\ \alpha_{21,\mathrm{N}} & \alpha_{22,\mathrm{N}} & \cdots & \alpha_{2K,\mathrm{N}} \\ \vdots & \vdots & \ddots & \vdots \\ \alpha_{K1,\mathrm{N}} & \alpha_{K2,\mathrm{N}} & \cdots & \alpha_{KK,\mathrm{N}} \end{bmatrix} \tag{5.12}$$

where $\alpha_{ij,C}$ and $\alpha_{ij,N}$ indicate the quantitative protein association ability between protein i and protein j for CPAN and NPAN, respectively, and K is the number of proteins in the protein association network. The resulting CPAN and NPAN constitute the network-based biomarker used for identifying significant proteins in lung carcinogenesis through the diagnostic evaluation

$$Y_\mathrm{C} = C \cdot Y_\mathrm{C} + E_\mathrm{C}, \quad Y_\mathrm{N} = N \cdot Y_\mathrm{N} + E_\mathrm{N} \tag{5.13}$$

where $Y_C = [y_{1,C(n)} \cdots y_{K,C(n)}]^T$, $Y_N = [y_{1,N(n)} \cdots y_{K,N(n)}]^T$ denotes the vectors of expression levels, and E_C and E_N indicate the noise vectors in cancer and non-cancer cases, respectively.

A matrix indicating the difference between two protein association networks is defined as $C - N$ *[37]*, i.e.,

$$D = \begin{bmatrix} d_{11} & d_{12} & \cdots & d_{1K} \\ d_{21} & d_{22} & \cdots & d_{2K} \\ \vdots & \vdots & \ddots & \vdots \\ d_{K1} & d_{K2} & \cdots & d_{KK} \end{bmatrix} = \begin{bmatrix} \alpha_{11,\mathrm{C}} - \alpha_{11,\mathrm{N}} & \alpha_{12,\mathrm{C}} - \alpha_{12,\mathrm{N}} & \cdots & \alpha_{1K,\mathrm{C}} - \alpha_{1K,\mathrm{N}} \\ \alpha_{21,\mathrm{C}} - \alpha_{21,\mathrm{N}} & \alpha_{22,\mathrm{C}} - \alpha_{22,\mathrm{N}} & \cdots & \alpha_{2K,\mathrm{C}} - \alpha_{2K,\mathrm{N}} \\ \vdots & \vdots & \ddots & \vdots \\ \alpha_{K1,\mathrm{C}} - \alpha_{K1,\mathrm{N}} & \alpha_{K2,\mathrm{C}} - \alpha_{K2,\mathrm{N}} & \cdots & \alpha_{KK,\mathrm{C}} - \alpha_{KK,\mathrm{N}} \end{bmatrix} \tag{5.14}$$

where d_{ij} denotes the difference in protein association ability between CPAN and NPAN among proteins i and j. Using matrix D to represent the difference in network structure between CPAN and NPAN, a carcinogenesis relevance value (CRV) was derived to quantify the correlation of each protein significant to lung carcinogenesis.

Table 5.2. Identified significant proteins in lung carcinogenesis

Protein symbol†	CRV	*P*-value	Functional annotation*		
			Cell growth	Cell survival	Cell migration
MAPK1	8.3418	< 1e-5	+	+	+
SMAD2	7.7901	< 1e-5	+	+	+
CREBBP	5.7870	0.00002	+		
EGFR	4.3635	0.00086	+	+	+
AR	4.0966	0.00159	+	+	+
UBC	4.0331	0.00180			
SRC	3.9446	0.00218	+	+	+
FGFR1	3.9227	0.00237	+		+
BRCA1	3.9049	0.00243	+	+	
ESR1	3.8409	0.00295	+	+	+
INSR	3.7946	0.00329	+		+
PTK2	3.6758	0.00432	+	+	+
HSP90AA1	3.6732	0.00436	+	+	+
CALM1	3.6363	0.00482		+	
POLR2A	3.5701	0.00547			
CSNK2A1	3.4128	0.00761	+	+	
PRKACA	3.3688	0.00856		+	
CTNNB1	3.2935	0.00994	+	+	+
SP1	3.2397	0.01133	+	+	
SMAD4	3.1947	0.01266	+	+	+
E2F1	3.1382	0.01407	+	+	
YWHAZ	3.1212	0.01467	+		
MEPCE	3.0968	0.01545			
AKT1	3.0193	0.01857	+	+	+
PLCG1	2.9654	0.02069			+
MYC	2.8987	0.02385	+	+	
MAPK3	2.8545	0.02654	+	+	+
NCOA6	2.8132	0.02892	+	+	
FYN	2.7833	0.03089	+		+
MAPK8IP3	2.7746	0.03141			+
YWHAQ	2.7582	0.03242	+		
TRAF6	2.7150	0.03535		+	
SMAD1	2.6940	0.03697	+	+	+
SMAD3	2.6815	0.03815	+	+	+
MAPK14	2.6727	0.03894	+	+	+
TP53	2.6522	0.04056	+	+	+
XRCC6	2.6270	0.04263		+	
EZR	2.6213	0.04314			+
TSC2	2.6116	0.04401	+	+	+
HGS	2.5730	0.04744	+		

† Full protein names according to UniProt database http://www.uniprot.org/website.

* Functional annotations taken from the Gene Ontology database http://www.geneontology.org/ website and literature.

To identify significant proteins, two important issues are taken into consideration. First, the magnitude of the association ability α_{ij} denotes the significance of association of one protein to another. A higher absolute value of α_{ij} implies that the two proteins are more tightly associated. Second, if a protein plays a more crucial role in lung carcinogenesis, then the difference in association numbers linked to the protein for CPAN and NPAN would be larger. For example, if one protein shares a strong association with many proteins in CPAN, but has weaker associations (no protein) in NPAN, then the protein in question is more likely to be involved in lung carcinogenesis. As a result, CRV is determined based on the difference in protein association abilities using the following equation [37]:

$$\mathrm{CRV}_i = \sum_{j=1}^{K} \left| d_{ij} \right| \tag{5.15}$$

For the i-th protein in the network-based biomarker, the implication of equation (5.15) is that the CRV quantifies the extent of protein associations that differentiate CPAN from NPAN.

The above network-based biomarker approach is applied to the molecular investigation and diagnosis of lung cancer. The primary data set of GSE4115 (79 smokers with lung cancer and 73 smokers without lung cancer; obtained from the GEO database, http://www.ncbi.nlm.nih.gov/geo/) was used for construction of the network-based biomarker. The CPAN and NPAN in equation (5.12), which consist of 399 and 393 protein associations respectively, constitute the network-based biomarker of lung cancer (Figure 5.5). The difference between CPAN and NPAN is shown in Figure 5.6. The CPAN indicates that 40 identified proteins play significant roles in lung carcinogenesis (Table 5.2).

The three functional subnetworks for lung cancer are shown in Figure 5.7. The mechanisms in cancinogenesis using significant proteins were further investigated through cell life cycle stages as follows:

(1) Cell Growth

Cancer is a complex disease of uncontrolled cell growth. Therefore, the proteins responsible for cell growth are likely to play critical roles in lung carcinogenesis. Among the 40 significant proteins identified based on the network-based biomarker, 30 proteins are annotated with cell growth (Table 5.2). The mitogen-activated protein kinase (MAPK) cascade is a highly conserved module that is identifiable in many cancers. Three MAPK-related proteins annotated with cell growth, MAPK1 (ERK2), MAPK3 (ERK1) and MAPK14 (p38a), were identified as significant proteins in lung carcinogenesis. Elevated expression of activated MAPK1/3 have been observed in NSCLC [39] and may play a role in lung metastasis [40]. MAPK14 (p38a) is well known for its important role in response to inflammation and environmental stress [41]; its protein expression is more than three times lower in human lung tumors compared with normal human lung tissue, suggesting that MAPK14 may function as a negative regulator of lung carcinogenesis [42]. Four proteins from the SMAD protein family (SMAD1, SMAD2, SMAD3 and SMAD4) were also identified as significant proteins in lung carcinogenesis. The SMAD proteins, which consist of three functional classes, are signal transducers and transcriptional modulators [43]. Mutations and altered expression for these four proteins were observed in human cancer [44,

45], highlighting the significant, albeit varied, roles of the SMAD protein family in carcinogenesis.

EGFR, FGFR1, and INSR are receptor tyrosine kinases (RTKs) bound by an epidermal growth factor, a fibroblast growth factor, and insulin, respectively. Receptor tyrosine kinases have been shown to not only perform a key regulatory role in normal cellular processes, but also become critically involved in the development and progression of human cancers [46]. EGFR is one of the most extensively studied proteins in carcinogenesis. It is overexpressed in NSCLC as well as in other common tumors, and its increased expression is associated with aggressive tumor growth and therapy resistance [47]. EGFR has also been found to be significantly mutated in lung adenocarcinoma [48, 49]. Because of its significance, EGFR becomes a popular therapeutic target for carcinogenesis. Gefitinib (Iressa) and erlotinib (Tarceva) are two targeted therapies that specifically inhibit EGFR tyrosine kinase [47, 49]. FGFR1 and INSR are both known to be involved in carcinogenesis and thus become novel, attractive targets for cancer therapeutic strategies [50, 51].

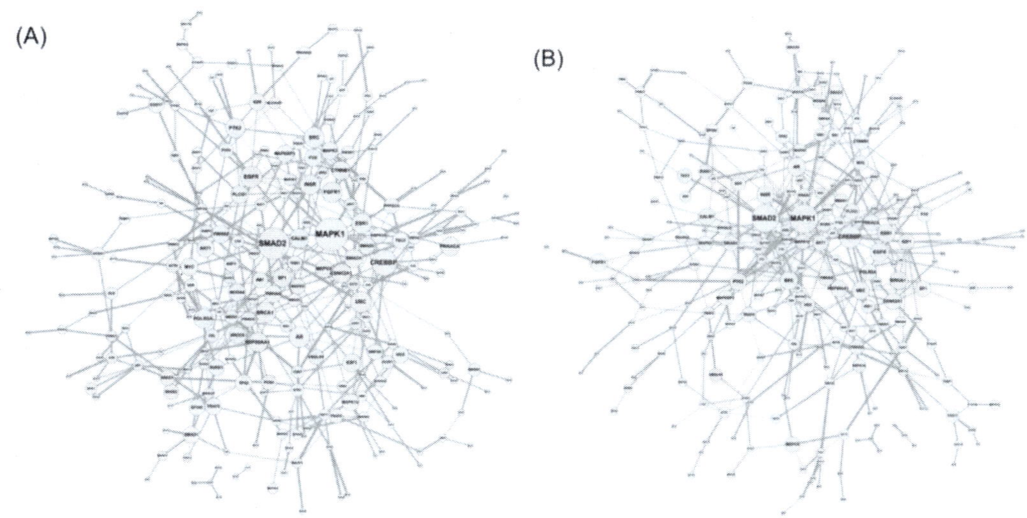

Figure 5.5. The constructed network-based biomarker. (A) Cancer protein association network (CPAN) obtained from C in (5.12) by maximum likelihood estimation (MLE), AIC selection, and Student's t-test. (B) Non-cancer protein association network (NPAN) obtained from N in equation (5.12) using the same criteria.

Many proteins acting as transcriptional regulators were also identified. CREBBP is a transcriptional co-activator downstream of the TGFb pathway and the mutations and deletions of the CREBBP gene are associated with lung cancer [52]. CTNNB1 (b-catenin) is one of the core components in the Wnt pathway. Mutation of b-catenin, which results in aberrant activation of the Wnt pathway, is a frequent cause of human cancer growth [53]. E2F1 is one of the significant proteins involved in the cell cycle. Its overexpression has been demonstrated in both NSCLC and SCLC and is induced by its upstream RB protein [54]. MYC is a nuclear phosphoprotein and functions as a transcription factor. It controls cell cycle progression by simulating G1/S transition and may result in loss of cell cycle arrest and uncontrolled tumor growth when dysfunctional [55] Other identified significant proteins such as AR, ESR1, SRC,

FYN, YWHAQ, YWHAZ, and HSP90AA1 were also shown to contribute to the process of carcinogenesis [56-62].

(2) Cell Survival

The ability of tumor cell populations to expand in number is determined not only by the rate of cell proliferation but also by the rate of cell death [63]. In addition, the acquired resistance of programmed cell death, apoptosis, is a hallmark of cancer. Consequently, proteins annotated with cell survival might be important in carcinogenesis. Twenty-seven significant proteins identified by the network-based biomarker were annotated with cell survival. TP53 (p53) is a well-studied tumor suppressor protein and plays important roles in anti-cancer mechanisms. Its activation is induced by a number of stress signals such as DNA damage, oxidative stress and activated oncogenes. Activated p53 induces cell cycle arrest, apoptosis and inhibition of angiogenesis and metastasis. Once damaged, tumor suppression is severely reduced, resulting in uncontrolled proliferation of the cell. Due to the importance in carcinogenesis, it is no surprise that p53 was found to be significantly mutated in lung adenocarcinoma as well as in squamous cell carcinoma and SCLC [48, 49, 64].

TRAF6 functions as a signal transducer in the NFB pathway that activates IKK, in response to proinflammatory cytokines. The identification of TRAF6 by the proposed network-based biomarker approach reinforces the linkage between inflammation and cancer [65, 66]. BRCA1 is a nuclear phosphoprotein that contributes to genomic stability. The mutant phenotype of BRCA1 predisposes to breast and ovarian cancer [67]. SP1 is a transcription factor downstream of the TGFb pathway and its overexpression contributes to malignant transformation [68]. Other protein kinases, AKT and CSNK2A1 (CK2), were also shown to participate in the carcinogenic process [69-71].

(3) Cell Migration

With the progression of cancer, the malignant tumor cells acquire the ability to migrate and metastasize to distant sites. As a result, the proteins that are relevant to the cell migration capability are crucial for the carcinogenic process. Twenty-three out of 40 significant proteins were annotated with cell migration. PTK2 (FAK), a protein tyrosine kinase in the RTK pathway, is an important mediator within the cell migration process, as well as in cell proliferation and cell survival. Substantial evidence has shown that activated PTK2 leads to tumor growth and metastasis [72], and the level of expression is substantially linked to the invasive potential of tumors [73]. High levels of TSC2 were correlated with increased metastasis and reduced survival in breast cancer patients, revealing a protumorigenic role for TSC2 [74]. The other two significant proteins, PLCG1 and EZR (ezrin), were demonstrated to play critical roles in the metastatic potential of cancer cells but not in primary tumor growth [75, 76].

For nine out of the 40 significant proteins identified, little is known about their roles in lung cancer (see Table 5.2). UBC is identified as a polyubiquitin precursor. Protein ubiquitination is a fundamental, regulatory posttranslational modification controlling intracellular signaling events. It has been associated with protein degradation, DNA repair, cell cycle regulation, endocytosis, and kinase modification [77]. Dysregulation of ubiquitin-mediated signaling is increasingly implicated in some human diseases. Therefore, UBC may be an important target for further characterization of lung carcinogenesis. CALM1 is calmodulin, which mediates the control of a large number of enzymes and other proteins by

Ca^{2+}. It is an essential regulator of cell cycle progression and cell survival. Further research is needed to examine its relation with carcinogenesis. PRKACA is a cAMPdependent protein kinase. The identification of PRKACA as a significant protein implies that cAMP signaling might also be involved in lung carcinogenesis. MAPK8IP3 functions as a scaffold protein in the RTK pathway; NCOA6, HGS, and XRCC6 are annotated with cell growth and/or cell survival. However, until now, no empirical evidence has linked their relevance to carcinogenesis, which makes them potential targets for further investigation into lung carcinogenesis.

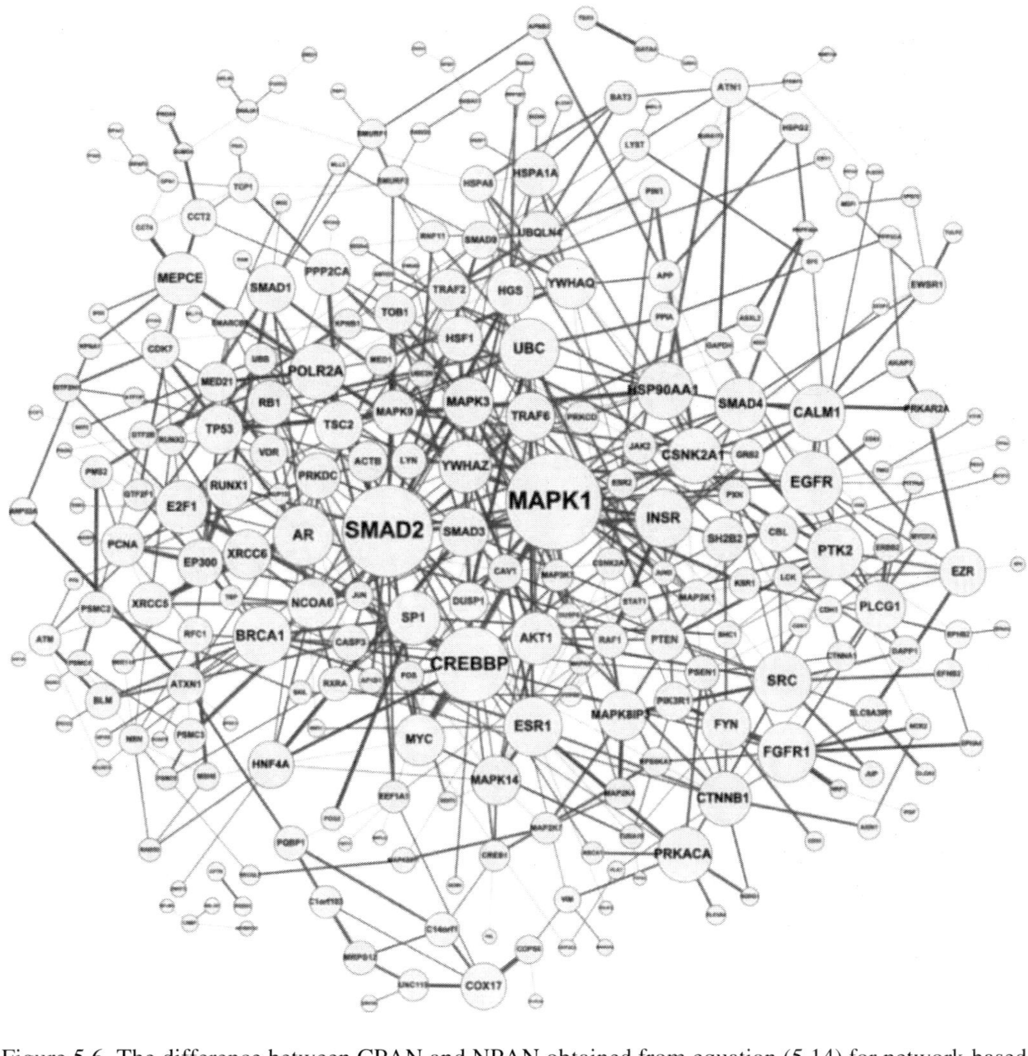

Figure 5.6. The difference between CPAN and NPAN obtained from equation (5.14) for network-based biomarkers for lung cancer. The significance of proteins (indicated by circle size) to the network-based marker is dependent on their CRVs in equation (5.15), which are listed in Table 5.2.

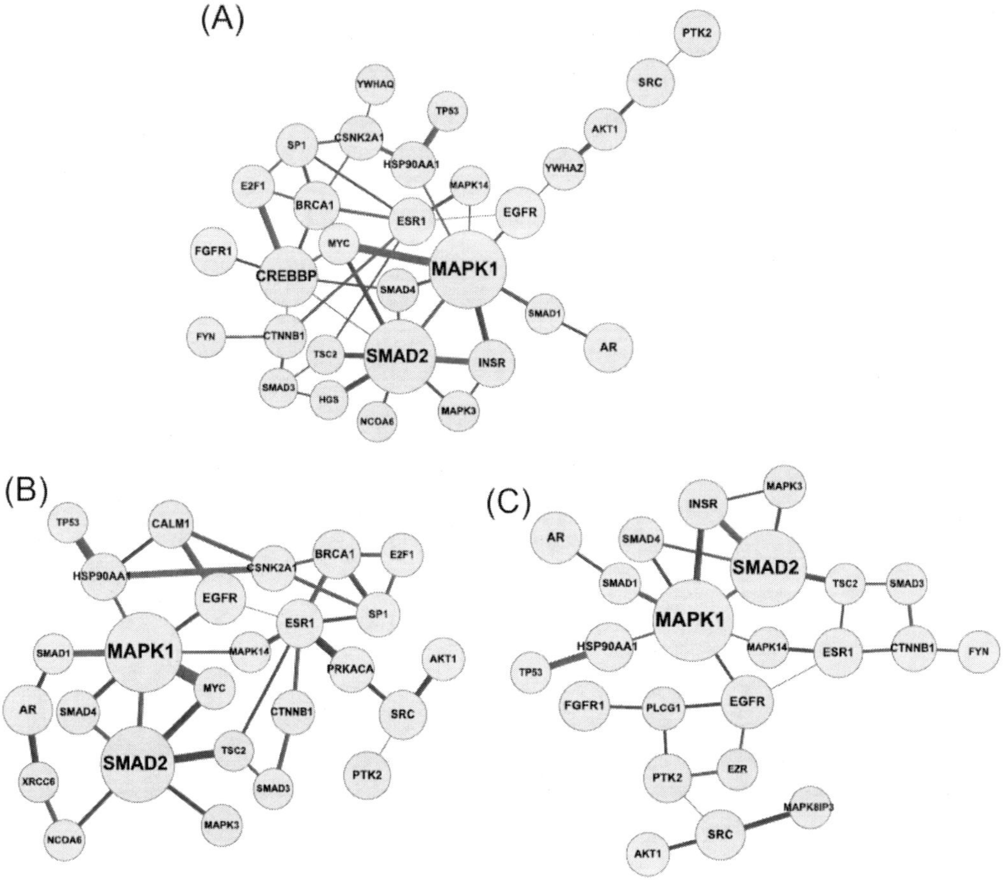

Figure 5.7. The functional subnetworks of significant proteins for lung cancer identified according to the network-based biomarker (A) Cell growth functional subnetwork (B) Cell survival functional subnetwork (C) Cell migration functional subnetwork. All the functional subnetworks are extracted from Figure 5.6.

5.4.2. Diagnostic Evaluation of Smokers Suspect of Lung Cancer Using the Network-based Biomarker

The network-based biomarker was constructed based on the primary data set of GSE4115. An independent data set (the prospective data set of GSE4115, 18 smokers with lung cancer and 17 smokers without lung cancer) was then used to evaluate the diagnostic performance of the proposed network-based biomarker. Among the 35samples, 26 were accurately classified, resulting in an accuracy of 74.29%. The sensitivity and specificity of the proposed approach were also evaluated. The network-based molecular diagnosis can identify smokers with or without cancer, using a high level of sensitivity of 83.33%, and a moderate specificity of 64.71%: this enables the proposed network-based biomarker to effectively diagnose the smokers with lung cancer. Further, this approach enables the network-based biomarker to act as a screening test, which, with the aid of other clinical diagnostic tools, both accelerates the process and improves the sensitivity of diagnosis.

The cause of the moderate specificity was further investigated. The difference between the specificity value and the sensitivity value can be attributed to a misclassification of a number of smokers without cancer in the cancer category. The misclassification may be due to similarities in molecular pattern, i.e., the gene expression profiles of smokers with cancer are similar to those of smokers without cancer. In order to validate the hypothesis, Pearson correlation coefficients [78] of gene expression profiles for both smokers with and without cancer were calculated. The mean correlation coefficient of smokers with cancer was 0.9616, whereas the mean correlation coefficient of smokers without cancer was 0.9441. In addition, the mean correlation coefficient of smokers both with and without cancer (pooled) was as high as 0.9437, suggesting that the molecular patterns shared among smokers with and without cancer are indeed highly similar. Because of the highly similar molecular patterns and the fact that cigarette smoking is the main investigator of lung cancer, it is likely smokers without cancer could one day develop lung cancer.

In order to validate the predictive performance of the proposed network-based biomarker, several comparisons were made. First, we tested the predictive performance without the information of protein-protein interactions. The 199 differentially expressed proteins selected by ANOVA using the primary data set were used for classification of the prospective data set. A simple hierarchical clustering was performed, illustrating that the use of gene expression alone cannot accurately classify the prospective data set (65.71% accuracy). The comparison shows that the integration of gene expression profiles and protein interaction information can improve lung cancer diagnosis. Second, we compared the predictive performance of the proposed method with randomly selected networks. The average accuracy for 100,000 randomly selected networks of 339 proteins was 48.44%, highlighting the significance of the proposed network-based biomarker. Third, in addition to evaluating the predictive performance using an independent data set, a 5-fold cross-validation was applied to the primary data set plus the prospective data set. The accuracy of the cross-validation is similar to the initial accuracy of 74.29% computed using the independent data set, illustrating the robustness and reproducibility of the proposed network-based biomarker approach.

5.5. CONCLUSION

In this chapter, we describe systematic methods to calculate some characteristics of biological networks such as network robustness to intrinsic perturbation, network sensitivity to external signals, and network-based biomarkers for significant PPI changes. From these system characteristics, the intrinsic system characteristics and systematic mechanisms of biological networks can be investigated, which may provide targets for drug design and therapy for cellular diseases.

5.6. Appendix

5.6.1. Proof of Proposition 5.1

Let us denote the Lyapunov (energy) function of perturbative gene regulatory network in equation (5.4) as $V(\tilde{X}(t))= \tilde{X}^T(t)P\tilde{X}(t)$, for some symmetric positive definite matrix $P=P^T>0$.

$$\begin{aligned}
\Delta V(\tilde{X}(t)) &= V\left(\tilde{X}(t+1)\right)-V(\tilde{X}(t)) \\
&= \tilde{X}^T(t+1)P\tilde{X}(t+1)-\tilde{X}^T(t)P\tilde{X}(t) \\
&= \tilde{X}^T(t)\left[\hat{A}(1+\eta)\right]^T P\left[\hat{A}(1+\eta)\right]\tilde{X}(t)-\tilde{X}^T(t)P\tilde{X}(t) \\
&\Rightarrow \tilde{X}^T(t)\left\{\left[\hat{A}(1+\eta)\right]^T P\left[\hat{A}(1+\eta)\right]-P\right\}\tilde{X}(t)\le 0 \\
&\Rightarrow \left[\hat{A}(1+\eta)\right]^T P\left[\hat{A}(1+\eta)\right]-P\le 0
\end{aligned}$$

Based on quadratic stability [6], if the LMI in equation (5.5) holds, then the perturbative network is robustly stable.

5.6.2. Proof of Proposition 5.2

Let us denote the Lyapunov function of gene regulatory network under external stimulation signals in equation (5.7) as $V(\tilde{X}(t))= \tilde{X}^T(t)P\tilde{X}(t)$, for some symmetric positive definite matrix $P=P^T>0$. We have

$$\begin{aligned}
&V\left(\tilde{X}(t+1)\right)=\tilde{X}^T(t+1)P\tilde{X}(t+1)=\left\{\hat{A}\tilde{X}(t)+U(t)\right\}^T P\left\{\hat{A}\tilde{X}(t)+U(t)\right\} \\
&=\tilde{X}^T(t)\hat{A}^T P\hat{A}\tilde{X}(t)+\tilde{X}^T(t)\hat{A}^T PU(t)+U^T(t)P\hat{A}\tilde{X}(t)+U^T(t)PU(t) \\
&=\tilde{X}^T(t)\hat{A}^T P\hat{A}\tilde{X}(t)+\tilde{X}^T(t)\hat{A}^T PU(t)+U^T(t)P\hat{A}\tilde{X}(t)+U^T(t)PU(t) \\
&+Y^T(t)Y(t)-\delta^2U^T(t)U(t)-Y^T(t)Y(t)+\delta^2U^T(t)U(t)-V\left(\tilde{X}(t)\right)+V\left(\tilde{X}(t)\right) \\
&=\tilde{X}^T(t)\hat{A}^T P\hat{A}\tilde{X}(t)+\tilde{X}^T(t)\hat{A}^T PU(t)+U^T(t)P\hat{A}\tilde{X}(t)+U^T(t)PU(t) \\
&+\tilde{X}^T(t)C^TC\tilde{X}(t)-\delta^2U^T(t)U(t)-\tilde{X}^T(t)P\tilde{X}(t)+V\left(\tilde{X}(t)\right)-Y^T(t)Y(t) \\
&+\delta^2U^T(t)U(t)
\end{aligned}$$

If the following inequality holds

$$\begin{aligned}&\tilde{X}^T(t)\hat{A}^T P\hat{A}\tilde{X}(t)+\tilde{X}^T(t)\hat{A}^T PU(t)+U^T(t)P\hat{A}\tilde{X}(t)+U^T(t)PU(t)\\&+\tilde{X}^T(t)C^T C\tilde{X}(t)-\delta^2 U^T(t)U(t)-\tilde{X}^T(t)P\tilde{X}(t)\le 0\end{aligned} \tag{5.16}$$

then, we get $V(\tilde{X}(t+1))\le V(\tilde{X}(t))-Y^T(t)Y(t)+\delta^2 U^T(t)U(t)$.

So we sum the above inequality from 0 to ∞ to get

$$\begin{aligned}&V\left(\tilde{X}(\infty)\right)\le V\left(\tilde{X}(0)\right)-\sum_{t=0}^{\infty}Y^T(t)Y(t)+\delta^2\sum_{t=0}^{\infty}U^T(t)U(t)\\&\Rightarrow V\left(\tilde{X}(\infty)\right)+\sum_{t=0}^{\infty}Y^T(t)Y(t)\le V\left(\tilde{X}(0)\right)+\delta^2\sum_{t=0}^{\infty}U^T(t)U(t)\\&\Rightarrow \sum_{t=0}^{\infty}Y^T(t)Y(t)\le V\left(\tilde{X}(\infty)\right)+\sum_{t=0}^{\infty}Y^T(t)Y(t)\\&\qquad\qquad\le \tilde{X}^T(0)P\tilde{X}(0)+\delta^2\sum_{t=0}^{\infty}U^T(t)U(t)\end{aligned} \tag{5.17}$$

which is the inequality of response ability in the case $\tilde{X}(0)\neq 0$. Or in the case $\tilde{X}(0)=0$:

$$\Rightarrow \sum_{t=0}^{\infty}Y^T(t)Y(t)\le\delta^2\sum_{t=0}^{\infty}U^T(t)U(t)\ \text{ or }\ \frac{\sum_{t=0}^{\infty}Y^T(t)Y(t)}{\sum_{t=0}^{\infty}U^T(t)U(t)}\le\delta^2$$

The above results hold only when the inequality (5.16) holds, i.e., the following inequality holds:

$$\begin{aligned}&\tilde{X}^T(t)\hat{A}^T P\hat{A}\tilde{X}(t)+\tilde{X}^T(t)\hat{A}^T PU(t)+U^T(t)P\hat{A}\tilde{X}(t)+U^T(t)PU(t)\\&+\tilde{X}^T(t)C^T C\tilde{X}(t)-\delta^2 U^T(t)U(t)-\tilde{X}^T(t)P\tilde{X}(t)\le 0\end{aligned}$$

The above inequality can be equivalent to the following inequality

$$\begin{bmatrix}\tilde{X}\\ U\end{bmatrix}^T\begin{bmatrix}\hat{A}^T P\hat{A}+C^T C-P & \hat{A}^T P\\ P\hat{A} & P-\delta^2 I\end{bmatrix}\begin{bmatrix}\tilde{X}\\ U\end{bmatrix}\le 0 \tag{5.18}$$

The inequality (5.18) holds for all $\tilde{X}(t)$ and $U(t)$ if and only if the following LMI holds.

$$\begin{bmatrix} \hat{A}^T P\hat{A} + C^T C - P & \hat{A}^T P \\ P\hat{A} & P - \delta^2 I \end{bmatrix} \le 0$$

REFERENCES

[1] Tu, CT; Chen, BS. "New measurement methods of network robustness and response ability via microarray data," *Plos One*, vol. 8, e55230, 2013.

[2] Tu, CT; Chen, BS. "On the Increase in Network Robustness and Decrease in Network Response Ability During the Aging Process: A Systems Biology Approach via Microarray Data," *IEEE/ACM Trans Comput Biol Bioinform*, Mar 19 2013.

[3] Chen, BS; Zhang, WH. "Stochastic H(2)/H(infinity) control with state-dependent noise," *IEEE Transactions on Automatic Control*, vol. 49, 45-57, Jan 2004.

[4] Zhang, WH; Chen, BS; Tseng, CS. "Robust H-infinity filtering for nonlinear stochastic systems," *IEEE Transactions on Signal Processing*, vol. 53, 589-598, Feb 2005.

[5] Zhang; WH; Chen, BS. "State feedback H(infinity) control for a class of nonlinear stochastic systems," *SIAM Journal on Control and Optimization*, vol. 44, 1973-1991, 2006.

[6] Boyd, SP. *Linear matrix inequalities in system and control theory*. Philadelphia: Society for Industrial and Applied Mathematics, 1994.

[7] Kitano, H. "Biological robustness," *Nat Rev Genet*, vol. 5, 826-37, Nov 2004.

[8] Charlesworth, B. "Patterns of age-specific means and genetic variances of mortality rates predicted by the mutation-accumulation theory of ageing," *Journal of Theoretical Biology*, vol. 210, 47-65, May 7 2001.

[9] Reynolds, RM; Temiyasathit, S; Reedy, MM; Ruedi, EA; Drnevich, JM; Leips, J; et al. "Age specificity of inbreeding load in Drosophila melanogaster and implications for the evolution of late-life mortality plateaus," *Genetics*, vol. 177, 587-595, Sep 2007.

[10] Harman, D. "The Aging Process," *Proceedings of the National Academy of Sciences of the United States of America-Biological Sciences*, vol. 78, 7124-7128, 1981.

[11] Tsukahara, T; Kannagi, M; Ohashi, T; Kato, H; Arai, M; Nunez, G et al. "Induction of Bcl-x(L) expression by human T-cell leukemia virus type 1 tax through NF-kappa B in apoptosis-resistant T-cell transfectants with tax," *Journal of Virology*, vol. 73, 7981-7987, Oct 1999.

[12] Modur, V; Nagarajan, R; Evers, BM; Milbrandt, J. "FOXO proteins regulate tumor necrosis factor-related apoptosis inducing ligand expression - Implications for PTEN mutation in prostate cancer," *Journal of Biological Chemistry*, vol. 277, 47928-47937, Dec 6 2002.

[13] Ramaswamy, S; Nakamura, N; Sansal, I; Bergeron, L; Sellers, WR. "A novel mechanism of gene regulation and tumor suppression by the transcription factor FKHR," *Cancer Cell*, vol. 2, 81-91, Jul 2002.

[14] Furukawa-Hibi, Y; Yoshida-Araki, K; Ohta, T; Ikeda, K; Motoyama, N. "FOXO forkhead transcription factors induce G(2)-M checkpoint in response to oxidative stress," *Journal of Biological Chemistry*, vol. 277, 26729-26732, Jul 26 2002.

[15] Kops, GJPL; Dansen, TB; Polderman, PE; Saarloos, I; Wirtz, KWA; Coffer, PJ; et al. "Forkhead transcription factor FOXO3a protects quiescent cells from oxidative stress," *Nature*, vol. 419, 316-321, Sep 19 2002.

[16] Sethi, G; Sung, B; Aggarwal, BB. "Nuclear factor-kB activation: From bench to bedside," *Experimental Biology and Medicine*, vol. 233, 21-31, Jan 2008.

[17] Swindell, WR. "Genes regulated by caloric restriction have unique roles within transcriptional networks," *Mechanisms of Ageing and Development*, vol. 129, 580-592, Oct 2008.

[18] Helenius, M; Hanninen, M; Lehtinen, SK; Salminen, A. "Aging induced up regulation of nuclear binding activities of oxidative stress responsive NF-kB transcription factor in mouse cardiac muscle," *Journal of Molecular and Cellular Cardiology*, vol. 28, 487-498, Mar 1996.

[19] Donato, AJ; Eskurza, I; Silver, AE; Levy, AS; Pierce, GL; Gates, PE. et al. "Direct evidence of endothelial oxidative stress with aging in humans - Relation to impaired endothelium-dependent dilation and upregulation of nuclear factor-kappa B," *Circulation Research*, vol. 100, 1659-1666, Jun 8 2007.

[20] Helenius, M; Kyrylenko, S; Vehvilainen, P; Salminen, A. "Characterization of aging-associated up-regulation of constitutive nuclear factor-kappa B binding activity," *Antioxidants & Redox Signaling*, vol. 3, 147-156, Feb 2001.

[21] Csiszar, A; Wang, MY; Lakatta, EG; Ungvari, Z. "Inflammation and endothelial dysfunction during aging: role of NF-kappa B," *Journal of Applied Physiology*, vol. 105, 1333-1341, Oct 2008.

[22] Gosselin, K; Abbadie, C. "Involvement of Rel/NF-kappa B transcription factors in senescence," *Exp Gerontol*, vol. 38, 1271-83, Nov-Dec 2003.

[23] Kim, HJ; Jung, KJ; Yu, BP; Cho, CG; Choi, JS. Chung, HY. "Modulation of redox-sensitive transcription factors by calorie restriction during aging," *Mechanisms of Ageing and Development*, vol. 123, 1589-1595, Nov 2002.

[24] Chung, HY; Sung, B; Jung, KJ; Zou, Y; Yu, BP. "The molecular inflammatory process in aging," *Antioxid Redox Signal*, vol. 8, 572-81, Mar-Apr 2006.

[25] Jobin, C; Bradham, CA; Russo, MP; Juma, B; Narula, AS; Brenner, DA; et al. "Curcumin blocks cytokine-mediated NF-kappa B activation and proinflammatory gene expression by inhibiting inhibitory factor I-kappa B kinase activity," *Journal of Immunology*, vol. 163, 3474-3483, Sep 15 1999.

[26] Johansson, R. *System modeling and identification*. Englewood Cliffs, NJ: Prentice Hall, 1993.

[27] Richmond, A. "Nf-kappa B, chemokine gene transcription and tumour growth," *Nat Rev Immunol*, vol. 2, 664-74, Sep 2002.

[28] Mukhopadhyay, A; Banerjee, S; Stafford, LJ; Xia, CZ; Liu, MY; Aggarwal, BB. "Curcumin-induced suppression of cell proliferation correlates with down-regulation of cyclin D1 expression and CDK4-mediated retinoblastoma protein phosphorylation," *Oncogene*, vol. 21, 8852-8861, Dec 12 2002.

[29] Vousden, KH; Lu, X. "Live or let die: The cell's response to p53," *Nature Reviews Cancer*, vol. 2, 594-604, Aug 2002.

[30] Tyson, JJ. "Another turn for p53," *Molecular Systems Biology*, vol. 2, 2006.

[31] Aranda-Anzaldo, A; Dent, MAR. "Reassessing the role of p53 in cancer and ageing from an evolutionary perspective," *Mechanisms of Ageing and Development*, vol. 128, 293-302, Apr 2007.

[32] Macleod, KF; Sherry, N; Hannon, G; Beach, D; Tokino, T; Kinzler, K; et al. "P53-Dependent and Independent Expression of P21 during Cell-Growth, Differentiation, and DNA-Damage," *Genes & Development*, vol. 9, 935-944, Apr 15 1995.

[33] Sherr, CJ; Roberts, JM. "CDK inhibitors: positive and negative regulators of G(1)-phase progression," *Genes & Development*, vol. 13, 1501-1512, Jun 15 1999.

[34] Abramovitch, S; Glaser, T; Ouchi, T; Werner, H. "BRCA1-Sp1 interactions in transcriptional regulation of the IGF-IR gene," *Febs Letters*, vol. 541, 149-154, Apr 24 2003.

[35] Werner, H; Karnieli, E; Rauscher, FJ; LeRoith, D; "Wild-type and mutant p53 differentially regulate transcription of the insulin-like growth factor I receptor gene," *Proceedings of the National Academy of Sciences of the United States of America*, vol. 93, 8318-8323, Aug 6 1996.

[36] van Heemst, D; Beekman, M; Mooijaart, SP; Heijmans, BT; Brandt, BW; Zwaan, BJ; et al. "Reduced insulin/IGF-1 signalling and human longevity," *Aging Cell*, vol. 4, 79-85, Apr 2005.

[37] Wang, YC; Chen, BS. "A network-based biomarker approach for molecular investigation and diagnosis of lung cancer," *BMC Medical Genomics*, vol. 4, Jan 6 2011.

[38] Wang YC; Chen, BS. "Integrated cellular network of transcription regulations and protein-protein interactions," *BMC Systems Biology*, vol. 4, Mar 8 2010.

[39] Vicent, S; Lopez-Picazo, JM; Toledo, G; Lozano, MD; Torre, W; Garcia-Corchon, C; et al. "ERK1/2 is activated in non-small-cell lung cancer and associated with advanced tumours," *British Journal of Cancer*, vol. 90, 1047-1052, Mar 8 2004.

[40] Hu, K; Gan, YH; Li, SL; Gao, Y; Wu, DC; Wang, CY; et al. "Relationship of activated extracellular signal-regulated kinase 1/2 with lung metastasis in salivary adenoid cystic carcinoma," *Oncology Reports*, vol. 21, 137-143, Jan 2009.

[41] Kyriakis JM; Avruch, J. "Mammalian mitogen-activated protein kinase signal transduction pathways activated by stress and inflammation," *Physiological Reviews*, vol. 81, 807-869, Apr 2001.

[42] Ventura, JJ; Tenbaum, S; Perdiguero, E; Huth, M; Guerra, C; Barbacid, M; et al. "p38 alpha MAP kinase is essential in lung stem and progenitor cell proliferation and differentiation," *Nature Genetics*, vol. 39, 750-758, Jun 2007.

[43] Shi, YG; Massague, J. "Mechanisms of TGF-beta signaling from cell membrane to the nucleus," *Cell*, vol. 113, 685-700, Jun 13 2003.

[44] Bierie, B; Moses, HL. "TGF-beta and cancer," *Cytokine & Growth Factor Reviews*, vol. 17, 29-40, Feb-Apr 2006.

[45] Langenfeld, EM; Kong, Y; Langenfeld, J. "Bone morphogenetic protein 2 stimulation of tumor growth involves the activation of Smad-1/5," *Oncogene*, vol. 25, 685-692, Feb 2006.

[46] Zwick, E; Bange, J; Ullrich, A. "Receptor tyrosine kinase signalling as a target for cancer intervention strategies," *Endocrine-Related Cancer*, vol. 8, 161-173, Sep 2001.

[47] Langer, CJ. "Emerging role of epidermal growth factor receptor inhibition in therapy for advanced malignancy: Focus on NSCLC," *International Journal of Radiation Oncology Biology Physics*, vol. 58, 991-1002, Mar 1 2004.

[48] Ding, L; Getz, G; Wheeler, DA; Mardis, ER; McLellan, MD; Cibulskis, K; et al. "Somatic mutations affect key pathways in lung adenocarcinoma," *Nature*, vol. 455, 1069-1075, Oct 23 2008.

[49] Herbst, RS; Heymach, JV; Lippman, SM. "Molecular origins of cancer: Lung cancer," *New England Journal of Medicine*, vol. 359, 1367-1380, Sep 25 2008.

[50] Behrens, C; Lin, HY; Lee, JJ; Raso, MG; Hong, WK; Wistuba, II; et al. "Immunohistochemical Expression of Basic Fibroblast Growth Factor and Fibroblast Growth Factor Receptors 1 and 2 in the Pathogenesis of Lung Cancer," *Clinical Cancer Research*, vol. 14, 6014-6022, Oct 1 2008.

[51] Cox, ME; Gleave, ME; Zakikhani, M; Bell, RH; Piura, E; Vickers, E; et al. "Insulin Receptor Expression by Human Prostate Cancers," *Prostate*, vol. 69, 33-40, Jan 1 2009.

[52] Kishimoto, M; Kohno, T; Okudela, K; Otsuka, A; Sasaki, H; Tanabe, C; et al. "Mutations and deletions of the CBP gene in human lung cancer," *Clinical Cancer Research*, vol. 11, 512-519, Jan 15 2005.

[53] Barker, N; Clevers, H. "Mining the Wnt pathway for cancer therapeutics," *Nature Reviews Drug Discovery*, vol. 5, 997-1014, Dec 2006.

[54] Imai, MA; Oda, Y; Oda, M; Nakanishi, I; Kawahara, E. "Overexpression of E2F1 associated with LOH at RB locus and hyperphosphorylation of RB in non-small cell lung carcinoma," *Journal of Cancer Research and Clinical Oncology*, vol. 130, 320-326, Jun 2004.

[55] Zajac-Kaye, M. "Myc oncogene: a key component in cell cycle regulation and its implication for lung cancer," *Lung Cancer*, vol. 34, S43-S46, Dec 2001.

[56] Chen, ZY; Cai, L; Bie, P; Wang, SG; Jiang, Y; Dong, JH; et al. "Roles of Fyn in pancreatic cancer metastasis," *Journal of Gastroenterology and Hepatology*, vol. 25, 293-301, Feb 2010.

[57] Vander Griend, DJ; D'Antonio, J; Gurel, B; Antony, L; DeMarzo, AM; Isaacs, JT. "Cell-Autonomous Intracellular Androgen Receptor Signaling Drives the Growth of Human Prostate Cancer Initiating Cells," *Prostate*, vol. 70, 90-99, Jan 1 2010.

[58] Henderson, BE; Feigelson, HS. "Hormonal carcinogenesis," *Carcinogenesis*, vol. 21, 427-433, Mar 2000.

[59] Byers, LA; Sen, B; Saigal, B; Diao, LX; Wang, J; Nanjundan, M; et al. "Reciprocal Regulation of c-Src and STAT3 in Non-Small Cell Lung Cancer," *Clinical Cancer Research*, vol. 15, 6852-6861, Nov 15 2009.

[60] Li, ZG; Zhao, J; Du, YH; Park, HR; Sun, SY; Bernal-Mizrachi, L; et al. "Down-regulation of 14-3-3 zeta suppresses anchorage-independent growth of lung cancer cells through anoikis activation," *Proceedings of the National Academy of Sciences of the United States of America*, vol. 105, 162-167, Jan 8 2008.

[61] Wang, B; Liu, K; Lin, HY; Bellam, N; Ling, SY; Lin, WC. "14-3-3 tau Regulates Ubiquitin-Independent Proteasomal Degradation of p21, a Novel Mechanism of p21 Downregulation in Breast Cancer," *Molecular and Cellular Biology*, vol. 30, 1508-1527, Mar 15 2010.

[62] Ruiz, MIG; Floor, K; Roepman, P; Rodriguez, JA; Meijer, GA; Mooi, WJ; et al. "Integration of Gene Dosage and Gene Expression in Non-Small Cell Lung Cancer, Identification of HSP90 as Potential Target," *Plos One*, vol. 3, Mar 5 2008.

[63] Hanahan, D; Weinberg, RA."The hallmarks of cancer," *Cell*, vol. 100, 57-70, Jan 7 2000.

[64] Hollstein, M; Sidransky, D; Vogelstein, B; Harris, CC. "P53 Mutations in Human Cancers," *Science*, vol. 253, 49-53, Jul 5 1991.

[65] Coussens, LM; Werb, Z. "Inflammation and cancer," *Nature*, vol. 420, 860-867, Dec 26 2002.

[66] Inoue, J; Gohda, J; Akiyama, T; Semba, K. "NF-kappa B activation in development and progression of cancer," *Cancer Science*, vol. 98, 268-274, Mar 2007.

[67] Welcsh, PL; Owens, KN; King, MC. "Insights into the functions of BRCA1 and BRCA2," *Trends in Genetics*, vol. 16, 69-74, Feb 2000.

[68] Deniaud, E; Baguet, J; Mathieu, AL; Pages, G; Marvel, J; Leverrier, Y; "Overexpression of Sp1 transcription factor induces apoptosis," *Oncogene*, vol. 25, 7096-7105, Nov 9 2006.

[69] Potter, CJ; Pedraza, LG; Xu, T. "Akt regulates growth by directly phosphorylating Tsc2," *Nature Cell Biology*, vol. 4, 658-665, Sep 2002.

[70] Wang, HM; Davis, A; Yu, SH; Ahmed, K. "Response of cancer cells to molecular interruption of the CK2 signal," *Molecular and Cellular Biochemistry*, vol. 227, 167-174, 2001.

[71] West, KA; Linnoila, IR; Belinsky, SA; Harris, CC; Dennis, PA. "Tobacco carcinogen-induced cellular transformation increases activation of the phosphatidylinositol 3 '-Kinase/Akt pathway in vitro and in vivo," *Cancer Research*, vol. 64, 446-451, Jan 15 2004.

[72] Mitra, SK; Schlaepfer, DD. "Integrin-regulated FAK-Src signaling in normal and cancer cells," *Current Opinion in Cell Biology*, vol. 18, 516-523, Oct 2006.

[73] McLean, GW; Carragher, NO; Avizienyte, E; Evans, J; Brunton, VG; Frame, MC; "The role of focal-adhesion kinase in cancer. A ew therapeutic opportunity," *Nature Reviews Cancer*, vol. 5, 505-515, Jul 2005.

[74] Liu, H; Radisky, DC; Nelson, CM; Zhang, H; Fata, JE; Roth, RA. et al. "Mechanism of Akt1 inhibition of breast cancer cell invasion reveals a protumorigenic role for TSC2," *Proceedings of the National Academy of Sciences of the United States of America*, vol. 103, 4134-4139, Mar 14 2006.

[75] Sala, G; Dituri, F; Raimondi, C; Previdi, S; Maffucci, T; Mazzoletti, M; et al. "Phospholipase C gamma 1 Is Required for Metastasis Development and Progression," *Cancer Research*, vol. 68, 10187-10196, Dec 15 2008.

[76] Deng, XY; Tannehill-Gregg, SH; Nadella, MVP; He, GC; Levine, A; Cao, Y; et al. "Parathyroid hormone-related protein and ezrin are up-regulated in human lung cancer bone metastases," *Clinical & Experimental Metastasis*, vol. 24, 107-119, Apr 2007.

[77] Komander, D. "The emerging complexity of protein ubiquitination," *Biochemical Society Transactions*, vol. 37, 937-953, Oct 2009.

[78] Pagano M; Gauvreau, K. *Principles of biostatistics*, 2nd ed. Pacific Grove, CA: Duxbury, 2000.

Chapter 6

NETWORK ROBUSTNESS, FILTERING, AND TRANSDUCTION ABILITY ANALYSIS OF STOCHASTIC BIOLOGICAL NETWORKS

ABSTRACT

Robust stabilization, environmental noise filtering, and signal transduction are ubiquitous systematic properties observed at different levels in biological networks. The underlying principles for these properties are present in both complex biological systems and sophisticated engineering systems. In many biological networks, network robustness should be sufficient to confer both intrinsic robustness for tolerating intrinsic parameter fluctuations and environmental robustness for filtering environmental noises. In this chapter, a unifying mathematical framework is developed to investigate the principles of both network robustness and environmental noise filtering in stochastic biological networks. It was determined that the phenotype robustness criterion for biological networks at different levels relies on the sum of intrinsic robustness and environmental robustness totaling less than or equal the network robustness. Transduction ability can be also measured based on filtering ability for signal transduction or metabolic pathways.

INTRODUCTION

Molecular noise plays many roles in GRN cellular functioning. These include noise-driven divergence of cell fates and population heterogeneity, noise-induced amplification of signals, generation of errors in DNA replication leading to mutation and evolution, and maintenance of quantitative individuality of cells [1]. Other cellular processes influenced by noise include ion-channel gating, neural firing, developmental modules, cytoskeleton dynamics, and motors [1]. Phase variation in pathogenic bacteria, in which cells alternate randomly between expressing certain genes and silencing others, is thought to be a form of cultivated noise [2]. These molecular-level noisy phenomena are deeply rooted in the statistical mechanical behavior of so-called nanoscale chemical systems, where concentrations of reactive molecule species are extremely low and, consequently, fluctuations (noises) in reaction rates are large [1]. Even though molecular fluctuations leading to phase variation seem random in the individual, regulatory factors in GRNs tune the variations through filtration or attenuation to ensure mean levels of heterogeneity for the population [2].

Gene expression is best viewed as a stochastic process, since cellular molecular events are subject to significant thermal fluctuations and noisy processes of transcriptional control, alternative splicing, translation, diffusion, and chemical modification reactions. Many observations suggest that molecular events underlying cellular physiology are subject to random fluctuations; these observations have led to the proposal of a stochastic model for gene expression and cellular functions [2]. Noise filtering can be considered from a signal processing perspective [3]. From this perspective, a pathway is viewed as an analog filter in terms of its response frequency. In terms of signal processing, these cellular pathways function as biological filters that transduce molecular signals and filter molecular noise [2].

In this chapter, we show how, based on stochastic Lyapunov theory, the network robustness of nonlinear stochastic biological networks can be obtained by solving a Hamilton-Jacobi inequality (HJI). Using H_∞ noise attenuation, the filtering ability of biological networks can then be calculated by solving an HJI-constrained optimization problem. Finally, based on the input/output system gain method, the transduction ability measure requires solving an HJI-constrained optimization problem. In order to avoid solving the difficult HJI for network robustness, network filtering, and transduction ability of biological networks, the T-S fuzzy method can be employed to interpolate some local stochastic linear networks to approximate a nonlinear stochastic network. In this situation, the HJI can be replaced by a set of LMIs, which are easily solvable with the help of the LMI toolbox in Matlab.

6.1. Network Robustness Analysis of Stochastic GRNs

For the convenience of illustration, the following linear biochemical dynamics of n genes GRN in Figure 6.1 [2] is simply considered initially:

$$\frac{dx(t)}{dt} = Nx(t) \tag{6.1}$$

where the concentration vector $x(t)$ and stoichiometric matrix N are given by

$$x(t) = \begin{bmatrix} x_1(t) \\ x_2(t) \\ \vdots \\ x_n(t) \end{bmatrix}, \quad N = \begin{bmatrix} N_{11} & \cdots & N_{1n} \\ \vdots & N_{ij} & \vdots \\ N_{n1} & \cdots & N_{nn} \end{bmatrix}$$

where $x_i(t)$ denotes the concentration of the i-th gene, and N_{ij} denotes the interaction between genes j and i.

Suppose the linear GRN suffers intrinsic molecular fluctuations so that the stoichiometric matrix N is perturbed as $N + \Delta N$,

$$\Delta N = \begin{bmatrix} \Delta N_{11}(t) & \cdots & \Delta N_{1n}(t) \\ \vdots & \Delta N_{ij}(t) & \vdots \\ \Delta N_{n1}(t) & \cdots & \Delta N_{nn}(t) \end{bmatrix} = \begin{bmatrix} M_{11} & \cdots & M_{1n} \\ \vdots & M_{ij} & \vdots \\ M_{n1} & \cdots & M_{nn} \end{bmatrix} n(t) = Mn(t)$$

where ΔN_{ij} denotes the random parametric fluctuation of N_{ij}; M_{ij} denotes the deterministic part (amplitude) of fluctuation; and $n(t)$ is white Gaussian noise with zero mean and unit variance, and denotes the stochastic part of fluctuation, i.e., the stochastic part of fluctuation is absorbed to $n(t)$.

The GRN with random parametric fluctuation can then be represented by

$$\begin{aligned} \frac{dx(t)}{dt} &= \left(N + \Delta N\right) x(t) \\ &= Nx(t) + Mx(t)n(t) \end{aligned} \tag{6.2}$$

For the convenience of analysis by the stochastic process, the above stochastic GRN can be represented as follows [3, 4]:

$$dx(t) = Nx(t)dt + Mx(t)d\omega\left(t\right) \tag{6.3}$$

where $d\omega(t) = n(t)dt$. $\omega(t)$ is called the stochastic Wiener process or Brownian motion [5].

The stochastic dynamic equations of GRNs are always nonlinear. In order to conform to nonlinear stochastic regulatory networks, equation (6.2) should be generalized as the following Langevin equation [2]

$$dx(t) = N(x)dt + M\left(x\right)d\omega\left(t\right) \tag{6.4}$$

where $N(x)$ denotes the nonlinear interaction equation of the nonlinear GRN, and $M(x)d\omega(t)$ is due to nonlinear intrinsic random fluctuation.

Based on stochastic Lyapunov theory, let $V(x) > 0$ with $V(0) = 0$ denote the Lyapunov power-like function, then the stochastic GRN in equation (6.2) or (6.3) is stochastically stable if $E[dV(x(t))/dt] \leq 0$ [5]. With the choice of $V(x) = x^T(t)Px(t)$ for some positive definite matrix P, the following result is derived.

Proposition 6.1 [2]: The linear GRN with stochastic perturbation in equation (6.2) or (6.3) is stochastically stable if the following LMI

$$PN + N^T P + M^T PM \leq 0 \tag{6.5}$$

has a symmetric positive definite solution $P > 0$, i.e., the phenotype of the linear stochastic GRN is maintained under intrinsic stochastic fluctuation.

Proof: See 6.5.1

Remark 6.1: (i) In the intrinsic noise-free case, the stable condition in (6.5) is reduced to $PN + N^T P \leq 0$, i.e., the eigenvalues of system matrix N should be on the left-hand side of the complex domain. Obviously, if the LMI in equation (6.5) has a positive solution $P > 0$, then the eigenvalues of N should be located on the far left-hand side of the complex domain with more negative real values in order to overcome the additional term $M^T PM$ due to intrinsic random noise. (ii) If some eigenvalues of system interaction matrix N are near the $j\omega$ axis, then intrinsic random molecular fluctuations across the $j\omega$ axis perturb these modes more easily such that the linear GRN becomes unstable. The LMI in equation (6.5) can be rearranged to

$$\underbrace{M^T PM}_{\text{intrinsic robustness}} \leq \underbrace{-(PN + N^T P)}_{\text{network robustness}} \tag{6.6}$$

that is, $-(PN + N^T P)$ in equation (6.6) can be taken as a measure of network robustness, and $M^T PM$ due to the random parametric fluctuation can be taken as a measure of intrinsic robustness. The physical interpretation of equation (6.6) is that if the network robustness can confer enough intrinsic robustness for tolerating intrinsic random parameter fluctuation, then the phenotype of the GRN is maintained.

For the nonlinear stochastic GRN in equation (6.4), the following result is also derived.

Proposition 6.2 [2]: The nonlinear stochastic GRN in equation (6.4) is still stochastically stable if the following Hamilton–Jacobi inequality (HJI) has a positive Lyapunov solution $V(x) > 0$ with $V(0) = 0$

$$\left(\frac{\partial V(x)}{\partial x}\right)^T N(x) + \frac{1}{2} M^T(x) \frac{\partial^2 V(x)}{\partial x^2} M(x) \leq 0 \tag{6.7}$$

That is, the phenotype of the nonlinear stochastic GRN in equation (6.4) is maintained under intrinsic stochastic fluctuation. Similarly, the HJI in equation (6.7) can be rearranged as the following phenotype robustness criterion:

$$\underbrace{\frac{1}{2} M^T(x) \frac{\partial^2 V(x)}{\partial x^2} M(x)}_{\text{intrinsic robustness}} \leq \underbrace{-\left(\frac{\partial V(x)}{\partial x}\right)^T N(x)}_{\text{network robustness}} \tag{6.8}$$

That is, if the network robustness of the nonlinear stochastic GRN in equation (6.4) can confer enough intrinsic robustness to tolerate intrinsic stochastic fluctuation, then the phenotype of the GRN is maintained.

Proof: See 6.5.2

6.2. FILTERING ANALYSIS OF STOCHASTIC GRNS

After the robust stability of GRN is guaranteed under intrinsic biochemical stochastic fluctuation, the effect of environmental random molecular noises on the GRN may be discussed. If the linear GRN in equation (6.3) also suffers from environmental molecular noises $v(t) = [v_1(t) \cdots v_n(t)]^T$ outside the network (see Figure 6.1), then

$$\begin{aligned} dx(t) &= \left(Nx(t) + Hv(t)\right)dt + Mx(t)d\omega(t) \\ Z(t) &= Cx(t) \end{aligned} \tag{6.9}$$

where H is a coupling matrix denoting the influence of environmental molecular signals $v(t)$ on the GRN. $Z(t)$ denotes the concentration of specific genes of interest. For example, if we want to examine the effect of molecular noises of $\omega(t)$ and $v(t)$ on gene i (i.e., $x_i(t)$), then we let $C = \mathrm{diag}(0\ldots010\ldots0)$. That is, every element of C is zero except for the i-th element. To investigate the effect of molecular noises on the whole GRN, then $C = I$, the identity matrix. The positive value ρ in the following inequality is then called the effect of environmental noises (or signals) on $Z(t)$ in the stochastic GRN in equation (6.9) with $x(0) = 0$

$$\frac{E\int_0^\infty Z^T(t)Z(t)dt}{E\int_0^\infty v^T(t)v(t)dt} \leq \rho^2 \tag{6.10}$$

That is, ρ is the upper bound of the effect of all environmental molecular signals or noises $v(t)$ on the GRN. It is called the response level of the GRN.

From this is derived the following response level result of the GRN in equation (6.9).

Proposition 6.3 [2]: The response level ρ of the linear stochastic GRN in equation (6.9) is guaranteed if the following matrix inequality has a positive solution $P > 0$

$$PN + N^T P + M^T PM + C^T C + \frac{1}{\rho^2} PHH^T P \leq 0 \tag{6.11}$$

By Shur complement [6], the above inequality is equivalent to the following LMI:

$$\begin{bmatrix} PN + N^T P + C^T C + M^T PM & PH \\ (PH)^T & -\rho^2 \end{bmatrix} \leq 0 \tag{6.12}$$

That is, if the above LMI has a positive solution $P > 0$, then the linear stochastic GRN in equation (6.9) is robustly stable under random intrinsic molecular fluctuation and has a response level ρ to environmental molecular noises.

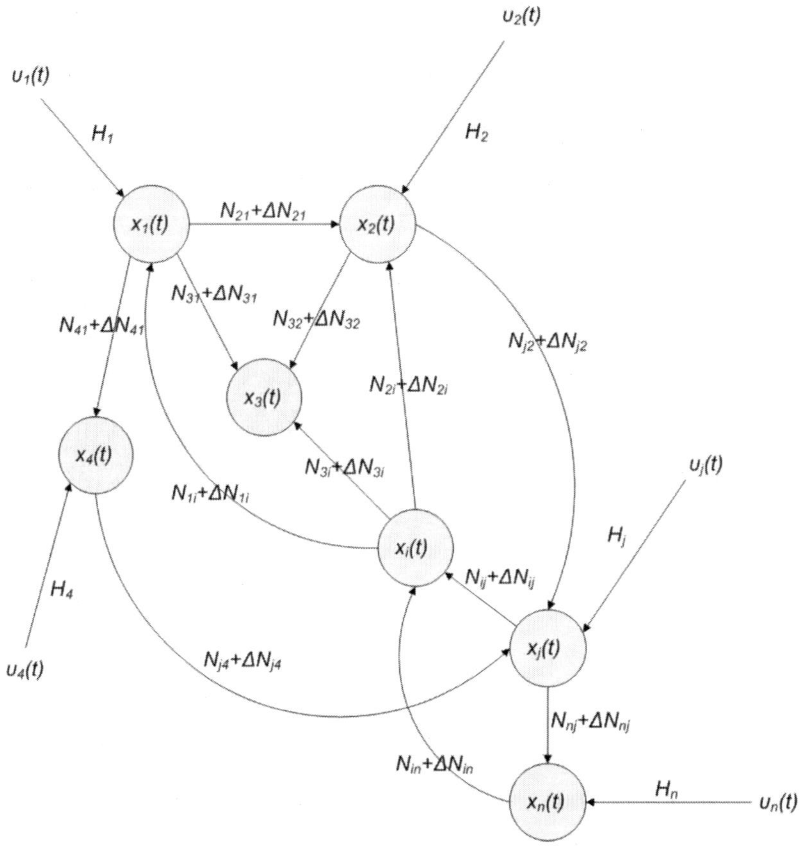

Figure 6.1. The linear n genes GRN with interaction N_{ij}, **intrinsic fluctuation** ΔN_{ij}, gene expression $x_i(t)$, and extrinsic fluctuation $\upsilon_i(t)$.

Proof: see 6.5.3

The optimal response level ρ_0 of the linear stochastic GRN (equation (6.9)) can be obtained by solving the following constrained optimization

$$\begin{aligned} &\rho_0 = \min \rho \\ &\text{subject to } \rho > 0, P > 0 \text{ and LMI in (6.12)} \end{aligned} \tag{6.13}$$

Remark 6.2: (i) If $\rho_0 < 1$, then environmental molecular noises $\upsilon(t)$ are attenuated by the GRN and ρ_0 is called the filtering ability of the GRN, i.e., the GRN is less sensitive to environmental noises. If $\rho_0 > 1$, then environmental molecular noises are amplified by the GRN, i.e., the GRN is more sensitive to environmental noises. (ii) Substituting ρ_0 for ρ in equation (6.11) and rearranging, the following phenotype robustness criterion of the stochastic GRN is derived

$$\underbrace{M^T PM}_{\text{intrinsic robustness}} + \underbrace{C^T C + \frac{1}{\rho_0^{\ 2}} PHH^T P}_{\text{environmental robustness}} \leq \underbrace{-\left(PN + N^T P\right)}_{\text{network robustness}} \tag{6.14}$$

The phenotype robustness criterion in (6.14) for the stochastic GRN in (6.9) can be denoted as "intrinsic robustness + environmental robustness ≤ network robustness". In other words, if the sum of intrinsic robustness and environmental robustness is less than network robustness, then the phenotype of the stochastic GRN remains robust under the influence of stochastic intrinsic fluctuation and environmental noises. In order to maintain the phenotype robustness criterion, GRNs need negative feedback loops to improve the network robustness on the right-hand side of equation (6.14). Parallel loops and modular and redundant structures are also required to reduce the effect of intrinsic parameter variations on the GRN and to resist environmental disturbance, i.e., to provide more intrinsic robustness and environmental robustness on the left hand side of equation (6.14). This is why feedback loops, parallel loops, and modular and redundant structures are abundant in GRNs as they contribute to phenotype robustness and favor natural selection. (iii) If the network robustness in equation (6.14) is not large enough to confer intrinsic robustness and environmental robustness simultaneously to maintain the phenotype of the GRN, some negative feedback gene loops are implemented as follows:

$$\begin{aligned} dx(t) &= \big((N+K)x(t) + Hv(t)\big)dt + Mx(t)d\omega(t) \\ Z(t) &= Cx(t) \end{aligned} \tag{6.15}$$

where K denotes specific negative feedback loops that improve the network robustness of the stochastic GRN. In this situation, the phenotype criterion in equation (6.14) is modified as

$$\underbrace{M^T P M}_{\text{intrinsic robustness}} + \underbrace{C^T C + \frac{1}{\rho_0^2} P H H^T P}_{\text{environmental robustness}} \le \underbrace{-\left[P(N+K) + (N+K)^T P\right]}_{\text{network robustness}} \tag{6.16}$$

It can be seen from the phenotype robustness criterion in equation (6.16) that an adequate K can improve the network robustness of the stochastic GRN such that it better tolerates intrinsic stochastic fluctuation and is less responsive to environmental noise (i.e., smaller ρ_0).

Similarly, the nonlinear stochastic GRN under environmental molecular noises (equation (6.4)) should be modified as follows:

$$\begin{aligned} dx(t) &= \big(N(x) + H(x)v(t)\big)dt + M(x)d\omega(t) \\ Z(t) &= Cx(t) \end{aligned} \tag{6.17}$$

The phenotype robustness criterion in equation (6.14) is then modified as follows [2]:

$$\underbrace{x^T C^T C x + \frac{1}{4\rho_0^2}\left(\frac{\partial V(x)}{\partial x}\right)^T H H^T \frac{\partial V(x)}{\partial x}}_{\text{environmental robustness}} + \underbrace{\frac{1}{2} M^T(x) \frac{\partial^2 V(x)}{\partial x^2} M}_{\text{intrinsic robustness}} \le \underbrace{-\left(\frac{\partial V(x)}{\partial x}\right)^T N(x)}_{\text{network robustness}} \tag{6.18}$$

In general, it is very difficult to solve the HJI in equations (6.8) and (6.18). Global linearization techniques [6, 7] or Takagi-Sugeno (T-S) fuzzy methods [8, 9] can be employed to interpolate several local linearized systems to approximate the nonlinear stochastic system in equation (6.17) as follows:

$$\begin{aligned} dx(t) &= \sum_{i=1}^{L}\left(h_i(z)N_i x(t) + H_i v(t)\right)dt + M_i x d\omega(t) \\ Z(t) &= Cx(t) \end{aligned} \quad (6.19)$$

where $h_i(z)$, $i = 1,\cdots, L$ denotes the interpolation functions of global linearization or fuzzy bases with $h_i(z) \geq 0$, $\sum_{i=1}^{L} h_i(z) = 1$, and $dx = (N_i x + H_i v)dt + M_i x d\omega$ is the i-th local linearized GRN.

In this situation, the network robustness criterion in equation (6.18) can be modified accordingly [2]

$$\underbrace{C^T C + \frac{1}{\rho_0^2} P H_i H_i^T P}_{\text{local environmental robustness}} + \underbrace{M_i^T P M_i}_{\text{local intrinsic robustness}} \leq \underbrace{-\left(P N_i + N_i^T P\right)}_{\text{local network robustness}},\ i = 1, 2, \cdots, L \quad (6.20)$$

From the phenotype robustness criterion in equation (6.20), it is seen that if local network robustness is large enough for every local linearized GRN to simultaneously provide enough local intrinsic robustness for tolerating local random parameter fluctuations and enough local environmental robustness for resisting local environmental disturbance, then the phenotype of nonlinear stochastic GRN is maintained in spite of intrinsic stochastic parameter fluctuations and environmental molecular noises. If the phenotype robustness criterion in equation (6.20) cannot be maintained, then negative feedback loops can be engineered to improve network robustness. These feedback loops may have potential applications for some types of therapy and drug target selection.

Systematic methodologies for the analysis of system characteristics of GRNs and PPINs detailed above are useful for the systematic designs of systems synthetic biology discussed in the following chapter.

6.3. Transduction Ability Analysis of Signal Transduction Pathways

The major function of signal transduction pathways in cells is to sense signals from the environment and process the information through signaling molecules in order to regulate the activity of transcription factors. On the molecular level, the information transmitted by a small number of signal molecules is amplified in the internal signaling pathway through enzyme catalysis, molecular modification and via the activation or inhibition of interactions. However, the dynamic system behavior of a signaling pathway can be complex and, despite knowledge of the pathway components and interactions, it is still a challenge to interpret the pathways behavior. Therefore, a systematic method is proposed in this chapter to quantify the signal transduction ability.

Signal transduction pathways are one of the fastest information processing networks. They enable cells to sense and transduce extracellular signals through inter- and intra- cellular communication. Signal transduction pathways consist of interactions between signaling proteins, where different external changes or events stimulate specific signaling pathways. Typical external signal stimuli are hormones, pheromones, heat, cold, light, osmotic pressure and concentration change of substances such as, K^+, Ca^{2+} or cAMP [10-12]. Metabolic processes, such as homeostasis [10], the growth [13] and differentiation of tissues through endocrine signaling [14], the synthesis and secretion of proteins, such as in blood clotting and immune-responses and the composition of intracellular and extracellular fluids, such as in the processes of photoreception and olfaction, are all closely regulated by specific signaling pathways [15]. Hence the proper function of cellular signaling pathways is essential for life.

On the molecular level, signal transduction involves the production or degradation of substances, molecular modification and the activation or inhibition of chemical reactions. Many previous studies have focused on signaling molecules, cell-surface receptors and the second messengers of the signaling pathway. Aside from molecular transfer, signaling pathways can also be viewed as the information processing and transferring systems of cells [16]. Many studies have investigated the properties of signaling pathways, such as amplification [10], specificity [17], adaption, ultrasensitivity [18, 19], oscillation [19] and synchronization. However, due to the complex behavior of signaling pathways, knowledge of the components of a pathway and their interactions is often not enough to interpret the dynamic system behavior of the pathway. To the best of our knowledge, the intension to express the signal transduction ability into mathematical formula appeared in 1982 [15]. It focused on the sensitivity amplification, which is defined as the ratio of the percent change in the output response to the percent change in the input stimulus, i.e., the relative change of the system output with respect to a specific input. Then the MAPK signaling pathway in linear form is studied and signal amplitude is defined as the ratio of the total response to the signal duration, which depends on the input [20]. Next, a stochastic differential model of MAPK signaling pathway is considered and the signal amplification is defined as signal gain, which exists at least one maximum with respect to the ratio of inputs to outputs [21]. And recently, Marhl, et al. compared the signaling pathways in biological systems with the cascade mechanisms in electrical engineering systems and then also defined the maximal amplification as the ratio of the maximum of output to the maximum of input, i.e., $output_{max}/input_{max}$ [22]. Obviously, these mentioned techniques defined the signal transduction ability on case-by-case basis, that is, the results are affected not only by the structure of the system but also the input to the system. This chapter specifically investigates signal transduction ability, a dynamic system behavior of signaling pathways.

As a system for processing and transferring information, signaling pathways require an amplification mechanism to reduce the expense associated with the production of signaling molecules and to ensure rapid responses. From the viewpoint of energy expenditure, signal amplification can effectively reduce the quantity of initial key substances necessary to ensure upstream sensing as well as having significant effects on downstream signal transduction. Also, if a deleterious stress impacts a cell, the cell is required to ensure a response is induced as soon as possible through signaling pathways. Accordingly, there are numerous studies indicating the importance and the functional role of signal amplification. An example is the osmoadaptation of yeast under osmotic stress, where an integrated mathematical model is provided for the signaling pathway to demonstrate the relationship between the rate constants of kinases and phosphatases and signal amplification [23]. For liver glycogenolysis and the release of glucose, a level of 10^{-10}M of epinephrine in blood can generate intracellular concentrations of 10^{-6}M cAMP with a signal amplification of 10^4 [24].

6.3.1. Dynamic Model of Signaling Pathways

Mathematical modeling of metabolic networks has been studied extensively and is useful for describing experimental data, deducing regulatory principles, and understanding more complex dynamic phenomena, such as oscillation [19, 25, 26]. However, in modeling signaling pathways, few examples have been studied due to the unavailability of experimental data. As the structure of signaling pathways consists of consecutive activations and deactivations of proteins, the signaling pathways can be modeled as a series of chemical reactions, expressed as a system of rate equations. The appropriate choice of rate expressions is a matter of a prior knowledge of the whole system and modeling intuition. For instance, linear and bilinear expressions or Michaelis-Menten kinetics have previously been used to model signaling pathways [16, 26].

Independent or coupled kinase cascades participate in a variety of different intracellular signaling pathways that cover the spectrum of cellular processes, including intracellular cell growth, differentiation, transformation and apoptosis. For example, mitogen-activated protein kinase (MAPK) cascades transduce signals from the cell membrane to the nucleus in response to a wide range of stimuli (Figure 6.2) [24]. The pathway consists of several levels, where the activated kinase at each level phosphorylates the kinase at the next level in the cascade (Figure 6.2). The dynamic modeling of a MAPK cascade may be represented by the system in (6.21) with Michaelis-Menten kinetics formulation [19].

$$\frac{d}{dt}x_1 = \frac{-k_1 \cdot x_1 \cdot u}{K_{m,1} + x_1} + \frac{V_{\max,2} \cdot x_2}{K_{m,2} + x_2}$$

$$\frac{d}{dt}x_2 = \frac{k_1 \cdot x_1 \cdot u}{K_{m,1} + x_1} - \frac{V_{\max,2} \cdot x_2}{K_{m,2} + x_2}$$

$$\frac{d}{dt}x_3 = \frac{-k_3 \cdot x_2 \cdot x_3}{K_{m,3} + x_3} + \frac{V_{\max,6} \cdot x_4}{K_{m,6} + x_4}$$

$$\frac{d}{dt}x_4 = \frac{k_3 \cdot x_2 \cdot x_3}{K_{m,3} + x_3} - \frac{V_{\max,6} \cdot x_4}{K_{m,6} + x_4} - \frac{k_4 \cdot x_2 \cdot x_4}{K_{m,4} + x_4} + \frac{V_{\max,5} \cdot x_5}{K_{m,5} + x_5}$$

$$\frac{d}{dt}x_5 = \frac{k_4 \cdot x_2 \cdot x_4}{K_{m,4} + x_4} - \frac{V_{\max,5} \cdot x_5}{K_{m,5} + x_5}$$

$$\frac{d}{dt}x_6 = \frac{-k_7 \cdot x_5 \cdot x_6}{K_{m,7} + x_6} + \frac{V_{\max,10} \cdot x_7}{K_{m,10} + x_7}$$

$$\frac{d}{dt}x_7 = \frac{k_7 \cdot x_5 \cdot x_6}{K_{m,7} + x_6} - \frac{V_{\max,10} \cdot x_7}{K_{m,10} + x_7} - \frac{k_8 \cdot x_5 \cdot x_7}{K_{m,8} + x_7} + \frac{V_{\max,9} \cdot x_8}{K_{m,9} + x_8}$$

$$\frac{d}{dt}x_8 = \frac{k_8 \cdot x_5 \cdot x_7}{K_{m,8} + x_7} - \frac{V_{\max,9} \cdot x_8}{K_{m,9} + x_8} \qquad (6.21)$$
$$y = x_8$$

Let us denote the state variable vector $x(t)$, protein interaction vector $f(x)$, input signal $u(t)$, output signal $y(t)$, input coupling vector $g(x)$ and output coupling matrix C in (6.22).

$$x(t) = \begin{bmatrix} x_1(t) \\ x_2(t) \\ x_3(t) \\ x_4(t) \\ x_5(t) \\ x_6(t) \\ x_7(t) \\ x_8(t) \end{bmatrix} = \begin{bmatrix} \text{MAPKKK}(t) \\ \text{MAPKKK-p}(t) \\ \text{MAPKK}(t) \\ \text{MAPKK-p}(t) \\ \text{MAPKK-pp}(t) \\ \text{MAPK}(t) \\ \text{MAPK-p}(t) \\ \text{MAPK-pp}(t) \end{bmatrix}$$

$$f(x) = \begin{bmatrix} \dfrac{V_{\max,2} \cdot x_2}{K_{m,2} + x_2} \\ -\dfrac{V_{\max,2} \cdot x_2}{K_{m,2} + x_2} \\ \dfrac{-k_3 \cdot x_2 \cdot x_3}{K_{m,3} + x_3} + \dfrac{V_{\max,6} \cdot x_4}{K_{m,6} + x_4} \\ \dfrac{k_3 \cdot x_2 \cdot x_3}{K_{m,3} + x_3} - \dfrac{V_{\max,6} \cdot x_4}{K_{m,6} + x_4} - \dfrac{k_4 \cdot x_2 \cdot x_4}{K_{m,4} + x_4} + \dfrac{V_{\max,5} \cdot x_5}{K_{m,5} + x_5} \\ \dfrac{k_4 \cdot x_2 \cdot x_4}{K_{m,4} + x_4} - \dfrac{V_{\max,5} \cdot x_5}{K_{m,5} + x_5} \\ \dfrac{-k_7 \cdot x_5 \cdot x_6}{K_{m,7} + x_6} + \dfrac{V_{\max,10} \cdot x_7}{K_{m,10} + x_7} \\ \dfrac{k_7 \cdot x_5 \cdot x_6}{K_{m,7} + x_6} - \dfrac{V_{\max,10} \cdot x_7}{K_{m,10} + x_7} - \dfrac{k_8 \cdot x_5 \cdot x_7}{K_{m,8} + x_7} + \dfrac{V_{\max,9} \cdot x_8}{K_{m,9} + x_8} \\ \dfrac{k_8 \cdot x_5 \cdot x_7}{K_{m,8} + x_7} - \dfrac{V_{\max,9} \cdot x_8}{K_{m,9} + x_8} \end{bmatrix}, \qquad (6.22)$$

$$u(t) = \text{MAPKKKK}(t), y(t) = \text{MAPK-pp}(t),$$

$$g(x) = \begin{bmatrix} \frac{-k_1 \cdot x_1}{K_{m,1} + x_1} & \frac{k_1 \cdot x_1}{K_{m,1} + x_1} & 0 & 0 & 0 & 0 & 0 & 0 \end{bmatrix}^T,$$

$$C = \begin{bmatrix} 0 & 0 & 0 & 0 & 0 & 0 & 0 & 1 \end{bmatrix}.$$

The dynamic system of the MAPK cascade signal transduction in (6.21) can then be represented in (6.23)

$$\begin{aligned} \dot{x}(t) &= f(x) + g(x)u(t) \\ y(t) &= Cx(t) \end{aligned} \tag{6.23}$$

where $x(t)$, $f(x)$, $g(x)$, and C are defined in (6.22).

In general, a signal transduction pathway may consist of m input signals, $u(t)=[u_1(t) \cdots u_m(t)]^T$, n proteins, $x(t)=[x_1(t) \cdots x_n(t)]^T$ in a signal transduction cascade, $f(x)$ and p output signals, $y(t)=[y_1(t) \cdots y_p(t)]^T$, where $u_i(t)$ denotes the i-th input signal, $x_i(t)$ denotes the concentration of the i-th protein in the signal cascade, and $y_i(t)$ denotes the concentration of the i-th output signal. In this situation, $f(x)$, $g(x)$, and C in (6.22) should be modified accordingly. For example, if $x_n(t)$ is the output signal to be measured then $C=[0 \cdots 0\ 1]$, i.e., C is the row vector of n elements with 0 expressed at all elements except 1 at the last entry.

System knowledge resides in models and data. A model is a construction, usually validated by data. Data validate models, and models give system meaning of data. They are opposite of the same coin. Actually, in this chapter, we assume the system models of signaling pathways have been identified by experimental data. Then, we use the system models of signaling pathways to investigate their signal transduction abilities. Since the signal transduction ability is a system property, it is not easy to investigate from input/output data directly. Further, from the signal transduction ability, we could get insight into the system meaning of signaling pathways. In this chapter, we assume the system dynamic model of signaling pathways in (6.23) has been well identified by rich input/output signals.

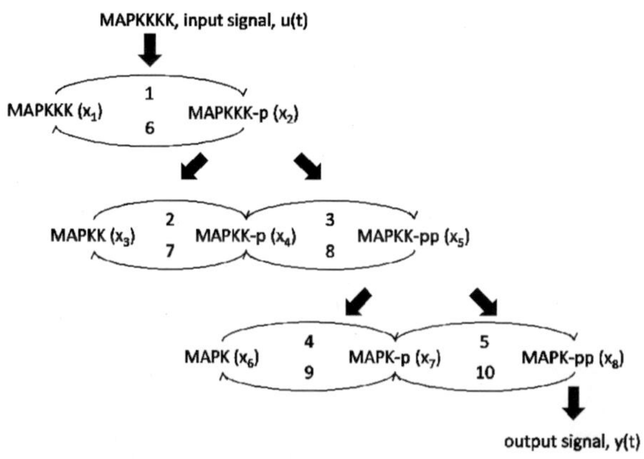

Figure 6.2. The scheme of the MAPK signal transduction pathway.

6.3.2. Signal Transduction Ability Measurement

Many properties of signaling pathways are investigated, including signaling time, signal duration or signal amplitude [20]. The focus of this chapter is on the signal transduction ability of signaling pathways. The transduction ability is defined from the system gain viewpoint as follows [6]:

$$\rho_a = \sup_{u(t)\in L_2[0,t_f]} \frac{\left(\int_0^{t_f} y^T(t)y(t)dt \Big/ t_f\right)^{1/2}}{\left(\int_0^{t_f} u^T(t)u(t)dt \Big/ t_f\right)^{1/2}} = \sup_{u(t)\in L_2[0,t_f]} \frac{\|y(t)\|_2}{\|u(t)\|_2} \tag{6.24}$$

where $u(t)$ and $y(t)$ are the input and output signals of the signaling pathway, respectively. The physical meaning of transduction ability is the maximum ratio of the output signal to the input signal from the root mean square sense for all of the possible bounded input signals within the time interval $[0,t_f]$, where $L_2[0,t_f]$ denotes the set of all possible bounded signals in $[0,t_f]$. The reason for employing the maximum root mean square ratio of the output signal to all possible input signals is that input signals may change under different conditions, causing the signal energy ratio to vary. However, transduction ability should be based on the maximum effect of all the possible input signals on the output signals and exhibit greater dependency on the characteristics of the dynamic system, in accordance with the system gain viewpoint. Theoretically, the transduction ability can be measured directly by testing all possible bounded input signals to get the maximum root mean square ratio. In general, the maximization problem described in (6.24) for measuring the signal transduction ability of a signaling pathway cannot be solved directly. It can only be solved indirectly by applying a systematic analysis method on the signal transduction dynamic in (6.23), as described in the following expressions.

Let us denote an upper bound of transduction ability as follows:

$$\|y(t)\|_2^2 \Big/ \|u(t)\|_2^2 \le \rho^2, \forall u(t) \in L_2[0,t_f] \tag{6.25}$$

or

$$\|y(t)\|_2^2 \le \rho^2 \|u(t)\|_2^2, \forall u(t) \in L_2[0,t_f] \tag{6.26}$$

In (6.25) or (6.26), if ρ^2 is the upper bound of $\|y(t)\|_2^2 / \|u(t)\|_2^2$ for every $u(t)\in L_2[0, t_f]$, then ρ^2 is also the upper bound of ρ_a^2 in (6.24), as every $u(t)\in L_2[0, t_f]$ should include the input signal which leads to the maximum mean square ratio ρ_a in (6.24). Therefore, we can find an upper bound ρ for the signal transduction ability ρ_a and then minimize the upper bound to approach the transduction ability, i.e., the signal transduction ability ρ_a is estimated by minimizing the upper bound ρ.

Proposition 6.4 [27]: If a positive function $V(x)$ with $V(0)=0$ is able to satisfy the following Hamilton-Jacobi inequality (HJI):

$$\left(\frac{\partial V(x)}{\partial x}\right)^T f(x)+x^T C^T Cx+\frac{1}{4\rho^2}\left(\frac{\partial V(x)}{\partial x}\right)^T g(x)g^T(x)\frac{\partial V(x)}{\partial x}\leq 0 \quad (6.27)$$

then the signal transduction ability of the signal pathway is less than or equal to ρ.

Proof: See 6.5.4

Since ρ is the upper bound of transduction ability ρ_a, the transduction ability can be obtained by solving the following constrained optimization problem:

$$\begin{gathered}\rho_a = \min \rho \\ \text{subject to HJI in (6.27) and } V(x)>0\end{gathered} \quad (6.28)$$

It means to make ρ as small as possible without violating HJI constraint in (6.27). After solving the transduction ability ρ_a from the HJI constraint optimization expression (8) and substituting ρ_a for ρ in (6.27), we find:

$$\left(\frac{\partial V(x)}{\partial x}\right)^T f(x)+x^T C^T Cx+\frac{1}{4\rho_a^2}\left(\frac{\partial V(x)}{\partial x}\right)^T g(x)g(x)^T\frac{\partial V(x)}{\partial x}\leq 0 \quad (6.29)$$

If the transduction ability ρ_a is small, the last term in (6.29) is positively large. In this situation, the first term $(\partial V(x)/\partial x)^T f(x)$ should be more negative (more stable) so that the HJI in (6.29) still holds. On the other hand, if $(\partial V(x)/\partial x)^T f(x)$ is less negative (less stable), then the transduction ability ρ_a should be large enough to satisfy the HJI in (6.29). From the HJI in (6.29), it can be seen that the signal transduction ability is more closely related to the system characteristics $f(x)$, $g(x)$, and C of signaling pathway than to the external stimuli. This is a similar case to a low-pass filter having greater dependence on the system characteristic of the filter than the incoming noise.

In general, it is still very difficult to solve the nonlinear HJI-constrained optimization problem for the measurement of signal transduction ability described in (6.28). Thus, a fuzzy interpolation method is employed to overcome the nonlinear difficulties associated with solving the HJI problem in (6.27) and to simplify the constrained optimization procedures required to solve the transduction ability measurement problem in (6.28).

6.3.3. Signal Transduction Measurement via Fuzzy Method

Fuzzy dynamic modeling has been widely employed to approximate nonlinear dynamic systems and to simplify nonlinear control problems by interpolating several local linear dynamic systems [9, 28]. The Takagi-Sugeno (T-S) fuzzy dynamic model [29] is described by fuzzy "If-Then" rules and employed here to solve the transduction ability measurement problem for the signaling pathway. The i-th rule of the fuzzy model for the nonlinear signal transduction pathway in (6.23) is proposed as follows [9, 28, 30, 31]:

$$\text{Rule } i: \text{If } x_1(t) \text{ is } F_{i1},\, x_2(t) \text{ is } F_{i2},\, \ldots,\, x_l(t) \text{ is } F_{il},$$
$$\text{Then } \begin{cases} \dot{x} = A_i x(t) + B_i u(t) \\ y = Cx(t) \end{cases} \tag{6.30}$$

where i=1,2,⋯,L. $x_i(t)$ is the premise variable, F_{ij} is the fuzzy set, A_i and B_i are the local system matrices, L is the number of fuzzy rules, and l is the number of premise variables. If all state variables are used as premise variables then l=n. The physical meaning of fuzzy rule i in (6.30) is that if the state variables $x_1(t)$, $x_2(t)$, ⋯, $x_l(t)$ are with the local fuzzy sets F_{i1}, F_{i2}, ⋯, F_{il}, then the nonlinear signaling pathway in (6.23) could be represented by the linear system in the "Then" part of (6.30). The fuzzy inference system of (6.30) can then be described as follows [28, 29, 32]:

$$\dot{x} = \frac{\sum_{i=1}^{L} \mu_i(x(t))\left[A_i x(t) + B_i u(t)\right]}{\sum_{i=1}^{L} \mu_i(x(t))} = \sum_{i=1}^{L} m_i(x(t))\left[A_i x(t) + B_i u(t)\right]$$
$$y = Cx(t) \tag{6.31}$$

where $m_i(x(t))=\mu_i(x(t))/\sum_{i=1}^{L} \mu_i(x(t))$ and $\mu_i(x(t))= \prod_{i=1}^{L} F_{ij}(x_i(t))$,with $F_{ij}(x_i(t))$ describing the grade of membership of $x_i(t)$ in F_{ij} and $m_i(x(t))$, i=1,2,⋯,L describing the fuzzy bases. The denominator $\sum_{i=1}^{L} \mu_i(x(t))$ is applied to enable normalization so that the total sum of the fuzzy bases $\sum_{i=1}^{L} m_i(x(t))=1$. The physical meaning of the fuzzy inference system in (6.31) is that L local linear signaling pathways are interpolated through nonlinear fuzzy bases $m_i(x(t))$ to approximate the nonlinear signaling pathway in (6.23). Aside from the fuzzy bases, other interpolation bases could also be employed to interpolate several linear signaling pathways to approximate a nonlinear signaling pathway.

By fuzzy approximation, the nonlinear signaling pathway can be represented by the following fuzzy interpolated signaling pathway:

$$\dot{x} = f(x) + g(x)u = \sum_{i=1}^{L} m_i(x(t))\left[A_i x(t) + B_i u(t)\right] \tag{6.32}$$

There are many system identification methods available to determine the local system matrixes A_i and B_i for the fuzzy model [29], such as the fuzzy toolbox in MATLAB.

Proposition 6.5 [27]: For the nonlinear signaling pathway in (6.32), if a positive definite symmetric matrix P exists, satisfying the following Riccati-like inequalities:

$$PA_i + A_i^T P + C^T C + \frac{1}{\rho^2} PB_i B_i^T P \le 0, i = 1, 2, \cdots, L \tag{6.33}$$

then the signal transduction ability is bounded by ρ.

Proof: see 6.5.5

The inequality in (6.33) can be considered as the local linearization of HJI in (6.27). In order to solve the above Riccati-like inequalities by the conventional LMI method, the positive definite symmetric matrix $Q=P^{-1}$ is multiplied at both sides of (6.33). The Riccati-like inequalities in (6.33) can be described as:

$$A_i Q + QA_i^T + \frac{1}{\rho^2} B_i B_i^T + QC^T CQ \le 0, i = 1, 2, \ldots, L \tag{6.34}$$

By the Schur complement method [6], the Riccati-like inequalities in (6.34) are then equivalent to the following LMIs:

$$\begin{bmatrix} A_i Q + QA_i^T + \frac{1}{\rho^2} B_i B_i^T & QC^T \\ CQ & -1 \end{bmatrix} \le 0, i = 1, 2, \ldots, L \tag{6.35}$$

If the LMIs in (6.35) hold for $Q>0$, the signal transduction ability of nonlinear signaling pathway is less than ρ and ρ is the upper bound of the signal transduction ability.

Using the fuzzy approximation method, the HJI in (6.27) is interpolated by a set of Riccati-like algebraic inequalities in (6.33) or (6.34). Therefore, the signal transduction ability of the nonlinear signaling pathway could be obtained by solving the following LMIs-constrained optimization problem [27]:

$$\begin{gathered} \rho_a = \min \rho \\ \text{subject to LMIs in (6.35) and } Q > 0 \end{gathered} \tag{6.36}$$

which can then be solved by decreasing ρ until no positive definite symmetric matrix Q is able to satisfy the inequalities in (6.33) or (6.34).

Remark 6.3: The Riccati-like inequalities in (6.33) or (6.34) are an approximation of HJI with L local linear systems. Similarly, the LMIs-constrained optimization problem in (6.36), describing the signal transduction ability measurement of a nonlinear signaling pathway, is also an efficient approach based on a fuzzy interpolation method to replace the very difficult HJI-constrained optimization problem in (6.28). The LMIs-constrained optimization problem for the signal transduction ability measurement is known as an "eigenvalue problem", which can be efficiently solved by the LMI toolbox in MATLAB.

Remark 6.4: After substituting ρ_a for ρ in (6.33), we get

$$PA_i + A_i^T P + C^T C + \frac{1}{\rho_a^2} PB_i B_i^T P \leq 0, i = 1, 2, \cdots, L \tag{6.37}$$

If ρ_a is smaller, the last term will be larger so that the eigenvalues of A_i would be in the far left hand complex plane (i.e., with more negative real part or more stable) and (6.37) would hold. If ρ_a is larger, the last term in (6.37) will be smaller, meaning that the eigenvalues of A_i would be closer to the imaginary axis (i.e., with smaller negative real part or less stable). In general, the signaling pathways of the cell amplify the input signals (i.e., ρ_a>1). In this situation, the eigenvalues of A_i would approach the imaginary axis (less stable) in order to amplify the extracellular signals so as to enable the downstream activation of the corresponding genes within the cell. The signal transduction ability of the signaling pathway is more dependent on the system characteristics A_i, B_i, and C than on the external signal $u(t)$, just as a low-pass filter is more dependent on the low-pass characteristic (poles and zeros) of the system than on the filtered noise.

Remark 6.5: In this chapter, we assume the system dynamic model of signaling pathways in (6.23) has been well identified. However, there are some parameter estimation errors due to the incorrect structure, stochastic noises [4, 5], infeasible parameter region and solution method. The effect of these parameter estimation errors may lead to an overestimate of signal transduction ability ρ_a (see 6.5.6 in Appendix).

6.3.4. MAPK Signal Transduction Pathway

The MAPK signal transduction pathway is one of the most well-known and well-studied signaling pathways. The pathway is highly conserved in eukaryotes, including humans, *Drosophila*, and yeast. A wide variety of membrane receptors, including RTK and cytokine receptors, utilize the MAPK cascade to transfer external upstream signals to the nucleus [24]. As soon as MAPK is activated by the signaling pathway, it is translocated into the nucleus, promoting the transcription of hundreds of genes. Based on these properties, the MAPK cascade was chosen as the first simulation example.

In mammalian cells, the MAPK cascade is operated in a sequential fashion downstream from an activated Ras protein. Active Ras binds to a Raf protein (MAPKKK), thereby activating it. The active Raf phosphorylates and thereby activates MEK (MAPKK). Active MEK then phosphorylates and activates MAP kinase. Finally MAP kinase phosphorylates many different proteins, including nuclear transcription factors that mediate cellular responses [33]. In summary, activation of Ras by RTK induces a kinase cascade that includes Raf, MEK and MAPK (RTK → Ras → Raf → MEK → MAPK) [24].

The dynamics of a MAPK cascade can be represented by the ODE system in (6.21). The input signal $u(t)$ is MAPKKKK and the output signal $y(t)$ is MAPK-pp. The values for the rate constants, the Michaelis-Menten constants and enzyme rates are adopted from [19], which were obtained from fitting the experimental data. The performance of the MAPK kinase cascade is measured in the ability to transmit and amplify the signal and then to notably enhance the concentration of double-phosphorylated MAPK which is dependent on the dynamic system of the signaling pathway and their constants. The simulation procedure was initiated by selecting a fuzzy model to approximate the nonlinear model. Selection of the model required the approximation of the parameters of the fuzzy model, A_i and B_i in (6.30) or (6.31), which can be found by the least mean square parameter estimation algorithm based on a system identification method [29]. The LMIs in (6.35) can then be solved via the LMI toolbox in MATLAB. Finally, the signal transduction ability in (6.36) can be obtained by decreasing ρ until no positive definite symmetric matrix Q is able to satisfy (6.35).

Using the least-squares parameter estimation method [29], the parameters A_i and B_i of the fuzzy model can be identified to achieve the minimum square error (simulation results were shown in Figure 6.3). Since there are 8 state variables in the model, and 2 fuzzy sets for each state variable, a total of 256 fuzzy rules are applied to the fuzzy model. The fuzzy approximation error of the MAPK signaling pathway in Figure 6.3 is about 10^{-10} with 256 fuzzy rules. Solving the optimization problem in (6.36), the signal transduction ability of the fuzzy signaling pathway is measured as ρ_a=2.73×10^6 with the corresponding positive definite symmetric matrix Q. This value implies the best ability of the MAPK signaling pathway to transduce input signals, that is, to amplify input signals. We have calculated ratio of the output signal to the input signal from the root mean square sense with a few kinds of input signals. From eqs.(6.24)–(6.26), it is seen that ρ_a is the least upper bound of these ratios. However, it seems not an easy task to find an adequate input signals to achieve the transduction ability.

In this example, we focused on the sequential phosphorylation of MAPKKK, MAPKK, and MAPK to investigate the transduction ability of on the MAPK signal transduction (Figure 6.2). The whole signaling pathway can be divided into three levels: (i) MAPKKK →MAPKKK-p, (ii) MAPKK→MAPKK-pp and (iii) MAPK→MAPK-pp. According to our measurements, the signal transduction ability of the entire signaling pathway is approximately 10^6, which closely resembles the value source from the literature [24]. Moreover, the measurement of transduction abilities of the above three stages by the proposed method were found to be about 80, 120, and 120, respectively. The results from measuring the MAPK signaling pathway as a whole or as separate pathways are similar, i.e., $10^6 \fallingdotseq 80\times120\times120$ through three separate sub-pathways. This finding implies that the measurement for the signaling pathway can be obtained as a whole or by measuring the separate pathways if the size of signaling pathway is very large and the computation of signal transduction ability in (6.36) becomes difficult by a whole calculation way [34].

6.3.5. NF-κB Signal Transduction Pathway

According to research on the evolution of the NF-κB pathway, it has existed for millions of years. The NF-κB pathway is referred to as the master of transcription factors for immune responses [35]. It is rapidly activated in mammalian immune-system cells in response to infection, inflammation, and a number of other stressful situations, such as ionizing radiation [33]. In contrast to the previous MAPK pathway, which is a reversible signaling pathway, the NF-κB is an irreversible pathway where a component is proteolytically cleaved. The NF-κB pathway in Figure 6.4 with dynamic model in (6.38) is investigated as the second simulation example to demonstrate our proposed measurement method.

$$
\begin{aligned}
&d\mathrm{IKKK_a}(t)/dt = k_a \times u(t) \times \left(K_N - \mathrm{IKKK_a}(t)\right) - k_i \times \mathrm{IKKK_a}(t) \\
&d\mathrm{IKK_n}(t)/dt = k_4 \times \left(K_{NN} - \mathrm{IKK_n} - \mathrm{IKK_a} - \mathrm{IKK_i}\right) - k_1 \times \mathrm{IKKK_a^2}(t) \times \mathrm{IKK_n}(t) \\
&d\mathrm{IKK_a}(t)/dt = k_1 \times \mathrm{IKKK_a^2}(t) \times \mathrm{IKK_n}(t) - k_3 \times \mathrm{IKK_a}(t) \\
&d\mathrm{IKK_i}(t)/dt = k_3 \times \mathrm{IKK_a}(t) - k_4 \times \mathrm{IKK_i}(t)
\end{aligned} \tag{6.38}
$$

In resting cells, the dimeric transcription factor NF-κB, composed of p50 and p65, is sequestered in the cytosol and bound to the inhibitor I-κB [36, 37]. Stimulation by TNF-α or IL-1 induces activation of TAK1 kinase (IKKK), leading to activation of the trimeric I-κB kinase (IKK). Ionizing radiation and other stresses are able to directly activate I-κB. Following phosphorylation of I-κB by IKK and the binding of ubiquitin ligase, the polyubiquitination of I-κB leads to degradation by proteasomes. The removal of I-κB unmasks the nuclear-localization signals in both subunits (p65 and p50) of NF-κB, allowing their translocation into the nucleus.

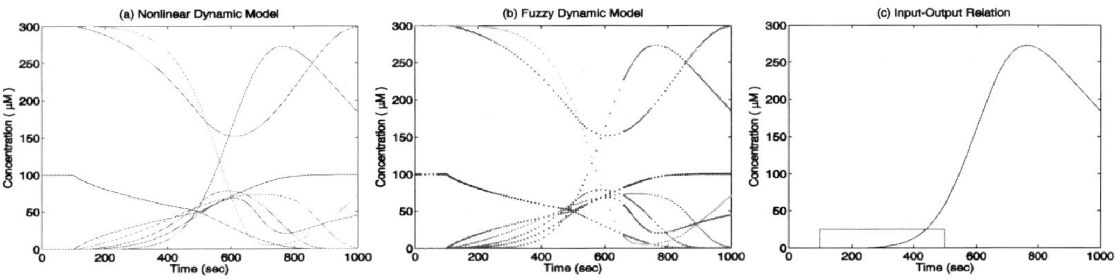

Figure 6.3. The simulation example for the MAPK signaling pathway. (a) The time profile of the state variables (MAPKKK:blue, MAPKK-p:dark green, MAPKK:red, MAPKK-p:cyan, MAPKK-pp:magenta, MAPK:gold, MAPK-p:grey, MAPK-pp:brown) for the nonlinear dynamic model of the MAPK signal transduction pathway in (6.21). (b) The time profile of the state variables for the fuzzy dynamic model of the MAPK signal transduction pathway in (6.31). (c) An input (red) and output (blue) signal of the MAPK signal transduction pathway.

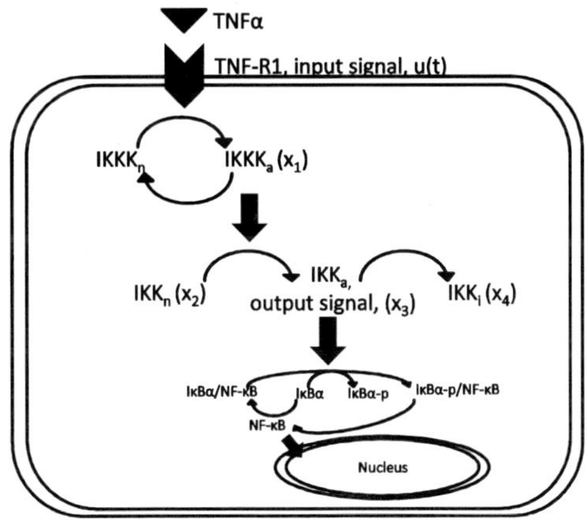

Figure 6.4. The scheme of NF-κB signal transduction pathway.

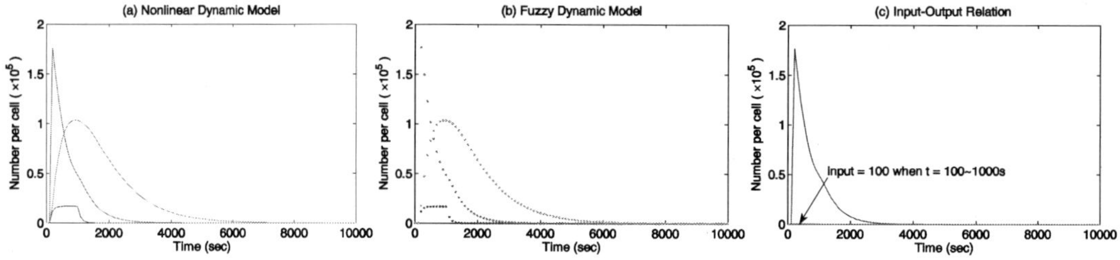

Figure 6.5. The simulation example of the NF-κB signaling pathway. (a) The state variables ($IKKK_a$:blue, IKK_n:green, IKK_a:red, IKK_i:cyan)of the nonlinear dynamic model of the NF-κB signal transduction pathway in (6.37). (b) The state variables of the fuzzy dynamic model of the NF-κB signal transduction pathway in the same form of (6.31). (c) An input (red) and output (blue) signal of the NF-κB signal transduction pathway.

Here NF-κB activates the transcription of numerous target genes, including the gene encoding the alpha subunit of I-κB, which acts to terminate signaling [38]. In this case, we adopted the mathematical model in (6.38) [35], and considered the IKK_a as the output, which directly affects the transcription factors. (Figure 6.4). Using similar procedures employed in the previous example, we first constructed the fuzzy model and then solved the corresponding LMIs (simulation results are shown in Figure 6.5). We found the signal transduction ability of the fuzzy signaling pathway to be ρ_a=6.48×10^5.

6.3.6. Discussion

In order to measure the signal transduction ability in biological experiments, it is necessary to test a multitude of input signals and then compute the ratio of the output to the input signals. This process requires numerous experiments and is not a realistic method of obtaining data. Therefore, a method that is independent of the measurement of output and input signals, which can be used to quantify the signal transduction ability based on the system dynamic model, is required. According to the concept of a system gain, we proposed a new method to measure the signal transduction ability of a signaling pathway, which is dependent only on the structure of the signaling pathway and the kinetic parameters of the components involved. Thus, the signal transduction ability can be viewed as a quantity to characterize a signaling pathway. Considering the important role of the MAPK and the NF-κB signaling pathways in a cell, an immediate and strong enough response to environmental changes or stresses is necessary. The simulation results describing a large system gain in concentration are unsurprising.

Here, we extend the concept of signal amplitude to the nonlinear signaling pathway through a fuzzy modeling method, compared to the linear case [20]. Moreover, the most of previous studies considered the relationship between the output and the inputs on case-by-case basis [15, 20-22]. But from the system point of view, the transduction ability is a systematic characteristic of a signaling pathway, which should be invariant unless the system has been changed. The value can be referred to as the best ability of the signaling pathway to transduce input signals, that is, to amplify input signals. However, the simulated signal transduction ability might be not achieved easily without testing all possible inputs for the least upper bound ratio ρ_a in (6.24). That is, infinite L_2-bounded input signals are required to measure the transduction ability of a signaling pathway, which could be an open question.

The simulation results also show that the proposed measurement method can be applied in two different ways. A signaling pathway can be viewed as a composition of several sub-signaling pathways. The total signal transduction ability of a signaling pathway is equal to the product of the signal transduction abilities of each of the sub-signaling pathways. Each sub-signaling pathway can be referred to as a basic element of the general signaling pathway. Hence, if the signal transduction abilities of sub-signaling pathway are known, we can design a signaling pathway with the desired signal transduction ability. A similar concept is utilized in synthetic biology [25, 39, 40], which designate DNA sequences as the bricks to construct an artificial function in a cell. It is possible, through the analysis of signal transduction ability, to analyze and calculate each sub-signaling pathway so that a synthetic signal pathway can be engineered with the desired signal transduction ability. In this situation, greater biological transduction insights can be obtained from the linear system viewpoint. In the future, characterization of signal transduction systems by the proposed transduction ability measurement may be useful for biotechnological and therapeutic applications as well as the development of drug design.

6.4. Conclusion

In this chapter, network robustness, network filtering and network transduction are discussed from the systematic perspective for stochastic nonlinear biological networks under intrinsic parameter fluctuation, genetic variation, and environmental noise. Phenotype robustness criteria of biological networks are investigated from the robust stabilization and noise filtering point of view. It is found that if the sum of intrinsic robustness and environmental robustness is less than or equal to network robustness, then the phenotype will be robust in the biological network. Measurement of network transduction ability for signal and metabolic pathways is also investigated. These system characteristics can be expected to be very useful in future system design in systems and synthetic biology.

6.5. Appendix

6.5.1. Proof of Proposition 6.1

Let us denote the Lyapunov (energy) function of linear stochastic regulatory network in equation (6.3) as $V(x) = x^T(t)Px(t)$, for some symmetric positive definite matrix $P^T = P > 0$. Then the change of Lyapunov function is obtained by following Itô derivative [5, 41]

$$E\left\{\frac{d}{dt}V(x)\right\} = E\left\{\left(\frac{\partial}{\partial x}V(x)\right)^T Nx(t) + \frac{1}{2}x^T(t)M^T\frac{\partial^2 V(x)}{\partial x^2}Mx(t)\right\}$$
$$= E\left\{x^T(t)PNx(t) + x^T(t)N^TPx(t) + x^T(t)M^TPMx(t)\right\} \quad (6.39)$$
$$= E\left\{x^T(t)\left(PN + N^TP + M^TPM\right)x(t)\right\}$$

From the inequality (6.5), we get

$$E\left\{\frac{d}{dt}V(x)\right\} \le 0 \quad (6.40)$$

Since the change of Lyapunov (energy) function is nonpositive, the linear stochastic genetic regulatory network is stable in probability.

6.5.2. Proof of Proposition 6.2

Let us denote the Lyapunov function $V(x)$ of the perturbative nonlinear genetic regulatory network. By following Itô derivative, we get [3, 4]

$$E\left\{\frac{d}{dt}V(x)\right\} = E\left\{\left(\frac{\partial}{\partial x}V(x)\right)^T N(x(t)) + \frac{1}{2}M^T(x(t))\frac{\partial^2 V(x)}{\partial x^2}M(x(t))\right\} \quad (6.41)$$

By the inequality (6.7), we get

$$E\left\{\frac{d}{dt}V(x)\right\} \le 0 \quad (6.42)$$

Then the nonlinear perturbative stochastic regulatory network in equation (6.4) is stable in probability.

6.5.3. Proof of Proposition 6.3

Consider the following equality,

$$\int_0^\infty Z^T(t)Z(t)dt = V(x(0)) - V(x(\infty)) + \int_0^\infty\left(Z^T(t)Z(t) + \frac{dV(x)}{dt}\right)dt \quad (6.43)$$

By Itô formula for stochastic equation (6.9), we get

$$E\left\{\frac{dV(x)}{dt}\right\}=E\left\{\left(\frac{\partial}{\partial x}V(x)\right)^T(Nx(t)+Hv(t))+\frac{1}{2}x^T(t)M^T\frac{\partial^2V(x)}{\partial x^2}Mx(t)\right\} \tag{6.44}$$

Let us choose the Lyapunov function as $V(x) = x^T(t)Px(t)$. Then

$$\begin{aligned}&E\left\{\frac{dV(x)}{dt}\right\}\\&=E\left\{2x^T(t)P(Nx(t)+Hv(t))+x^T(t)M^TPMx(t)\right\}\\&=E\left\{x^T(t)P(Nx(t)+Hv(t))+(Nx(t)+Hv(t))^TPx(t)+x^T(t)M^TPMx(t)\right\}\end{aligned} \tag{6.45}$$

Taking expectation to both sides of (6.43) and substituting equation (6.45) into equation (6.43), we get

$$\begin{aligned}&E\left\{\int_0^\infty Z^T(t)Z(t)dt\right\}\\&=E\{V(x(0))\}-E\{V(x(\infty))\}\\&+E\left\{\int_0^\infty x^T(t)C^TCx(t)+x^T(t)PNx(t)+x^T(t)PHv(t)+v^T(t)H^TPx(t)\right.\\&\left.+x^T(t)NPx(t)+x^T(t)M^TPMx(t)dt\right\}\end{aligned} \tag{6.46}$$

By the fact that

$$X^TY+Y^TX\le\rho^2X^TX+\rho^{-2}Y^TY,\ \text{for any } \rho^2>0 \tag{6.47}$$

and $E\{V(x(\infty))\}\geq 0$, we get

$$\begin{aligned}&E\left\{\int_0^\infty Z^T(t)Z(t)dt\right\}\\&\le E\{V(x(0))\}\\&+E\left\{\int_0^\infty x^T(t)\left(C^TC+PN+N^TP+M^TPM+\frac{1}{\rho^2}PHH^TP\right)x(t)+\rho^2v^T(t)v(t)dt\right\}\end{aligned} \tag{6.48}$$

By the inequality (6.11), we get

$$E\left\{\int_0^\infty Z^T(t)Z(t)dt\right\}\le E\{V(x(0))\}+\rho^2E\left\{\int_0^\infty v^T(t)v(t)dt\right\} \tag{6.49}$$

or

$$\|Z(t)\|_2^2 \le E\{V(x(0))\} + \rho^2 \|v(t)\|_2^2 \tag{6.50}$$

6.5.4. Proof of Proposition 6.4

By integrating the norm of output signal from 0 to t_f, we have

$$\int_0^{t_f} y^T(t)y(t)dt = V(x(0)) - V(x(t_f)) + \int_0^{t_f}\left(x^T(t)C^TCx(t) + \frac{d}{dt}V(x)\right)dt \tag{6.51}$$

By the fact that $V(x(t_f))>0$ and by chain rule

$$\begin{aligned}\int_0^{t_f} y^T(t)y(t)dt &\le V(x(0)) + \int_0^{t_f}\left(x^T(t)C^TCx(t) + \left(\frac{dV}{dx}\right)^T(f(x)+g(x)u(t))\right)dt \\ &= V(x(0)) + \int_0^{t_f}\left(x^T(t)C^TCx(t) + \left(\frac{dV}{dx}\right)^T f(x) + \left(\frac{dV}{dx}\right)^T g(x)u(t)\right)dt\end{aligned} \tag{6.52}$$

By the fact, $2a^Tb \le a^Ta + b^Tb$ for any vector a and b, we get

$$\begin{aligned}\left(\frac{dV}{dx}\right)^T g(x)u(t) &= 2\left(\frac{1}{2\rho}\left(\frac{dV}{dx}\right)^T g(x)\right)\rho u(t) \\ &\le \frac{1}{4\rho^2}\left(\frac{dV}{dx}\right)^T g(x)g^T(x)\frac{dV}{dx} + \rho^2 u^T(t)u(t)\end{aligned} \tag{6.53}$$

By substituting $(dV/dx)^T g(x)u(t)$ with above inequality, we have

$$\begin{aligned}&\int_0^{t_f} y^T(t)y(t)dt \\ &\le V(x(0)) \\ &+\int_0^{t_f}\left(x^T(t)C^TCx(t) + \left(\frac{dV}{dx}\right)^T f(x) + \rho^2 u^T(t)u(t) + \frac{1}{4\rho^2}\left(\frac{dV}{dx}\right)^T g(x)g^T(x)\frac{dV}{dx}\right)dt\end{aligned} \tag{6.54}$$

By the HJI in (6.27), we get

$$\int_0^{t_f} y^T(t)y(t)dt \le V(x(0)) + \rho^2\int_0^{t_f} u^T(t)u(t)dt \tag{6.55}$$

i.e.,

$$\|y(t)\|_2^2 \le V(x(0)) + \rho^2\|u(t)\|_2^2 \tag{6.56}$$

When $x(0) = 0$ and $V(0) = 0$, the above inequality becomes the inequality (6.25) or (6.26).

6.5.5. Proof of Proposition 6.5

By the fact that

$$V(x) = x^T(t)Px(t), f(x) = \sum_{i=1}^{L} m_i(x(t))A_i x, \text{ and } g(x) = \sum_{i=1}^{L} m_i(x(t))B_i$$

Then (6.51) becomes

$$\begin{aligned}\int_0^{t_f} y^T(t)y(t)dt &\le V(x(0)) \\ &+\int_0^{t_f}\left(x^T C^T Cx + \left(\frac{dV}{dx}\right)^T \sum_{i=1}^{L} m_i(x(t))A_i x + \left(\frac{dV}{dx}\right)^T \sum_{i=1}^{L} m_i(x(t))B_i u(t)\right)dt\end{aligned} \tag{6.57}$$

By the fact: $dV/dx = 2Px$

$$\begin{aligned}\left(\frac{dV}{dx}\right)^T \sum_{i=1}^{L} m_i(x(t))A_i x &= \sum_{i=1}^{L} m_i(x(t))x^T\left(PA_i + A_i^T P\right)x \\ \left(\frac{dV}{dx}\right)^T \sum_{i=1}^{L} m_i(x(t))B_i u(t) &= \sum_{i=1}^{L} m_i(x(t))\left(x^T PB_i u(t) + u^T(t)B_i^T Px\right) \\ &\le \sum_{i=1}^{L} m_i(x(t))\left(\frac{1}{\rho^2}x^T PB_i B_i^T Px + \rho^2 u^T(t)u(t)\right)\end{aligned} \tag{6.58}$$

Then substitute the above equality and inequality to get

$$\begin{aligned}&\int_0^{t_f} y^T(t)y(t)dt \\ &\le V(x(0)) \\ &+\int_0^{t_f}\left[\sum_{i=1}^{L} m_i(x(t))x^T\left(C^T C + PA_i + A_i^T P + \frac{1}{\rho^2}PB_i B_i^T P\right)x + \rho^2 u^T(t)u(t)\right]dt\end{aligned} \tag{6.59}$$

If MI in (6.33) holds, then we get

$$\int_0^{t_f} y^T(t)y(t)dt \le V(x(0)) + \rho^2\int_0^{t_f} u^T(t)u(t)dt \tag{6.60}$$

With assumption $x(0)=0$ and $V(0)=0$, then we get

$$\int_0^{t_f} y^T(t)y(t)dt \le \rho^2 \int_0^{t_f} u^T(t)u(t)dt \tag{6.61}$$

i.e.,

$$\|y(t)\|_2^2 \le V(x(0)) + \rho^2 \|u(t)\|_2^2 \tag{6.62}$$

When $x(0) = 0$ and $V(0) = 0$, the above inequality becomes the inequality (6.25) or (6.26).

6.5.6. The Effect of Parameter Estimation Errors May Lead to an Overestimate of Transduction Ability ρ_a

The measurement of signal transduction ability in this chapter is based on the assumption that the system dynamic model of signaling pathways in (6.23) has been well identified. However, there are at least three possible error sources in the parameter estimation process due to (a) incorrect model, (b) infeasible parameter region, and (c) solution method. Further, different kinds of experimental data may also lead to different kinds of parameter estimates due to stochastic noises. In general, these parameter estimation errors are stochastic. Therefore, these parameter estimation errors should be considered in the parameter estimation process. Therefore, in the real system, the model system in (6.23) should be modified as follows:

$$\begin{aligned} \dot{x}(t) &= f(x) + g(x)u(t) + \sum_{i=1}^{M} h_i(x)n_i(t) \\ y(t) &= Cx(t) \end{aligned} \tag{6.63}$$

where the stochastic process $\Sigma_{i=1}^{M} h_i(x)n_i(t)$ denotes the modeling error due to M possible parameter estimation error sources as mentioned above, $n_i(t)$ is a zero mean white noise with variance σ_i^2 to denote the i-th error source and $h_i(t)$ denotes the effect of the i-th error source on the system model. For the convenience analysis the stochastic system in (6.63) is changed to an equivalent Ito system [4, 5]

$$\begin{aligned} dx(t) &= (f(x) + g(x)u(t))dt + \sum_{i=1}^{M} h_i(x)dw_i(t) \\ y(t) &= Cx(t) \end{aligned} \tag{6.64}$$

where $dw_i(t)=n_i(t)dt$, and $w_i(t)$ denotes a Wiener process (or Brownian motion). Since we use the stochastic system (6.64) to replace the dynamic system in (6.23), the transduction ability in (6.24) needs to be modified as:

$$\rho_a = \sup_{u \in L_2[0,t_f]} \frac{E(\int_0^{t_f} y^T(t)y(t)dt)^{1/2}}{E(\int_0^{t_f} u^T(t)u(t)dt)^{1/2}} \tag{6.65}$$

where $E(*)$ denotes the expectation of $(*)$. Let us denote the upper bound of ρ_a as follows:

$$\frac{E\int_0^{t_f} y^T(t)y(t)dt}{E\int_0^{t_f} u^T(t)u(t)dt} \le \rho^2 \tag{6.66}$$

Then we get the following result.

Proposition 6.6

If a positive function $V(x)$ with $V(0)=0$ is able to satisfy the following HJI:

$$x^T C^T Cx + (\frac{\partial V(x)}{\partial x})^T f(x) + \frac{1}{4\rho^2}(\frac{\partial V(x)}{\partial x})^T g(x)g^T(x)\frac{\partial V(x)}{\partial x} + \frac{1}{2}\sum_{i=1}^{M}\sigma_i^2 h_i^T(x)\frac{\partial^2 V(x)}{\partial x^2}h_i(x) \le 0 \tag{6.67}$$

then the signal transduction ability of signaling pathway is less than or equal to ρ.

Proof:

By the equality of two sides, we get

$$E\int_0^{t_f} y^T(t)y(t)dt = EV(x(0)) - EV(x(t_f)) + E\int_0^{t_f} (x^T(t)C^T Cx(t) + \frac{dV(x(t))}{dt})dt \tag{6.68}$$

By Ito formula [4, 5] , we get

$$dV(x) = (\frac{\partial V(x)}{\partial x})^T[(f(x)+g(x)u(t))dt + \sum_{i=1}^{M} h_i(x)dw_i(t)] + \frac{1}{2}\sum_{i=1}^{M}\sigma_i^2 h_i^T(x)\frac{\partial^2 V(x)}{\partial x^2}h_i(x)dt \tag{6.69}$$

Substituting (6.69) into (6.68) and by the fact $EV(x(t_f))>0$ and $Edw_i(t)=0$, we get

$$\begin{aligned} E\int_0^{t_f} y^T(t)y(t)dt &\le EV(x(0)) \\ &+E\int_0^{t_f} (x^T(t)C^T Cx(t) + (\frac{\partial V(x)}{\partial x})^T[(f(x)+g(x)u(t))] + \frac{1}{2}\sum_{i=1}^{M}\sigma_i^2 h_i^T(x)\frac{\partial^2 V(x)}{\partial x^2}h_i(x))dt \end{aligned} \tag{6.70}$$

By the fact

$$(\frac{\partial V(x)}{\partial x})^T g(x)u(t) \le \frac{1}{4\rho^2}(\frac{\partial V(x)}{\partial x})^T g(x)g^T(x)(\frac{\partial V(x)}{\partial x}) + \rho^2 u^T(t)u(t) \tag{6.71}$$

then we get

$$\begin{aligned} &E\int_0^{t_f} y^T(t)y(t)dt \\ &\le EV(x(0)) + E\int_0^{t_f}[x^T(t)C^TCx(t) + (\frac{\partial V(x)}{\partial x})^T f(x) + \frac{1}{4\rho^2}(\frac{\partial V(x)}{\partial x})^T g(x)g^T(x)(\frac{\partial V(x)}{\partial x}) \\ &\qquad + \rho^2 u^T(t)u(t) + \frac{1}{2}\sum_{i=1}^{M}\sigma_i^2 h_i^T(x)\frac{\partial^2 V(x)}{\partial x^2}h_i(x)]dt \end{aligned} \tag{6.72}$$

By the HJI in (6.67) and with $x(0)=0$, we get

$$E\int_0^{t_f} y^T(t)y(t)dt \le \rho^2 E\int_0^{t_f} u^T(t)u(t)dt \tag{6.73}$$

which is (6.66)

Q.E.D.

Therefore the transduction ability ρ_a for the stochastic signaling pathway in (6.65) is obtained by solving the following constrained optimization problem:

$$\begin{gathered} \rho_a = \min \rho \\ \text{subject to HJI in (6.67)} \end{gathered} \tag{6.74}$$

Substituting ρ_a into ρ in (6.67), we get

$$\begin{aligned} &(\frac{\partial V(x)}{\partial x})^T f(x) + x^T C^T Cx + \frac{1}{4\rho_a^2}(\frac{\partial V(x)}{\partial x})^T g(x)g^T(x)\frac{\partial V(x)}{\partial x} \\ &\qquad + \frac{1}{2}\sum_{i=1}^{M}\sigma_i^2 h_i^T(x)\frac{\partial^2 V(x)}{\partial x^2}h_i(x) \le 0 \end{aligned} \tag{6.75}$$

Comparing HJI in (6.75) with HJI in (6.29), the last term in (6.75) is due to the modeling error in the parameter estimation process. If this term becomes large, then ρ_a should also become large in order to compensate the last term in (6.75). It implies the modeling error in the parameter estimation process will lead to an overestimate of the signal transduction ability. Similarly, the fuzzy approximation of the signaling pathway in (6.32) can be modified

$$\dot{x}(t) = f(x) + g(x)u(t) + \sum_{i=1}^{M} h_i(x)n_i(t) = \sum_{i=1}^{L} m_i(x)[A_i x(t) + B_i u(t) + \sum_{j=1}^{M} H_{ij}x(t)n_j(t)] \tag{6.76}$$

where $\sum_{i=1}^{L}\sum_{j=1}^{M} m_i(x)H_{ij}x(t)n_j(t)$ is the fuzzy approximation of $\sum_{i=1}^{M} h_i(x)n_i(t)$. In this situation, the Riccati-like inequality in (6.33) can be modified as:

$$PA_i + A_i^T P + C^T C + \frac{1}{\rho_a^2} PB_i B_i^T P + \sum_{j=1}^{M} \sigma_j^2 H_{ij}^T PH_{ij} \le 0, i = 1, \cdots, L \tag{6.77}$$

The last term of (6.77) denotes the effect of modeling error in the parameter estimation process on the estimation of ρ_a and also leads to an overestimate of the signal transduction ability.

An example is given in the following to illustrate the overestimate of ρ_a due to these modeling errors in the parameter estimation process. Suppose the following three modeling errors occur in the parameter estimation process of MAPK signaling pathway

$$h_1(x)n_1(t) = \begin{bmatrix} 0.4018x_1 \\ 0.1233x_2 \\ 0.4173x_3 \\ 0.9448x_4 \\ 0.3377x_5 \\ 0.1112x_6 \\ 0.2417x_7 \\ 0.1320x_8 \end{bmatrix} n_1(t), h_2(x)n_2(t) = \begin{bmatrix} 0.0760x_1 \\ 0.1839x_2 \\ 0.0497x_3 \\ 0.4909x_4 \\ 0.9001x_5 \\ 0.7803x_6 \\ 0.4039x_7 \\ 0.9421x_8 \end{bmatrix} n_2(t), h_3(x)n_3(t) = \begin{bmatrix} 0.2399x_1 \\ 0.2400x_2 \\ 0.9027x_3 \\ 0.4893x_4 \\ 0.3692x_5 \\ 0.3897x_6 \\ 0.0965x_7 \\ 0.9561x_8 \end{bmatrix} n_3(t) \tag{6.78}$$

where $n_1(t)$, $n_2(t)$, and $n_3(t)$ are zero mean white noise with variance $\sigma_1^2=0.1^2$, $\sigma_2^2=0.2^2$, and $\sigma_3^2=0.2^2$, respectively, and the coefficients of $h_i(x)$ are randomly generated from the uniform distribution between 0 and 1. Then, from (6.77), we estimate $\rho_a=6.965\times10^6$, which is large than 2.73×10^6 in the perfect model case. Obviously, the modeling error in the parameter estimation process will lead to an overestimate of signal transduction ability in our method.

REFERENCES

[1] McAdams, HH; Arkin, A. "It's a noisy business! Genetic regulation at the nanomolar scale," *Trends in Genetics*, vol. 15, 65-69, Feb 1999.

[2] Chen, BS; Wang, YC. "On the attenuation and amplification of molecular noise in genetic regulatory networks," *BMC Bioinformatics*, vol. 7, Feb 2 2006.

[3] Zhang, WH; Chen, BS; Tseng, CS. "Robust H-infinity filtering for nonlinear stochastic systems," *IEEE Transactions on Signal Processing*, vol. 53, 589-598, Feb 2005.

[4] Zhang, WH; Chen, BS. "State feedback H(infinity) control for a class of nonlinear stochastic systems," *SIAM Journal on Control and Optimization*, vol. 44, 1973-1991, 2006.

[5] Chen, BS; Zhang, WH. "Stochastic H(2)/H(infinity) control with state-dependent noise," *IEEE Transactions on Automatic Control*, vol. 49, 45-57, Jan 2004.

[6] Boyd, SP; *Linear matrix inequalities in system and control theory*. Philadelphia: Society for Industrial and Applied Mathematics, 1994.

[7] Chen, BS; Chen, WH; Wu, HL. "Robust H-2/H-∞ Global Linearization Filter Design for Nonlinear Stochastic Systems," *IEEE Transactions on Circuits and Systems I-Regular Papers*, vol. 56, 1441-1454, Jul 2009.

[8] Chen, BS; Lee, TS; Feng, JH. "A Nonlinear H-Infinity Control Design in Robotic Systems under Parameter Perturbation and External Disturbance," *International Journal of Control*, vol. 59, 439-461, Feb 1994.

[9] Chen, BS; Tseng, CS; Uang, HJ. "Mixed H-2/H-infinity fuzzy output feedback control design for nonlinear dynamic systems: An LMI approach," *IEEE Transactions on Fuzzy Systems*, vol. 8, 249-265, Jun 2000.

[10] Hohmann, S. "Osmotic stress signaling and osmoadaptation in Yeasts," *Microbiology and Molecular Biology Reviews*, vol. 66, 300-+, Jun 2002.

[11] Teraoka, H; Maruyama, Y; Takehana, K; Iwanaga, T; Hiraga, T; Fujita, S; et al. "Ca2+ signaling in porcine duodenal glands by muscarinic receptor activation," *American Journal of Physiology-Gastrointestinal and Liver Physiology*, vol. 280, G729-G737, Apr 2001.

[12] Xie, ZJ; Askari, A. "Na+/K+-ATPase as a signal transducer," *European Journal of Biochemistry*, vol. 269, 2434-2439, May 2002.

[13] Aplin, AE; Juliano, RL. "Integrin and cytoskeletal regulation of growth factor signaling to the MAP kinase pathway," *Journal of Cell Science*, vol. 112, 695-706, Mar 1999.

[14] Herrington, J; Carter-Su, C. "Signaling pathways activated by the growth hormone receptor," *Trends in Endocrinology and Metabolism*, vol. 12, 252-257, Aug 2001.

[15] Koshland, DE; Goldbeter, A; Stock, JB. "Amplification and Adaptation in Regulatory and Sensory Systems," *Science*, vol. 217, 220-225, 1982.

[16] Klipp, E. *Systems biology in practice : concepts, implementation and application*. Weinheim: Wiley-VCH, 2005.

[17] Zou, XF; Peng, T; Pan, ZS. "Modeling specificity in the yeast MAPK signaling networks," *Journal of Theoretical Biology*, vol. 250, 139-155, Jan 7 2008.

[18] Huang, CYF; Ferrell, JE. "Ultrasensitivity in the mitogen-activated protein kinase cascade," *Proceedings of the National Academy of Sciences of the United States of America*, vol. 93, 10078-10083, Sep 17 1996.

[19] Kholodenko, BN. "Negative feedback and ultrasensitivity can bring about oscillations in the mitogen-activated protein kinase cascades," *European Journal of Biochemistry*, vol. 267, 1583-1588, Mar 2000.

[20] Heinrich, R; Neel, BG; Rapoport, TA. "Mathematical models of protein kinase signal transduction," *Molecular Cell*, vol. 9, 957-970, May 2002.

[21] Samoilov, M; Plyasunov, S; Arkin, AP. "Stochastic amplification and signaling in enzymatic futile cycles through noise-induced bistability with oscillations," *Proceedings of the National Academy of Sciences of the United States of America*, vol. 102, 2310-2315, Feb 15 2005.

[22] Grubelnik, V; Dugonik, B; Osebik, D; Marhl, M. "Signal amplification in biological and electrical engineering systems Universal role of cascades," *Biophysical Chemistry*, vol. 143, 132-138, Aug 2009.

[23] Klipp, E; Nordlander, B; Kruger, R; Gennemark, P; Hohmann, S. "Integrative model of the response of yeast to osmotic shock," *Nature Biotechnology*, vol. 23, 975-982, Aug 2005.

[24] Lodish, HF. *Molecular cell biology*, 4th ed. New York: W.H. Freeman, 2000.

[25] Hasty, J; McMillen, D; Collins, JJ. "Engineered gene circuits," *Nature*, vol. 420, 224-230, Nov 14 2002.

[26] Lin, CL; Liu, YW; Chuang, CH. "Control design for signal transduction networks," *Bioinform Biol Insights*, vol. 3, 1-14, 2009.

[27] Chen, BS; Wu, CC. "On the calculation of signal transduction ability of signaling transduction pathways in intracellular communication: systematic approach," *Bioinformatics*, vol. 28, 1604-1611, Jun 15 2012.

[28] Chen, BS; Tseng, CS; Uang, HJ. "Robustness design of nonlinear dynamic systems via fuzzy linear control," *IEEE Transactions on Fuzzy Systems*, vol. 7, 571-585, Oct 1999.

[29] Takagi, T; Sugeno, M. "Fuzzy Identification of Systems and Its Applications to Modeling and Control," *IEEE Transactions on Systems Man and Cybernetics*, vol. 15, 116-132, 1985.

[30] Tseng, CS; Chen, BS. "H-infinity decentralized fuzzy model reference tracking control design for nonlinear interconnected systems," *IEEE Transactions on Fuzzy Systems*, vol. 9, 795-809, Dec 2001.

[31] Tseng, CS; Chen, BS; Uang, HJ. "Fuzzy tracking control design for nonlinear dynamic systems via T-S fuzzy model," *IEEE Transactions on Fuzzy Systems*, vol. 9, 381-392, Jun 2001.

[32] Sherr, CJ; Roberts, JM. "CDK inhibitors: positive and negative regulators of G(1)-phase progression," *Genes & Development*, vol. 13, 1501-1512, Jun 15 1999.

[33] Karin, M; Greten, FR. "NF kappa B: Linking inflammation and immunity to cancer development and progression," *Nature Reviews Immunology*, vol. 5, 749-759, Oct 2005.

[34] Zumsande, M; Gross, T. "Bifurcations and chaos in the MAPK signaling cascade," *Journal of Theoretical Biology*, vol. 265, 481-491, Aug 7 2010.

[35] Hoffmann, A; Levchenko, A; Scott, ML; Baltimore, D. "The I kappa B-NF-kappa B signaling module: Temporal control and selective gene activation," *Science*, vol. 298, 1241-1245, Nov 8 2002.

[36] Hacker, H; Karin, M. "Regulation and function of IKK and IKK-related kinases," *Sci STKE*, vol. 2006, p. re13, Oct 17 2006.

[37] Hoffmann, A; Baltimore, D. "Circuitry of nuclear factor kappa B signaling," *Immunological Reviews*, vol. 210, 171-186, Apr 2006.

[38] Hayden, MS; Ghosh, S. "Shared principles in NF-kappa B signaling," *Cell*, vol. 132, 344-362, Feb 8 2008.

[39] Kaern, M; Blake, WJ; Collins, JJ. "The engineering of gene regulatory networks," *Annual Review of Biomedical Engineering*, vol. 5, 179-206, 2003.

[40] Murphy, KF; Balazsi, G; Collins, JJ. "Combinatorial promoter design for engineering noisy gene expression," *Proceedings of the National Academy of Sciences of the United States of America*, vol. 104, 12726-12731, Jul 31 2007.

[41] Chen, G; Chen, G; Hsu, SH. *Linear stochastic control systems*. Boca Raton: CRC Press, 1995.

Chapter 7

Systems Synthetic Biology

Abstract

Synthetic biology brings engineers and biologists together to design and build novel biomolecular components, circuits, and pathways, and to use these constructs to rewrite and reprogram organisms. Re-engineered organisms will change lives in the near future, providing cheaper drugs, green fuel for our cars, and targeted therapies to combat diseases such as cancer. In this chapter, robust synthetic gene network designs based on robust design specifications or robust optimal reference tracking schemes are introduced that can tolerate intrinsic parameter perturbation and resist environmental disturbance. This is done by selecting device components that are adequate from the systems biology perspective.

Introduction

Biological systems and their underlying components show a number of functional parallels with engineered systems. For example, the exquisitely sensitive biological sensors of the olfactory system can detect single odorant molecules and decode them. Biological systems can send and receive signals rapidly and in a highly specific manner, and here are pathways to sense and respond to the environment. Plants and microbes can use sunlight as an energy source. However, biological systems are uniquely capable of self-replication and mutation, leading to evolutionary change. Synthetic biologists aim to take advantage of parallels between biological and artificial systems to develop engineering principles for their design and construction. It remains an open question whether or not we understand biological systems sufficiently to be able to fulfill design specifications.

Synthetic biology is expected to have important applications in biotechnology and medicine, and to contribute significantly to a better understanding of the functioning of complex biological systems. Synthetic biology is concerned with engineering biological systems that fulfill a particular purpose. This is achieved through transformative innovations that make it possible to build living machines from chemical ingredients. Synthetic biology seeks to redesign natural biological systems based on a set of powerful biomolecular techniques for greater efficiency and construction of practical and functional genetic circuits and metabolic pathways [1]. The construction of GRNs has recently demonstrated the feasibility of synthetic biology [2-5], although the design of robust and functional gene networks still presents a difficult challenge. Design failures are mainly due to intrinsic perturbations such as gene expression noises, splicing, mutation, evolutionary changes, and environmental disturbances such as changing extracellular environments [6, 7]. Designing robust synthetic gene networks that can tolerate intrinsic parameter perturbations, attenuate extrinsic disturbances, and function properly in a host cell is therefore an important topic in synthetic biology [8].

7.1. Design of Specifications-Based Systems Synthetic Biology

Analysis of the dynamic properties of gene networks has been previously implemented using sensitivity analysis, either by qualitative simulation of coarse-grained models or by extensive numerical simulations of nonlinear differential equation models or stochastic dynamic models [9]. For applications in systems synthetic biology, however, these approaches are not satisfactory since local sensitivity analysis can provide only a partial description of all possible behaviors of a nonlinear gene network. In particular, it cannot guarantee that a synthetic network behaves as expected for all uncertain initial conditions, external disturbances, and parameter variations in a given range. Recently, Kuepfer et al. developed an approach based on semidefinite programming for partitioning parameter spaces of polynomial differential equation models into "feasible" and "infeasible" regions [10]. In this approach, "feasible" simply refers to the existence of a desired steady state of the synthetic network. Another approach using robustness analysis and tuning of synthetic networks was proposed by Batt et al. to provide a means to assess the robustness of a synthetic gene network with respect to parameter variations [11]. This approach allows searching for parameter sets for which a given property is satisfied, using a publicly available tool called RoVerGeNe. Several gene circuit design methods have recently been introduced to insert or delete specific circuits in an existing gene network to modify its structure toward improved robust stability and filtering ability [12]. Robust synthetic gene network design, however, is a topic in itself, necessitating the design of a completely synthetic network with enough robust stability toward parameter fluctuations and with enough noise filtering ability to resist external disturbances, which allows it to function properly in a host cell.

A robust synthetic biological design incorporating molecular noises has been developed based on stochastic game theory [6]. However, the intrinsic parameter fluctuations of synthetic gene networks have not been considered in the design procedure. In a previous study [7] some system design specifications (engineering designs) had been provided by users so that the designer must engineer an artificial gene network to meet these design specifications. For the convenience of problem description, a simple design example is provided to give an overview of design problems for robust synthetic gene networks. A more general problem scenario is treated in preparation for this overview. First, consider the cross-inhibition network shown in Figure 7.1. This network is synthesized with two genes (a, b) that code for two repressor proteins (A, B). More specifically, protein A represses the expression of gene b, and, at higher concentration, the expression of its own gene. Protein degradation is not regulated. This synthetic system can be modeled by the following synthetic equations:

$$\dot{x}_a = k_a r_{a1}(x_b) r_{a2}(x_a) - \gamma_a x_a \tag{7.1}$$

$$\dot{x}_b = k_b r_b(x_a) - \gamma_b x_b \tag{7.2}$$

The state variables x_a and x_b denote the concentrations of proteins A and B, respectively. k_i and γ_i are the kinetic parameter and decay rate, respectively. r_i is the regulation function capturing the regulator effect of a transcriptional protein on gene expression, and has a smooth sigmoidal form (e.g., Hill function) [13].

The simple cross-inhibition network in (7.1) and (7.2) can be represented by the following stoichiometric matrix equation [14]

$$\begin{bmatrix} \dot{x}_a \\ \dot{x}_b \end{bmatrix} = \begin{bmatrix} k_a & 0 & -\gamma_a & 0 \\ 0 & k_b & 0 & -\gamma_b \end{bmatrix} \begin{bmatrix} r_{a1}(x_b) r_{a2}(x_a) \\ r_b(x_a) \\ x_a \\ x_b \end{bmatrix} \tag{7.3}$$

However, the stoichiometric matrix *in vivo* suffers from intrinsic parameter perturbations because of gene expression noises, splicing, mutation, evolutionary change, etc., as in [6]

$$\begin{array}{ll} k_a \to k_a + \Delta k_a n_a & \gamma_a \to \gamma_a + \Delta \gamma_a n_a \\ k_b \to k_b + \Delta k_b n_b & \gamma_b \to \gamma_b + \Delta \gamma_b n_b \end{array} \tag{7.4}$$

where Δk_i and $\Delta \gamma_i$ denote the amplitudes of fluctuations of the stochastic parameters and decay rates, respectively; and n_i is a random white noise with zero mean and unit variance.

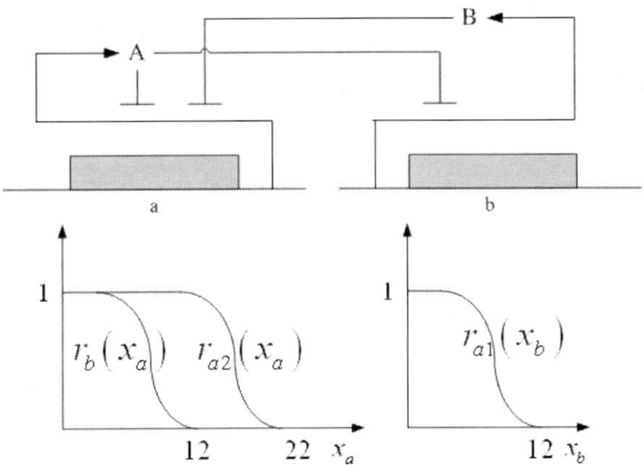

Figure 7.1. A simple two-gene cross-inhibition network. The network's regulation functions are given in equations (7.1) and (7.2).

Suppose the synthetic gene network also suffers from environmental disturbances because of changing extracellular environments and interactions with the cellular context in its host cell. The stochastic gene network can be then represented as

$$\begin{bmatrix} \dot{x}_a \\ \dot{x}_b \end{bmatrix} = \begin{bmatrix} k_a + \Delta k_a n_a & 0 & -\gamma_a - \Delta\gamma_a n_a & 0 \\ 0 & k_b + \Delta k_b n_b & 0 & -\gamma_b - \Delta\gamma_b n_b \end{bmatrix} \begin{bmatrix} r_{a1}(x_b) r_{a2}(x_a) \\ r_b(x_a) \\ x_a \\ x_b \end{bmatrix} + \begin{bmatrix} v_a \\ v_b \end{bmatrix}$$

$$= \begin{bmatrix} k_a & 0 & -\gamma_a & 0 \\ 0 & k_b & 0 & -\gamma_b \end{bmatrix} \begin{bmatrix} r_{a1}(x_b) r_{a2}(x_a) \\ r_b(x_a) \\ x_a \\ x_b \end{bmatrix} + \begin{bmatrix} \Delta k_a & -\Delta\gamma_a \\ 0 & 0 \end{bmatrix} \begin{bmatrix} r_{a1}(x_b) r_{a2}(x_a) \\ x_a \end{bmatrix} n_a \tag{7.5}$$

$$+ \begin{bmatrix} 0 & 0 \\ \Delta k_b & -\Delta\gamma_b \end{bmatrix} \begin{bmatrix} r_b(x_a) \\ x_b \end{bmatrix} n_b + \begin{bmatrix} v_a \\ v_b \end{bmatrix}$$

where $x = [x_a \; x_b]^T$ and $v = [v_a \; v_b]^T$ denote the state vector and environmental disturbance of the synthetic gene network in the host cell, respectively. These intrinsic parameter fluctuations and environmental disturbances may cause the engineered synthetic gene network to be dysfunctional in the host cell.

If a synthetic gene network consists of n genes, then the stochastic gene network of equation (7.5) in the host cell can be extended according to the following n-gene network dynamics [6]

$$\dot{x} = Nf(x) + \sum_{i=1}^{m} M_i g_i(x) n_i + v \tag{7.6}$$

where the state vector $x = [x_1 \cdots x_n]^T$ denotes the concentrations of proteins in the synthetic gene network, N denotes the corresponding stoichiometric matrix of the n-gene network, M_i; $i = 1,..., m$ denotes fluctuation matrices associated with independent random noise sources n_i, $i = 1,..., m$; and the elements of M_i denote the standard deviations of the corresponding parameter fluctuations. $v = [v_1 \cdots v_n]^T$ denotes the vector of external disturbances. This stochastic system is used to mimic the realistic dynamic behavior of a synthetic gene network of n genes in the host cell. As the network is subject to intrinsic parameter fluctuations and environmental disturbances in the context of the host cell, a robust synthetic gene network should be designed with the ability to not only tolerate parameter fluctuations and attenuate external disturbances, but also to achieve desired steady-state behaviors.

For convenience of analysis and design, the stochastic dynamic in equation (7.6) of a synthetic gene network in vivo can be represented by the following Ito stochastic differential equation [15, 16]

$$dx = \left(Nf(x) + v\right)dt + \sum_{i=1}^{m} M_i g_i(x) dW_i \tag{7.7}$$

where $W_i(t)$ is a standard Wiener process with $dW_i(t) = n_i dt$.

To ensure correct and efficient operation of the gene network, several systematic design specifications should be imposed on it from the systematic engineering point of view.

(i) The kinetic parameters and the decay rates in stoichiometric matrix should be chosen from a biologically feasible range

$$N \in [N_1 \ , \ N_2]$$

(ii) The network should tolerate the stochastic kinetic parameter and decay rate fluctuations with prescribed standard deviations in M_i in the state-dependent noise term

$$\sum_{i=1}^{m} M_i g_i(x) dW_i$$

(iii) The following desired steady state should be achieved

$$x \rightarrow x_d \text{ as } t \rightarrow \infty$$

(iv) The following prescribed attenuation level ρ of environmental disturbance should be achieved (i.e., the H_∞ filtering ability)

$$\frac{E\int_0^\infty (x-x_d)^T Q(x-x_d)dt}{E\int_0^\infty v^T v dt} \leq \rho^2 \tag{7.8}$$

for all bounded environmental disturbances $v(t)$, where $Q \geq 0$ is a symmetric weighting matrix and ρ is a prescribed attenuation level less than 1. That is, the effect of environmental disturbance v on the regulation error $x - x_d$ should be less than the attenuation level ρ from the average energy perspective.

In order to achieve the desired steady state x_d and for convenience of design, the origin of the nonlinear stochastic system in equation (7.7) should be shifted to x_d. Stabilizing the shifted nonlinear stochastic system at the origin would then also achieve x_d, simplifying the design procedure. Let $\tilde{x} = x - x_d$. The following shifted stochastic synthetic genetic system is then derived [7]:

$$d\tilde{x} = \left(Nf(\tilde{x}+x_d)+v\right)dt + \sum_{i=1}^{m} M_i g_i(\tilde{x}+x_d)dW_i \tag{7.9}$$

That is, the origin $\tilde{x} = 0$ of the stochastic system in equation (7.9) is at the desired steady state x_d of the original stochastic system in equation (7.7). $N \in [N_1,N_2]$ is then specified to tolerate the stochastic parameter fluctuation $\sum_{i=1}^{m} M_i g_i(\tilde{x}+x_d)dW_i$ and efficiently attenuate the environmental disturbance v to the prescribed level

$$\frac{E\int_0^\infty \tilde{x}^T Q\tilde{x}dt}{E\int_0^\infty v^T v dt} \le \rho^2 \tag{7.10}$$

From the stochastic synthetic gene network equation (7.9) *in vivo*, we obtain the following result.

Proposition 7.1 [7]: If design kinetic parameters and decay rates in $N \in [N_1,N_2]$ are chosen such that the following HJI has a positive solution $V(\tilde{x}) > 0$

$$\begin{aligned}&\left(\frac{\partial V(\tilde{x})}{\partial \tilde{x}}\right)^T Nf(\tilde{x}+x_d)+\tilde{x}^T Q\tilde{x}+\frac{1}{4\rho^2}\left(\frac{\partial V(\tilde{x})}{\partial \tilde{x}}\right)^T\left(\frac{\partial V(\tilde{x})}{\partial \tilde{x}}\right)\\&+\frac{1}{2}\sum_{i=1}^{m} g_i^T(\tilde{x}+x_d)M_i^T\frac{\partial^2 V(\tilde{x})}{\partial \tilde{x}^2}M_i g_i(\tilde{x}+x_d)<0\end{aligned} \tag{7.11}$$

then (a) the stochastic gene network in equation (7.9) can achieve both the robust stabilization necessary to tolerate the intrinsic stochastic parameter perturbations and the prescribed attenuation level ρ of environmental disturbances, i.e., design specifications (i), (ii), and (iv) are satisfied. (b) If the stochastic gene network is free of environmental disturbances ($v(t)$ = 0), then $\tilde{x}\to 0$ or $x\to x_d$ in probability, i.e., design specification (iii) is achieved.

Proof: See 7.4.1

It is generally very difficult to specify $N \in [N_1,N_2]$ to solve HJI in equation (7.11) for $V(\tilde{x}) > 0$ using a systematic method. If all global linearizations are bound by a polytope consisting of M vertices

$$\begin{bmatrix}\partial f(\tilde{x}+x_d)/\partial\tilde{x}\\ \partial g_1(\tilde{x}+x_d)/\partial\tilde{x}\\ \vdots\\ \partial g_m(\tilde{x}+x_d)/\partial\tilde{x}\end{bmatrix}\in \mathrm{Co}\left[\begin{pmatrix}F_1\\G_{11}\\ \vdots\\ G_{m1}\end{pmatrix},\cdots,\begin{pmatrix}F_M\\G_{1M}\\ \vdots\\ G_{mM}\end{pmatrix}\right],\quad \forall\tilde{x} \tag{7.12}$$

where Co denotes the convex hull of a polytope with M vertices defined in equation (7.12), then through the global linearization method [17, 18], the state trajectories $\tilde{x}(t)$ of the shifted gene network in equation (7.9) can be represented by the convex combination of M linearized gene networks as

$$d\tilde{x} = \sum_{j=1}^{M} \alpha_j(\tilde{x}) \left(NF_j \tilde{x} dt + \sum_{i=1}^{m} M_i G_{ij} \tilde{x} dW_i \right) + v dt \tag{7.13}$$

where the interpolation function $\alpha_j(\tilde{x})$ satisfies $0 \leq \alpha_j(\tilde{x}) \leq 1$ and $\sum \ \alpha_j(\tilde{x}) = 1$. The trajectory of the nonlinear stochastic synthetic gene network in equation (7.9) could thus be represented by the interpolated synthetic gene network in equation (7.13). This yields the following result.

Proposition 7.2 [7]: Assuming that design kinetic parameters and decay rates in $N \in [N_1, N_2]$ are chosen such that the following M LMIs have a common symmetric positive definite solution $P > 0$

$$\begin{bmatrix} PNF_j + F_j^T N^T P + \sum_{i=1}^{m} G_{ij}^T M_i^T P M_i G_{ij} + Q & P \\ P & -\rho^2 I \end{bmatrix} < 0, \ j = 1, \ldots, M \tag{7.14}$$

then there are two results: (a) The stochastic synthetic gene network in equation (7.7) is robustly stable toward intrinsic stochastic parameter perturbations and achieves a prescribed attenuation level ρ of environmental disturbances. That is, design specifications (i), (ii) and (iv) are satisfied. (b) If the synthetic gene network is free of environmental disturbance ($v(t) = 0$), then the synthetic gene network in equation (7.7) may asymptotically converge to the desired steady state.

Proof: see 7.4.2

An *in silico* design example for a robust synthetic gene network to confirm the robust stabilization and performance in environmental disturbance attenuation of the proposed design method for a synthetic gene network is introduced here. The synthetic gene network is shown in Figure 7.2. The goal is to synthesize a cascade loop of transcriptional inhibitions built in *E. coli* [11]. It consists of four genes, *tetR*, *lacI*, *cI*, and *eyfp*, which code for three repressor proteins, TetR, LacI, CI, and the fluorescent protein EYFP, respectively. The measured output is the fluorescence caused by the protein EYFP. The regulatory dynamic equations of the synthetic transcriptional cascade (Figure 7.2) are as follows [11].

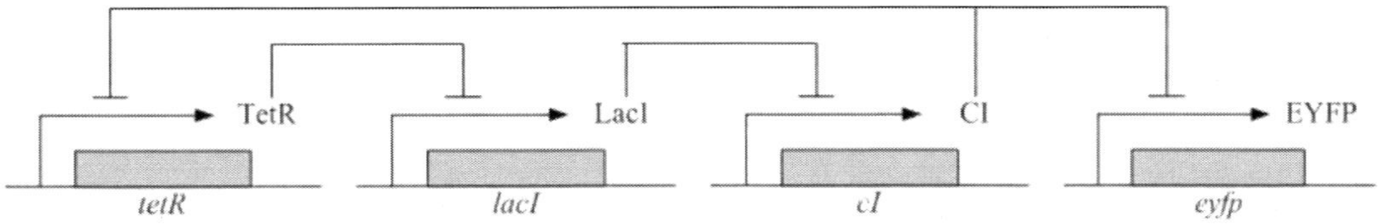

Figure 7.2. Synthetic transcriptional cascade loop. An in silico design example of a synthetic transcriptional cascade loop. TetR represses *lacI*, LacI represses *cI*, and CI represses *eyfp* and *tetR*. The fluorescent protein EYFP is the output. The regulatory dynamic equations of the synthetic transcriptional cascade are described in equation (7.15), and its stochastic model under random parameter fluctuations and environmental disturbance is described in equation (7.16).

$$\begin{aligned}
\dot{x}_{tetR} &= \kappa_{tetR,0} + \kappa_{tetR} r_{tetR}\left(x_{cI}\right) - \gamma_{tetR} x_{tetR} \\
\dot{x}_{lacI} &= \kappa_{lacI,0} + \kappa_{lacI} r_{lacI}\left(x_{tetR}\right) - \gamma_{lacI} x_{lacI} \\
\dot{x}_{cI} &= \kappa_{cI,0} + \kappa_{cI} r_{cI}\left(x_{lacI}\right) - \gamma_{cI} x_{cI} \\
\dot{x}_{eyfp} &= \kappa_{eyfp,0} + \kappa_{eyfp} r_{eyfp}\left(x_{cI}\right) - \gamma_{eyfp} x_{eyfp}
\end{aligned} \tag{7.15}$$

where $\kappa_{tetR,0}$, $\kappa_{lacI,0}$, $\kappa_{cI,0}$, and $\kappa_{eyfp,0}$ are the nominal generating ratios with stochastic parameter fluctuations for the corresponding proteins, which are assumed to be 150, 587, 210, and 3487, respectively. κ_{tetR}, κ_{lacI},, κ_{cI}, and κ_{eyfp} are the kinetic parameters, and γ_{tetR}, γ_{lacI}, γ_{cI}, and γ_{eyfp} are the decay rates of the corresponding proteins, which are also subject to parameter fluctuations in the host cell (i.e., *E. coli*) and are specified such that they meet the four design specifications. $r_{tetR}(x)$, $r_{lacI}(x)$, $r_{cI}(x)$, and $r_{eyfp}(x)$ are the Hill functions for the repressors. They have the form $r_i(x) = \beta/[1 + (x/k_i)^n]$, with $\beta = 1$, $n = 2$, $k_i = 1000$, and $i = tetR, lacI, cI, eyfp$ [13].

The stochastic synthetic gene network of equation (7.15) with four random parameter fluctuation sources *in vivo* in equation (7.7) can be represented by

$$\begin{aligned}
\begin{bmatrix} dx_{tetR} \\ dx_{lacI} \\ dx_{cI} \\ dx_{eyfp} \end{bmatrix} &= \left(\begin{bmatrix} \kappa_{tetR,0} & -\gamma_{tetR} & 0 & 0 & 0 & \kappa_{tetR} & 0 & 0 & 0 \\ \kappa_{lacI,0} & 0 & -\gamma_{lacI} & 0 & 0 & 0 & \kappa_{lacI} & 0 & 0 \\ \kappa_{cI,0} & 0 & 0 & -\gamma_{cI} & 0 & 0 & 0 & \kappa_{cI} & 0 \\ \kappa_{eyfp,0} & 0 & 0 & 0 & -\gamma_{eyfp} & 0 & 0 & 0 & \kappa_{eyfp} \end{bmatrix} \begin{bmatrix} 1 \\ x_{tetR} \\ x_{lacI} \\ x_{cI} \\ x_{eyfp} \\ r_{tetR}\left(x_{cI}\right) \\ r_{lacI}\left(x_{tetR}\right) \\ r_{cI}\left(x_{lacI}\right) \\ r_{eyfp}\left(x_{cI}\right) \end{bmatrix} + \begin{bmatrix} v_1 \\ v_2 \\ v_3 \\ v_4 \end{bmatrix} \right) dt \\
&+ \begin{bmatrix} \Delta\kappa_{tetR,0} & -\Delta\gamma_{tetR} & \Delta\kappa_{tetR} \\ 0 & 0 & 0 \\ 0 & 0 & 0 \\ 0 & 0 & 0 \end{bmatrix} \begin{bmatrix} 1 \\ x_{tetR} \\ r_{tetR}\left(x_{cI}\right) \end{bmatrix} dw_1 + \begin{bmatrix} 0 & 0 & 0 \\ \Delta\kappa_{lacI,0} & -\Delta\gamma_{lacI,0} & \Delta\kappa_{lacI} \\ 0 & 0 & 0 \\ 0 & 0 & 0 \end{bmatrix} \begin{bmatrix} 1 \\ x_{lacI} \\ r_{lacI}\left(x_{tetR}\right) \end{bmatrix} dw_2 \\
&+ \begin{bmatrix} 0 & 0 & 0 \\ 0 & 0 & 0 \\ \Delta\kappa_{cI,0} & -\Delta\gamma_{cI} & \Delta\kappa_{cI} \\ 0 & 0 & 0 \end{bmatrix} \begin{bmatrix} 1 \\ x_{cI} \\ r_{cI}\left(x_{lacI}\right) \end{bmatrix} dw_3 + \begin{bmatrix} 0 & 0 & 0 \\ 0 & 0 & 0 \\ 0 & 0 & 0 \\ \Delta\kappa_{eyfp,0} & -\Delta\gamma_{eyfp} & \Delta\kappa_{eyfp} \end{bmatrix} \begin{bmatrix} 1 \\ x_{eyfp} \\ r_{eyfp}\left(x_{cI}\right) \end{bmatrix} dw_4
\end{aligned} \tag{7.16}$$

Four kinetic parameters, κ_{tetR}, κ_{lacI}, κ_{cI}, and κ_{eyfp}, and four decay rates, γ_{tetR}, γ_{lacI}, γ_{cI}, and γ_{eyfp}, are designed to satisfy the following four design specifications.

(i) The biologically acceptable ranges of kinetic parameters and decay rates are given by [11]

$$\kappa_{tetR} \in [50,\ 5000],\ \kappa_{lacI} \in [70,\ 7000],\ \kappa_{cI} \in [75,\ 8000],\ \kappa_{eyfp} \in [30,\ 30000],$$
$$\gamma_{tetR} \in [0.2,\ 4.8],\ \gamma_{lacI} \in [0.02,\ 0.14],\ \gamma_{cI} \in [0.6,\ 0.8],\ \gamma_{eyfp} \in [0.1,\ 1]$$

(ii) The standard deviations of parameter fluctuations to be tolerated are expressed as

$$[\Delta\kappa_{tetR,0}, \Delta\gamma_{tetR}, \Delta\kappa_{tetR}] = [30,\ 0.3,\ 50], \quad [\Delta\kappa_{lacI,0}, \Delta\gamma_{lacI}, \Delta\kappa_{lacI}] = [50,\ 0.3,\ 200],$$
$$[\Delta\kappa_{cI,0}, \Delta\gamma_{cI}, \Delta\kappa_{cI}] = [30,\ 0.3,\ 50], \quad [\Delta\kappa_{eyfp,0}, \Delta\gamma_{eyfp}, \Delta\kappa_{eyfp}] = [50,\ 0.3,\ 200].$$

(iii) The desired steady state x_d is given by [11]

$$x_d = \begin{bmatrix} x_{tetR,d} \\ x_{lacI,d} \\ x_{cI,d} \\ x_{eyfp,d} \end{bmatrix} = \begin{bmatrix} 1000 \\ 30000 \\ 300 \\ 30000 \end{bmatrix}$$

(iv) The prescribed attenuation level of external disturbance is specified as $\rho = 0.3$.

By solving the LMIs in equation (7.14) under the constraints of the design specifications, we find that if the design kinetic parameters κ_i and decay rates γ_i of the synthetic gene network are specified by the following ranges

$$\begin{aligned} &\kappa_{tetR} \in [891,\ 4159], && \kappa_{lacI} \in [1248,\ 5822] \\ &\kappa_{cI} \in [1422,\ 6652], && \kappa_{eyfp} \in [5124, 24906] \\ &\gamma_{tetR} \in [0.97,\ 4.03], && \gamma_{lacI} \in [0.04,\ 0.08] \\ &\gamma_{cI} \in [0.65,\ 0.75], && \gamma_{eyfp} \in [0.27,\ 0.83] \end{aligned} \tag{7.17}$$

then the four design specifications (i)–(iv) are satisfied.

To confirm the performance of the proposed robust synthetic gene network, the network is designed by using the set of kinetic parameters κ_i and decay rates γ_i in the ranges in equation (7.17). This approach provides a test of the network's ability to achieve the desired steady state regardless of initial conditions, parameter fluctuations, and extrinsic disturbances. The following design parameters are chosen from the ranges given in equation (7.17).

$$\begin{aligned}(\kappa_{tetR}, \kappa_{lacI}, \kappa_{cI}, \kappa_{eyfp}) &= (2000,\ 2000,\ 2000,\ 15000) \\ (\gamma_{tetR}, \gamma_{lacI}, \gamma_{cI}, \gamma_{eyfp}) &= (1.98,\ 0.05,\ 0.7,\ 0.57)\end{aligned} \tag{7.18}$$

The desired steady states of the synthetic gene network *in vivo* can be achieved under intrinsic parameter fluctuations and environmental disturbances. From the *in silico* simulation in Figure 7.3(a) with $v(t) = [10n_1, 1000n_2, 10n_3, 1000n_4]$, where n_i, $i = 1,\ldots,4$ are independent Gaussian white noises with unit variance, the disturbance attenuation level of environmental disturbance, which is prescribed by $\rho = 0.3$, is estimated as

$$\frac{\left(E\int_0^{1000} \tilde{x}^T Q \tilde{x} dt\right)^{1/2}}{\left(E\int_0^{1000} v^T v dt\right)^{1/2}} = 0.2715 < 0.3 \tag{7.19}$$

The prescribed level of disturbance attenuation (filtering ability) is thus achieved by the proposed method. A synthetic gene network with parameters outside the ranges in equation (7.17) is also designed. For example, kinetic parameters κ_i and decay rates γ_i may be (150, 100, 500, 1500) and (0.5, 0.05, 0.5, 0.2), respectively, which are outside the regions specified in equation (7.17). The simulation is shown in Figure 7.3(b). The time response of the synthetic network clearly suffers from more external disturbances and cannot achieve the desired steady states. In this design case, the disturbance attenuation level of external disturbance is estimated as

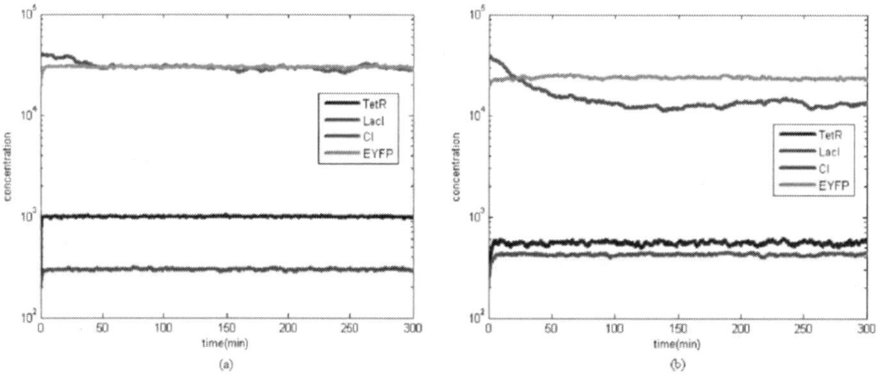

Figure 7.3. Simulation of the example of synthetic gene network design. To confirm the stability robustness and filtering ability of the synthetic gene network in the *in silico* example, the synthetic gene network is simulated with initial values [200, 40000, 200, 20000] and desired steady states [1000, 30000, 300, 30000]. (a) With design parameters $(\kappa_{tetR}, \kappa_{lacI}, \kappa_{cI}, \kappa_{eyfp}) = (2000, 2000, 2000, 15000)$ and $(\gamma_{tetR}, \gamma l_{acI}, \gamma_{cI}, \gamma_{eyfp}) = (1.98, 0.05, 0.7, 0.57)$ in the specified parameter range given in equation (7.17), the network shows sufficient stability and noise-filtering ability to achieve the desired steady state in spite of parameter fluctuations and disturbances in the host cell. (b) If the design parameters are outside the specified range, with $(\kappa_{tetR}, \kappa_{lacI}, \kappa_{cI}, \kappa_{eyfp}) = (150, 100, 500, 1500)$ and $(\gamma_{tetR}, \gamma_{lacI}, \gamma_{cI}, \gamma_{eyfp}) = (0.5, 0.05, 0.5, 0.2)$, then expression of the synthetic gene network shows greater fluctuation and cannot achieve the desired steady state under parameter fluctuations and environmental disturbances.

$$\frac{\left(E\int_0^{1000} \tilde{x}^T Q\tilde{x}dt\right)^{1/2}}{\left(E\int_0^{1000} v^T vdt\right)^{1/2}} = 2.8093>0.3 \tag{7.20}$$

The design specification for the filtering ability is significantly violated in this case.

Remark 7.1:

(i) Gene circuit design can now be implemented using recombination technology to insert or delete TF binding sites in the promoter region of a regulated gene with the aim of increasing or decreasing the value of the kinetic parameter κ_i (i.e., different levels of affinity) of the regulated gene [19]. By inserting strong or weak binding sites, large or small values can be obtained. For example, the binding site of $\kappa_i = 1$ is 10 times larger than that of $\kappa_i = 0.1$ at the promoter region of target gene i. Changes to the decay rate γ_i can be implemented by shortening the 3' polyadenylate tail (referred to as deadenylation), which primarily triggers decapping, resulting in 5' to 3' exonucleolysis. Alternatively, removal of the 3' polyadenylate tail can facilitate γ_i [7]. Therefore, by shortening or elongating the gene's 3' polyadenylate tail, we can increase or decrease the decay rate γ_i of gene i. Directed evolution methods are also useful in changing the elasticity (kinetic property of κ_i) and in designing biochemical circuits [7]. From a systems biology perspective, these advances in implementation techniques of κ_i and γ_i enable engineering of synthetic gene networks in the near future.

(ii) Because of the uncertain values of v and initial state $\tilde{x}(0)$ of the synthetic gene network (7.9) in the host cell, the following minimax design has also been considered to reset the uncertainty of $v(t)$ and $\tilde{x}(0)$ *in vivo* [6]. $v(t)$ and $\tilde{x}(0)$ are considered players maximizing their effects on regulation error $\tilde{x}(t)$ when the kinetic parameters are specified as another player minimizing the effects of v and $\tilde{x}(0)$ on $\tilde{x}(t)$.

$$\min_{\substack{k\in[k_1,k_2]\\ \gamma\in[\gamma_1,\gamma_2]}} \max_{\substack{\tilde{x}(0)\\ v}} \frac{E\int_0^{t_f} \tilde{x}^T Q\tilde{x}dx}{E\int_0^{t_f} v^T vdx} \tag{7.21}$$

A robust synthetic gene network based on dynamic game theory and fuzzy interpolation of local linearized linear systems can thus be efficiently designed [6]. It will be discussed in Chapter 10 in detail.

(iii) When the desired output of the synthetic gene network is not a steady state but a time-varying signal like a specific binary or periodic signal, and the network structure becomes complex, the above systematic analysis methods are difficult to apply. Because natural selection of traits suited for environmental change is an important evolutionary mechanism, kinetic parameters of synthetic gene networks can be tuned by genetic algorithms (GAs) or evolution algorithms (EAs) in Chapter 2 to optimally track desired biological functions [20, 21]. Based on the evolutionary network algorithm, the kinetic parameters may be tuned to maximize the fitness to some desired phenotype selected by natural selection. Consider the optimal tracking design of a synthetic gene network by network evolution shown in Figure 7.4. Let the tracking error be defined as

$$e(t) = y_d(t) - y(t)$$

The design purpose is then to tune the design parameter k_i using evolutionary (or genetic) algorithms such that the network can achieve the following optimal tracking

$$\min_{k_i \in \left[k_i^L, k_i^U\right]} E\int_0^{T_p} e^T(t)e(t)dt \tag{7.22}$$

where T_p denotes the present time. Let

$$J(k) = E\int_0^{T_p} e^T(t)e(t)dt \tag{7.23}$$

Let the fitness function $F(k)$ be defined as

$$F(k) = \frac{1}{J(k)} \tag{7.24}$$

Adapting a parameter vector (chromosome) $k_i \in [k_i^L, k_i^U]$ by EA or GA to minimize $J(k)$ for the desired network behavior tracking of the synthetic gene network is equivalent to maximizing the fitness function in equation (7.24) to meet the natural selection criteria in this evolutionary process for the synthetic gene network. A robust biological network design with a desired output behavior $y_d(t)$ is therefore equivalent to as solution to the following fitness maximization problem using an evolutionary network method [20, 21]

$$F(k^*) = \max_{k \in \left[k^L, k^U\right]} F(k) \tag{7.25}$$

The evolutionary network algorithm (or evolutionary systems biology algorithm [21]) is employed to solve the above fitness maximization problem via genetic operators such as selection, crossover, and mutation. It mimics natural selection in an evolutionary process to tune the kinetic parameter network of the synthetic gene network to an environmental change. A GA-based algorithm with binary coding of the chromosome has been proposed for the design of robust synthetic genetic oscillators with prescribed amplitude, period, and phase [20]. The oscillator is intended to allow protein concentrations to track desired periodic reference signals under intrinsic and environmental noises. Based on an evolutionary systems biology algorithm that encodes each chromosome in a real-valued vector, the design parameters of target gene circuits can evolve to specific values in order to robustly track a desired biological function in spite of such interferences [21].

7.2. Robust Synthetic Gene Network Design via Library-Based Search Methods

Over the past decade, synthetic biology has made significant progress in designing biological parts. Even synthetic gene networks may have been constructed using a variety of biological components to achieve desired behaviors. A limitation on the development of complex synthetic gene networks intended to track specific reference trajectories is the lack of an efficient method for selecting suitable biological parts from libraries; experimental data in promoter libraries cannot be directly used for selecting adequate promoter parts. Current promoter libraries therefore need to be redefined based on promoter activity. This would allow development of library-based search methods.

A dynamic model can be used in the indirect evaluation of the activities of promoter parts to help redefine existing promoter libraries. In the TetR-regulated promoter library shown in Figure 7.5, $x(c, t)$ and $X(c, t)$ respectively denote the concentrations of the mRNA and protein of gene *yegfp* (which is used to measure protein expression). The dynamic model of the promoter-regulation gene part is constructed with [9, 22]

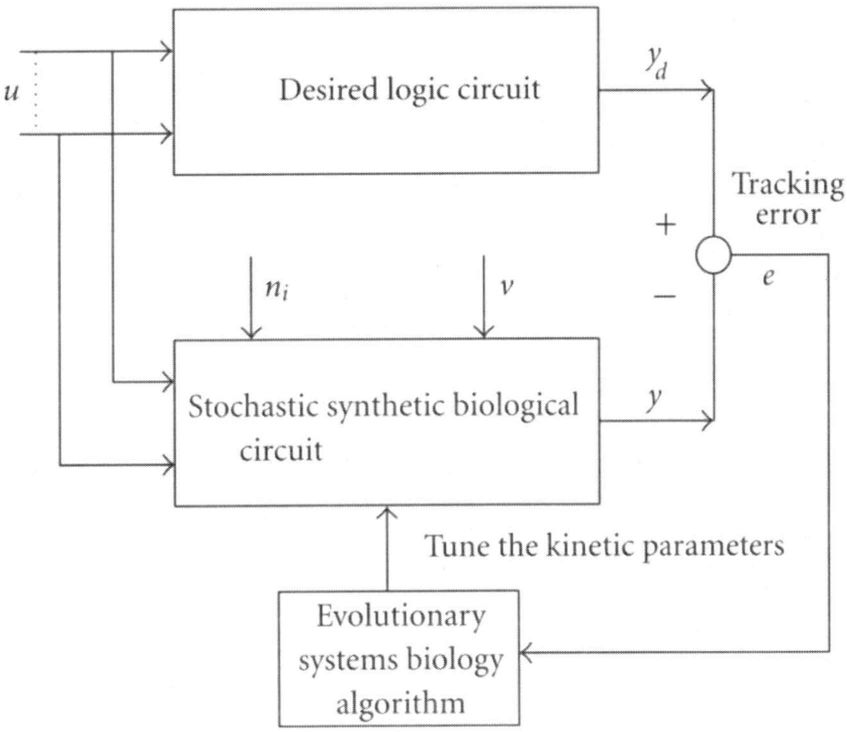

Figure 7.4. Block diagram of the optimal tracking scheme for synthetic biological circuit design using an evolutionary systems biology algorithm. Based on a network algorithm mimicking natural selection in an evolutionary process, the design parameters *k* of a synthetic biological circuit are tuned to minimize the tracking error between the desired logic circuit and the stochastic synthetic biological circuit, and to achieve the desired behavior tracking.

$$\begin{aligned}\dot{x}(c,t) &= P_{TetR}(c,r) - \beta x(c,t) \\ \dot{X}(c,t) &= \alpha x(c,t) - \gamma_{yEGFP} X(c,t)\end{aligned} \tag{7.26}$$

where β and γ_{yEGFP} denote the degradation rates of mRNA and protein yEGFP, respectively, and α denotes the translation rate. The promoter regulation function $P_{TetR}(c,r)$, which is dependent on repressor activity γ and the choice of promoter c, has the form

$$P_{TetR}(c,r) = c_r + \frac{c_s - c_r}{1 + (r/K_{TetR})^{n_{TetR}}} = c_r + (c_s - c_r) H_{TetR}(r), c = [c_r, c_s] \in Lib_{TetR} \tag{7.27}$$

where the promoter c has two promoter activities c_r and c_s (the minimum and maximum values of the promoter regulation function $P_{TetR}(c, r)$, respectively) for the TetR-regulated promoter library Lib_{TetR}. That is, $c = (c_r, c_s) \in Lib_{TetR}$. K_{TetR} and n_{TetR} denote the TetR–DNA binding affinity and binding cooperativity of regulatory protein TetR and DNA, respectively. $H_{TetR}(r)$ is a Hill function capturing the effect of a regulatory protein.

Based on the estimated promoter activities $c = (c_r, c_s)$ via maximum and minimum values of the steady state of protein expression data, some promoter libraries can be redefined in such a way that they can be efficiently selected from the design of the synthetic gene network (Table 7.1) [9, 22]. Since a synthetic gene network always consists of a set of promoter-regulation gene circuits (Figure 7.5), the design of complex synthetic gene network addresses how to select a set of promoters from the corresponding promoter libraries that have promoter activities adequate for achieving the design specifications. A well-known gene circuit topology, the simple toggle switch, is shown for illustration purposes in Figure 7.6. The toggle switch has two distinct stable states and can be reversibly switched between them by changing the inducers ATc and IPTG. Let $x_1(c_1, t)$, $x_2(c_2, t)$, and $x_3(c_3, t)$ denote the concentrations of mRNAs *tetR*, *lacI*, and *yegfp*, respectively; and let $X_1(c_1, t)$, $X_2(c_2, t)$, and $X_3(c_3, t)$ denote the concentrations of proteins TetR, LacI, and yEGFP, respectively. The dynamic model of the toggle switch gene network in Figure 7.6 is then modeled as

$$\begin{aligned}&\begin{cases}\dot{x}_1(c_1,t) = p_{LacI}(c_1,r_1) - \beta x_1(c_1,t) \\ \dot{X}_1(c_1,t) = \alpha x_1(c_1,t) - \gamma_1 X_1(c_1,t) \\ \dot{x}_2(c_2,t) = p_{TetR}(c_2,r_2) - \beta x_2(c_2,t) \\ \dot{X}_2(c_2,t) = \alpha x_2(c_2,t) - \gamma_2 X_2(c_2,t) \\ \dot{x}_3(c_3,t) = p_{TetR}(c_3,r_3) - \beta x_3(c_3,t) \\ \dot{X}_3(c_3,t) = \alpha x_3(c_3,t) - \gamma_3 X_3(c_3,t)\end{cases} \\ &y(c,t) = X_3(c_3,t) \\ &c = [c_1, c_2, c_3], c_1 \in Lib_{LacI}, c_2, c_3 \in Lib_{TetR}\end{aligned} \tag{7.28}$$

The promoter regulation functions $p_{LacI}(c_1, r_1)$, $p_{TetR}(c_2, r_2)$, and $p_{TetR}(c_3, r_3)$ are dependent on the selection of promoters c_1, c_2, and c_3 from the corresponding promoter libraries in Table 7.1. The output of interest $y(c, t)$ is dependent of the selected promoter set $c = [c_1, c_2, c_3]$ with adequate promoter activities from the corresponding promoter libraries in Table 7.1. This dynamic toggle switch gene network model consists of three interactive dynamic models of promoter-regulation gene parts, as shown in equation (7.26).

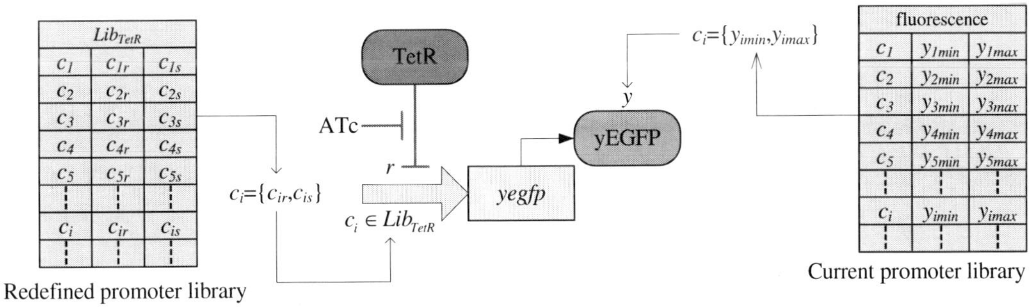

Figure 7.5. Single schematic diagram of the synthetic promoter-regulation gene circuit. The existing TetR-regulated promoter library contains the minimum and maximum values of fluorescence $[y_{imin}, y_{imax}]$ corresponding to with and without TetR (repressor) binding. Based on the promoter regulation function (7.27) and these values, the promoter library is redefined for the design of synthetic gene networks (Table 7.1).

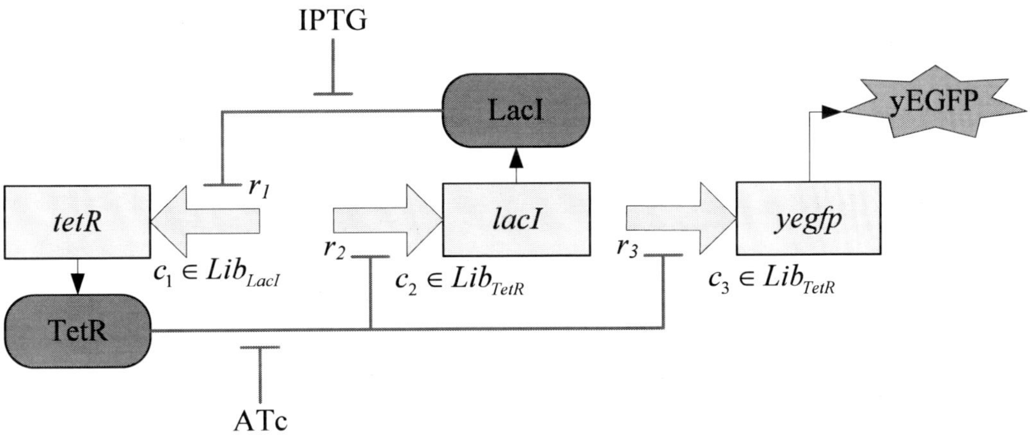

Figure 7.6. Synthetic gene circuit topology: simple toggle switch. The regulatory protein TetR, which is induced by ATc, inhibits the transcription of *lacI* by binding promoter c_2. TetR also inhibits transcription of *yegfp* by binding promoter c_3 to repress the expression of the fluorescent protein yEGFP. The protein LacI, which is induced by the inducer IPTG, inhibits the transcription of *tetR* by binding promoter c_1. The gene circuit has two distinct stable states, and can reversibly switch between them by changing the inducers ATc and IPTG. If an adequate promoter set $c = [c_1, c_2, c_3]$ is selected from corresponding promoter libraries, then yEGFP can be used to track the desired behaviors generated by a reference model. In the reference model, c_1 is selected from the LacI-regulated promoter library, and c_2 and c_3 are selected from the TetR-regulated promoter library in Table 7.1 (i.e., $c_1 \in Lib_{LacI}, c_2, c_3 \in Lib_{TetR}$).

The multiobjective design approach for the H_2/H_∞ synthetic gene network based on promoter libraries selects an adequate promoter set $c = [c_1, c_2, c_3]$ from corresponding promoter libraries such that the following two design objectives are achieved simultaneously [9]:

(i) H_∞ desired noise attenuation level ρ_d:

$$\frac{E\int_0^{t_f}\left(y(c,t)-y_r(t)\right)^T Q\left(y(c,t)-y_r(t)\right)dt}{E\int_0^{t_f} v^T(t)v(t)dt} \le \rho_d^2 \text{ for } c \in Lib_j,\ j=1,\ldots,m \tag{7.29}$$

(ii) H_2 optimal reference tracking:

$$\min_{\substack{c\in Lib_j \\ j=1,\ldots,m}} E\int_0^{t_f}\left(y(c,t)-y_r(t)\right)^T Q\left(y(c,t)-y_r(t)\right)dt \tag{7.30}$$

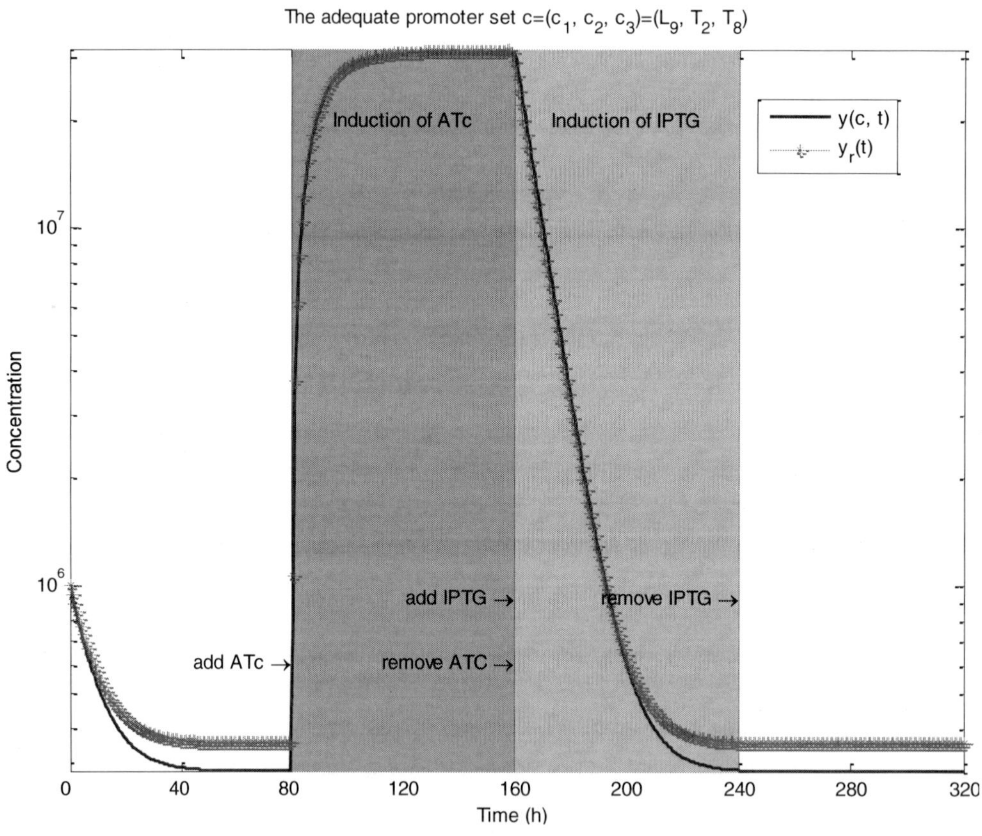

Figure 7.7. Simulation of toggle switch. By solving the LMI-constrained optimization problem of the H_2/H_∞ design objective equations (7.29) and (7.30) for the synthetic gene network in Figure 11 through the library searching method, an adequate promoter set $c = [c_1, c_2, c_3] = [L_9, T_2, L_8]$ is selected from the corresponding promoter libraries. The inducer ATc is added to the synthetic gene network at 80–160 hours to induce the gene network, and then the inducer IPTG is added at 160–240 hours. The output $y(c, t)$ clearly produces a robust track with the desired reference output $y_r(t)$.

By solving an optimization problem with two constraints [9], adequate promoters can be selected from the corresponding libraries in Table 7.1 to achieve the two design objectives in equations (7.29) and (7.30). Based on the synthetic toggle switch (Figure 7.6) and dynamic model (equation (7.28)) with intrinsic parameter variation and external disturbance, the adequate promoter set $c=[c_1, c_2, c_3]=[L_9, T_2, L_8]$ is selected from promoter libraries in Table 7.1 to achieve the multi-objective H_2/H_∞ reference tracking specified in equations (7.29) and (7.30). Simulation results with $v(t) = 10 \times [n_1 \cdots n_6]^T$ are shown in Figure 7.7.

7.3. Conclusion

In this chapter, we design a robust synthetic gene circuit based on four design specifications that tolerates genetic mutations, resists environmental noise, and functions properly in host cells. Evolutionary algorithm (EA) searching methods are used to select adequate genetic components from corresponding component libraries to match the input and output of a desired reference circuit. The two robust synthetic biology design methods proposed could help develop reliable means to realize predictable and robust behaviors in genetic networks.

7.4. Appendix

7.4.1. Proof of Proposition 7.1

Let us choose a Lyapunov function $V(\tilde{x}) > 0$ with $V(0) = 0$

$$E\int_0^\infty \tilde{x}^T Q\tilde{x}dt = E\left\{V(\tilde{x}(0)) - V(\tilde{x}(\infty)) + \int_0^\infty \left[\tilde{x}^T Q\tilde{x} + \frac{dV(\tilde{x})}{dt}\right]dt\right\} \tag{7.31}$$

By the Ito's formula [23] and $E\{dW_i/dt\}=E\{n_i\}=0$, we get

$$E\frac{dV(\tilde{x})}{dt}$$
$$= E\left\{\left(\frac{\partial V(\tilde{x})}{\partial \tilde{x}}\right)^T \left[Nf(\tilde{x}+x_d)+v\right] + \frac{1}{2}\sum_{i=1}^m g_i^T(\tilde{x}+x_d)M_i^T \frac{\partial^2 V(\tilde{x})}{d\tilde{x}^2} M_i g_i(\tilde{x}+x_d)\right\} \tag{7.32}$$

Substituting (7.32) into (7.31), we get

$$E\int_0^\infty \tilde{x}^T Q\tilde{x}dt$$
$$= E\left\{V(\tilde{x}(0)) - V(\tilde{x}(\infty)) + \int_0^\infty \left[\tilde{x}^T Q\tilde{x} + \left(\frac{\partial V(\tilde{x})}{\partial \tilde{x}}\right)^T Nf(\tilde{x}+x_d) + \left(\frac{\partial V(\tilde{x})}{\partial \tilde{x}}\right)^T v \right.\right. \tag{7.33}$$
$$\left.\left. + \frac{1}{2}\sum_{i=1}^{m} g_i^T(\tilde{x}+x_d) M_i^T \frac{\partial^2 V(\tilde{x})}{d\tilde{x}^2} M_i g_i(\tilde{x}+x_d)\right]dt\right\}$$

By the inequality (7.11), we have

$$E\int_0^\infty \tilde{x}^T Q\tilde{x}dt$$
$$< E\left\{V(\tilde{x}(0)) - V(\tilde{x}(\infty)) + \int_0^\infty \left[\left(\frac{\partial V(\tilde{x})}{\partial \tilde{x}}\right)^T v - \frac{1}{4\rho^2}\left(\frac{\partial V(\tilde{x})}{\partial \tilde{x}}\right)^T \frac{\partial V(\tilde{x})}{\partial \tilde{x}}\right]dt\right\}$$
$$\le E\left\{V(\tilde{x}(0)) + \int_0^\infty \left[\left(\frac{\partial V(\tilde{x})}{\partial \tilde{x}}\right)^T v - \frac{1}{4\rho^2}\left(\frac{\partial V(\tilde{x})}{\partial \tilde{x}}\right)^T \frac{\partial V(\tilde{x})}{\partial \tilde{x}}\right]dt\right\} \tag{7.34}$$
$$\le E\left\{V(\tilde{x}(0)) + \int_0^\infty \left[\rho^2 v^T v - \left(\rho v - \frac{1}{2\rho}\frac{\partial V(\tilde{x})}{\partial \tilde{x}}\right)^T \left(\rho v - \frac{1}{2\rho}\frac{\partial V(\tilde{x})}{\partial \tilde{x}}\right)\right]dt\right\}$$
$$\le E\left\{V(\tilde{x}(0)) + \int_0^\infty \rho^2 v^T v dt\right\}$$

If $V(\tilde{x}(0)) = 0$, then (7.34) will be reduced to (7.10).
If the synthetic gene network is free of the external disturbances,i.e., $v = 0$, then we have

$$E\int_0^\infty \tilde{x}^T Q\tilde{x}dt < V(\tilde{x}(0)) \tag{7.35}$$

For some positive constant $V(\tilde{x}(0))$, it means $\tilde{x} \to 0$ in probability as $t \to \infty$

7.4.2. Proof of Proposition 7.2

Now, we will derive the sufficient condition to ensure that the linearized synthetic gene network in (7.13) can attenuate the external disturbance below a prescribed attenuation level ρ in (7.10). By choosing a positive Lyapunov function as $V(\tilde{x}) = \tilde{x}^T P\tilde{x} > 0$, we have

$$E\int_0^\infty \tilde{x}^T Q\tilde{x}dt = E\left\{\tilde{x}^T(0) P\tilde{x}(0) - \tilde{x}^T(\infty) P\tilde{x}(\infty) + \int_0^\infty \left[\tilde{x}^T Q\tilde{x} + \frac{d}{dt}\tilde{x}^T P\tilde{x}\right]dt\right\} \tag{7.36}$$

By Ito's formula and $E\{dW_i/dt\}=0$, we have

$$
\begin{aligned}
&E\int_0^{\infty}\tilde{x}^T Q\tilde{x}dt \\
&= E\Big\{\tilde{x}^T(0)P\tilde{x}(0)-\tilde{x}^T(\infty)P\tilde{x}(\infty)+\int_0^{\infty}\Big[\tilde{x}^T Q\tilde{x}+\sum_{j=1}^{M}\alpha_j(\tilde{x})\big(NF_j\tilde{x}+v\big)^T P\tilde{x} \\
&+\sum_{j=1}^{M}\alpha_j(\tilde{x})\tilde{x}^T P\big(NF_j\tilde{x}+v\big)+\sum_{j=1}^{M}\alpha_j(\tilde{x})\sum_{i=1}^{m}\tilde{x}^T G_{ij}^T M_i^T PM_iG_{ij}\tilde{x}\Big]dt\Big\} \\
&\le E\Big\{\tilde{x}^T(0)P\tilde{x}(0)+\int_0^{\infty}\Big[\sum_{j=1}^{M}\alpha_j(\tilde{x})\tilde{x}^T\Big(F_j^T N^T P+PNF_j+Q+\sum_{i=1}^{m}G_{ij}^T M_i^T PM_iG_{ij}+\frac{1}{\rho^2}PP\Big)\tilde{x} \\
&+\sum_{j=1}^{M}\alpha_j(\tilde{x})\Big(\tilde{x}^T Pv+v^T P\tilde{x}-\frac{1}{\rho^2}\tilde{x}^T PP\tilde{x}-\rho^2 v^T v\Big)+\rho^2 v^T v\Big]dt\Big\}
\end{aligned}
\tag{7.37}
$$

Table 7.1. Redefined TetR- and LacI-regulated promoter libraries. The redefined TetR- and LacI-regulated promoter libraries (i.e., Lib_{TetR} and Lib_{LacI}) comprise different promoters (i.e., T_k and L_k, $k = 0,\ldots,20$) with their corresponding activities of c_s and c_r obtained from previous libraries of experimental data

TetR-regulated promoter library (Lib_{TetR})			LacI-regulated promoter library (Lib_{LacI})		
Promoter	Promoter activity		Promoter	Promoter activity	
	c_s	c_r		c_s	c_r
T_0	2121	0.1724	L_0	1657.5	0.3018
T_1	1604	0.7576	L_1	923.97	0.2567
T_2	1376.6	0.1936	L_2	860.87	0.2244
T_3	1169.8	0.4672	L_3	674.92	1.9189
T_4	974.52	0.0753	L_4	651.58	1.1680
T_5	942.77	0.2281	L_5	570.07	3.5062
T_6	967.17	0.1493	L_6	527.83	0.5497
T_7	738.57	0.0702	L_7	323.45	0.1248
T_8	641.74	0.7135	L_8	327.77	0.1772
T_9	564.24	0.2620	L_9	309.74	0.5439
T_{10}	501.35	0.0756	L_{10}	298.35	0.1146
T_{11}	469.35	0.0788	L_{11}	250.16	0.1326
T_{12}	466.16	0.1636	L_{12}	248.03	0.1171
T_{13}	356.88	0.0927	L_{13}	239.32	0.1010
T_{14}	348.95	0.1483	L_{14}	190.2	0.0959
T_{15}	274.79	0.1067	L_{15}	163.84	0.4813
T_{16}	250.04	0.0857	L_{16}	166.42	0.0989
T_{17}	188.77	0.1366	L_{17}	131.63	0.1190
T_{18}	119.57	0.0753	L_{18}	108.96	0.0903
T_{19}	111.57	0.1185	L_{19}	101.89	0.0982
T_{20}	70.909	0.1606	L_{20}	85.673	0.2174

By the inequality in (7.14), we have

$$\begin{aligned}
&E\int_0^{\infty} \tilde{x}^T Q\tilde{x}dt \\
&\leq E\left\{\tilde{x}^T(0)P\tilde{x}(0)+\int_0^{\infty}\rho^2 v^T v dt-\int_0^{\infty}\left[\sum_{j=1}^{M}\alpha_j(\tilde{x})\left(\frac{1}{\rho}\tilde{x}^T P-\rho v\right)^T\left(\frac{1}{\rho}\tilde{x}^T P-\rho v\right)\right]dt\right\} \qquad (7.38)\\
&\leq E\left\{\tilde{x}^T(0)P\tilde{x}(0)+\int_0^{\infty}\rho^2 v^T v dt\right\}
\end{aligned}$$

Then, the remainder of the proof is similar to the procedure in Proof of Proposition 7.1.

REFERENCES

[1] Endy, D. "Foundations for engineering biology," *Nature*, vol. 438, 449-53, Nov 24 2005.

[2] Stricker, J; Cookson, S; Bennett, MR; Mather, WH; Tsimring, LS; Hasty, J. "A fast, robust and tunable synthetic gene oscillator," *Nature*, vol. 456, 516-U39, Nov 27 2008.

[3] Tigges, M; Marquez-Lago, TT; Stelling, J; Fussenegger, M. "A tunable synthetic mammalian oscillator," *Nature*, vol. 457, 309-12, Jan 15 2009.

[4] Basu, S; Gerchman, Y; Collins, CH; Arnold, FH; Weiss, R. "A synthetic multicellular system for programmed pattern formation," *Nature*, vol. 434, 1130-4, Apr 28 2005.

[5] Tsai, TY; Choi, YS; Ma, W; Pomerening, JR; Tang, C; Ferrell, JE. Jr., "Robust, tunable biological oscillations from interlinked positive and negative feedback loops," *Science*, vol. 321, 126-9, Jul 4 2008.

[6] Chen, BS; Chang, CH; Lee, HC. "Robust synthetic biology design: stochastic game theory approach," *Bioinformatics*, vol. 25, 1822-1830, Jul 15 2009.

[7] Chen, BS; Wu, CH. "A systematic design method for robust synthetic biology to satisfy design specifications," *BMC Systems Biology*, vol. 3, Jun 30 2009.

[8] Chen, BS; Wu, CH. "Robust Optimal Reference-Tracking Design Method for Stochastic Synthetic Biology Systems: T-S Fuzzy Approach," *IEEE Transactions on Fuzzy Systems*, vol. 18, 1144-1159, Dec 2010.

[9] Wu, CH; Zhang, WH; Chen, BS. "Multiobjective H-2/H-infinity synthetic gene network design based on promoter libraries," *Mathematical Biosciences*, vol. 233, 111-125, Oct 2011.

[10] Kuepfer, L; Sauer, U; Parrilo, PA. "Efficient classification of complete parameter regions based on semidefinite programming," *BMC Bioinformatics*, vol. 8, Jan 15 2007.

[11] Batt, G; Yordanov, B; Weiss, R; Belta, C. "Robustness analysis and tuning of synthetic gene networks," *Bioinformatics*, vol. 23, 2415-2422, Sep 15 2007.

[12] Chen, BS; Chang, YT; Wang, YC. "Robust H infinity-stabilization design in gene networks under stochastic molecular noises: fuzzy-interpolation approach," *IEEE Trans Syst Man Cybern B Cybern*, vol. 38, 25-42, Feb 2008.

[13] Alon, U. *An introduction to systems biology : design principles of biological circuits*. Boca Raton, FL: Chapman & Hall/CRC, 2007.

[14] Voit, EO. *Computational analysis of biochemical systems : a practical guide for biochemists and molecular biologists*. New York: Cambridge University Press, 2000.

[15] Chen, BS; Zhang, WH. "Stochastic H(2)/H(infinity) control with state-dependent noise," *IEEE Transactions on Automatic Control*, vol. 49, 45-57, Jan 2004.

[16] Zhang, WH; Chen, BS; Tseng, CS. "Robust H-infinity filtering for nonlinear stochastic systems," *IEEE Transactions on Signal Processing*, vol. 53, 589-598, Feb 2005.

[17] Boyd, SP. *Linear matrix inequalities in system and control theory*. Philadelphia: Society for Industrial and Applied Mathematics, 1994.

[18] Chen, BS; Chen, WH; Wu, HL. "Robust H-2/H-∞ Global Linearization Filter Design for Nonlinear Stochastic Systems," *IEEE Transactions on Circuits and Systems I-Regular Papers*, vol. 56, 1441-1454, Jul 2009.

[19] Ellis, T; Wang, X; Collins, JJ. "Diversity-based, model-guided construction of synthetic gene networks with predicted functions," *Nature Biotechnology*, vol. 27, 465-471, May 2009.

[20] Chen, BS; Chen, PW. "GA-based Design Algorithms for the Robust Synthetic Genetic Oscillators with Prescribed Amplitude, Period and Phase," *Gene Regul Syst Bio*, vol. 4, 35-52, 2010.

[21] Chen, BS; Hsu, CY; Liou, JJ. "Robust Design of Biological Circuits: Evolutionary Systems Biology Approach," *Journal of Biomedicine and Biotechnology*, 2011.

[22] Wu, CH; Lee, HC; Chen, BS. "Robust synthetic gene network design via library-based search method," *Bioinformatics*, vol. 27, 2700-2706, Oct 1 2011.

[23] Chen, G; Chen, G; Hsu, SH. *Linear stochastic control systems*. Boca Raton: CRC Press, 1995.

Chapter 8

SYSTEMS METABOLIC ENGINEERING

ABSTRACT

Metabolic engineering aims to design biochemical networks that use engineered metabolic pathways for biotechnological applications. In this chapter, a robust control method for biochemical circuits is proposed based on steady-state analyses of the synergism and saturation system (S-system) model. Two design schemes are developed. One scheme improves the system's structural stability to tolerate larger kinetic parameter variations, while the other compensates for the kinetic parameter variations to eliminate their effect. In addition, a multi-objective biochemical circuit design scheme is introduced for both the robust and sensitivity designs. The proposed robust circuit design scheme provides a systematic method with potential applications in synthetic circuit design for metabolic engineering and drug design. First, a robust circuit design of the tricarboxylic acid cycle *in silico* is introduced to illustrate the design procedure and to confirm the performance of the proposed biochemical pathway. Then, a robust metabolic pathway is designed based on H_∞ robust stabilization of dynamic metabolic networks modeled by a discrete-time stochastic system under intrinsic fluctuations and environmental disturbances. Finally, an engineered isobutanol biosynthesis pathway in *E. coli* using systems synthetic biology technology is introduced for biofuel production, and integration of a temperature-controllable enzyme expression circuit in the biosynthetic isobutanol pathway is synthesized to improve yield in *E. coli*.

INTRODUCTION

Systems metabolic engineering aims to amplify or delete specific genes in metabolic pathways to engineer metabolisms within a systems biology framework. Regulatory gene networks, metabolic networks, and other cellular networks can thus be engineered in an integrative manner. Systems biological analysis with large-scale genome-wide analyses and computational bioinformatic tools can allow for rapid evaluation of the global physiology of a cell with respect to cellular regulation. This includes transcriptional and translational regulation as well as metabolic engineering distributions [1-4]. Results of this approach can be used to predict targets for metabolic engineering within the host strain. Furthermore, integration of high-throughput, large-scale, genome-wide analyses with in silico simulation results might provide additional information on cellular status at various hierarchical levels from genome to fluxome. Strain selection for improvement by systems metabolic engineering is divided into several phases [3]: (i) a base strain is allowed to develop; (ii) the base strain is further engineered using synthetic biological methods, based on results obtained from high-throughput genome-wide bioinformatic data and systems biology computational analyses; (iii) the performance of this preliminary production strain is evaluated in actual fermentation; (iv) the results are fed back to further develop the strain until the derivation of a superior strain with the desired performance.

Systems biology may allow us to reach an unprecedented level of understanding of cellular physiology and characteristics by combining results of multiple genome-wide and computational analyses [5]. This understanding can subsequently be used within a systems synthetic biology framework to design systems metabolic engineering strategies, such as strategies for strain improvement. However, novel computational methods such as large-scale data mining, multidimensional data integration, data-driven network inference, and deep curation schemes still need to be developed and applied to the integration and interpretation of heterogeneous large-scale genome-wide bioinformatic data, which are interconnected by complex regulatory and metabolic pathways [3, 6]. Several robust biochemical circuit designs have recently been proposed to improve the network robustness of metabolic pathways. The proposed design schemes provide methods within systems biology with potential applications in systems synthetic circuit design for systems metabolic engineering and drug design.

8.1. Robust Biochemical Circuit Design in Metabolic Networks

Broadly, a metabolic network is a collection of enzymatic reactions that process cellular and intercellular metabolites. In systems metabolic engineering, the rates of reactions or fluxes, which correspond directly to changes in concentrations of substrates, enzymes, factors, or products, are often measured. Such systematic changes in concentrations can be expressed in terms of dynamic differential equations. The following S-system synergism and saturation representation has been used for the last three decades as an efficient model for describing a dynamic metabolic network [7, 8]

$$
\begin{aligned}
\dot{X}_1 &= \alpha_1 \prod_{j=1}^{n+m} X_j^{g_{1j}} - \beta_1 \prod_{j=1}^{n+m} X_j^{h_{1j}} \\
&\vdots \\
\dot{X}_i &= \alpha_i \prod_{j=1}^{n+m} X_j^{g_{ij}} - \beta_i \prod_{j=1}^{n+m} X_j^{h_{ij}} = V_i - V_{-i}, \quad i=1,\ldots,n \qquad (8.1) \\
&\vdots \\
\dot{X}_n &= \alpha_n \prod_{j=1}^{n+m} X_j^{g_{nj}} - \beta_n \prod_{j=1}^{n+m} X_j^{h_{nj}}
\end{aligned}
$$

where $X_1,\cdots,X_{n+m}$ are metabolites, such as substrates, enzymes, factors, or products of a biochemical network, in which $X_1,\cdots,X_n$ denote n dependent variables and X_{n+1}, $\cdots,X_{n+m}$ denote the independent variables. In a metabolic network, intermediate metabolites and products are dependent variables, whereas substrates and enzymes are independent variables. The rate of change in X_i, $\dot{X}_i$ is equal to the difference between two terms, one for production or accumulation, and the other for degradation or clearance. Each term is the product of a positive rate constant α_i or β_i and all dependent and independent variables that directly affect production or degradation, respectively. Each variable X_j is raised to the power of a kinetic parameter g_{ij} and h_{ij}, which represents an activating effect of X_j on X_i when its value is positive, and an inhibitive effect when its value is negative. V_i and V_{-i} represent aggregate flux into and out of the X_i pool.

Construction of the S-system representation of a metabolic network and estimation of its kinetic parameters from experimental data are described in [8] and the literature references therein. The nonlinear parameter estimation problem of S-systems has recently been solved efficiently by evolution optimization methods [9]. It is generally difficult to study the network robustness or sensitivity of a nonlinear system such as equation (8.1). Fortunately, many important characteristics of an S-system at or close to the steady state can be analyzed by using simple algebraic methods. Since most metabolic networks in nature operate near the steady state, at which inputs and outputs are almost balanced, the following focuses on the network robustness of metabolic networks at the steady state [7, 8].

Consider the steady state of the metabolic network in equation (8.1),

$$\alpha_i \prod_{j=1}^{n+m} X_j^{g_{ij}} = \beta_i \prod_{j=1}^{n+m} X_j^{h_{ij}}, \quad i = 1, 2, \ldots, n \tag{8.2}$$

Taking the logarithm on both sides of equation (8.2) and making some rearrangements,

$$\sum_{j=1}^{n} (g_{ij} - h_{ij}) \ln X_j = \ln \beta_i - \ln \alpha_i - \sum_{j=n+1}^{n+m} (g_{ij} - h_{ij}) \ln X_j, \quad i = 1, 2, \ldots, n \tag{8.3}$$

Introducing new variables and coefficients,

$$y_j = \ln X_j, \quad a_{ij} = g_{ij} - h_{ij}, \quad b_i = \ln\left(\frac{\beta_i}{\alpha_i}\right) \tag{8.4}$$

the steady state of the metabolic network is obtained as

$$A_D Y_D = b - A_I Y_I \tag{8.5}$$

where

$$Y_D = \begin{bmatrix} y_1 \\ \vdots \\ y_n \end{bmatrix},\ b = \begin{bmatrix} b_1 \\ \vdots \\ b_n \end{bmatrix},\ Y_I = \begin{bmatrix} y_{n+1} \\ \vdots \\ y_{n+m} \end{bmatrix}, A_D = \begin{bmatrix} a_{11} & \cdots & a_{1n} \\ \vdots & \ddots & \vdots \\ a_{n1} & \cdots & a_{nn} \end{bmatrix},\ A_I = \begin{bmatrix} a_{1,n+1} & \cdots & a_{1,n+m} \\ \vdots & \ddots & \vdots \\ a_{n,n+1} & \cdots & a_{n,n+m} \end{bmatrix}$$

In equation (8.5), A_D denotes the system matrix of the interactions between dependent variables Y_D, and A_I indicates the interactions between dependent variables Y_D and independent variables Y_I. In the nominal parameter case, it is assumed that the inverse of A_D exists so that Y_D can be solved uniquely, i.e., the metabolic network results in only one steady state (phenotype). Therefore, the steady state of the biochemical system is given by

$$Y_D = A_D^{-1}(b - A_I Y_I) \tag{8.6}$$

Suppose that parameter variations due to mutation, thermal changes, or disease can alter the kinetic properties of the steady state of a metabolic network as follows [7]

$$(A_D + \Delta A_D)(Y_D + \Delta Y_D) = (b + \Delta b) - (A_I + \Delta A_I)(Y_I + \Delta Y_I) \tag{8.7}$$

Parameter perturbations are defined as

$$\Delta A_D = \begin{bmatrix} \Delta a_{11} & \cdots & \Delta a_{1n} \\ \vdots & \Delta a_{ij} & \vdots \\ \Delta a_{n1} & \cdots & \Delta a_{nn} \end{bmatrix} = \begin{bmatrix} \Delta g_{11} - \Delta h_{11} & \cdots & \Delta g_{1n} - \Delta h_{1n} \\ \vdots & \Delta g_{ij} - \Delta h_{ij} & \vdots \\ \Delta g_{n1} - \Delta h_{n1} & \cdots & \Delta g_{nn} - \Delta h_{nn} \end{bmatrix}$$

$$\Delta b = \begin{bmatrix} \Delta b_1 \\ \vdots \\ \Delta b_n \end{bmatrix},\ \Delta A_I = \begin{bmatrix} \Delta g_{1,n+1} - \Delta h_{1,n+1} & \cdots & \Delta g_{1,n+m} - \Delta h_{1,n+m} \\ \vdots & \Delta g_{i,n+j} - \Delta h_{i,n+j} & \vdots \\ \Delta g_{n,n+1} - \Delta h_{n,n+1} & \cdots & \Delta g_{n,n+m} - \Delta h_{n,n+m} \end{bmatrix}$$

where ΔA_D denotes perturbations due to kinetic parameter variations, Δb denotes perturbations due to rate constant variations, and ΔA_I denotes perturbations due to kinetic parameter variations between dependent and independent variables. ΔA_D can influence the existence of the steady state of the metabolic network.

From equation (8.7), we derive

$$A_D\left(I + A_D^{-1}\Delta A_D\right)(Y_D + \Delta Y_D) = (b + \Delta b) - (A_I + \Delta A_I)Y_I \tag{8.8}$$

If the following robustness condition holds [7]

$$\left\| A_D^{-1}\Delta A_D \right\|_2 < 1 \tag{8.9}$$

then the singular values of $I + A_D^{-1}\Delta A_D$ are free of zero and the inverse $(I + A_D^{-1}\Delta A_D)^{-1}$ exists. Therefore, the steady state of the perturbative metabolic network in equation (8.8) is uniquely solved as:

$$Y_D + \Delta Y_D = \left(I + A_D^{-1}\Delta A_D\right)^{-1} A_D^{-1}\left[(b+\Delta b)-(A_I + \Delta A_I)Y_I\right] \quad (8.10)$$

The above analysis implies that if the robustness condition in equation (8.9) holds, then the steady state of a metabolic system is preserved under parameter variations ΔA_D, i.e., $Y_D + \Delta Y_D$ in equation (8.10) has a small difference ΔY_D from the nominal equation in (8.6) under small perturbation. However, if the condition in equation (8.9) does not hold, then individual values of $I + A_D^{-1}\Delta A_D$ may be zero, the inverse $(I + A_D^{-1}\Delta A_D)^{-1}$ may not exist, and the steady state $Y_D + \Delta Y_D$ may cease to exist under the parameter perturbation ΔA_D. As an example, consider the singular value decomposition:

$$A_D = U^T \Sigma V = \sum_{i=1}^{n} \sigma_i u_i^T v_i \quad (8.11)$$

where σ_i denotes the i-th singular value and u_i, $v_i \in R^n$ denote the corresponding left and right singular vectors, respectively. Therefore, if a parameter variation is specified as follows:

$$\Delta A_D = -\sigma_i u_i^T v_i, \quad i = 1, 2, \ldots, n \quad (8.12)$$

then

$$A_D + \Delta A_D = U^T \begin{bmatrix} \sigma_1 & & & & & & 0 \\ & \ddots & & & & & \\ & & \sigma_{i-1} & & & & \\ & & & 0 & & & \\ & & & & \sigma_{i+1} & & \\ & & & & & \ddots & \\ 0 & & & & & & \sigma_n \end{bmatrix} V \quad (8.13)$$

Obviously, the inverse $(A_D+\Delta A_D)^{-1}$ or $(I+A_D^{-1}\Delta A_D)^{-1}$ does not exist under the parameter perturbation in equation (8.12).

Remark 8.1: The parameter perturbations in the direction of singular vectors like equation (8.12) are the network's points of fragility. The robustness prevents this kind of parameter variation and guarantees the existence of the steady state of the metabolic networks. When unexpected perturbations like equation (8.12) are encountered, a catastrophic failure of the network follows. Robust circuit design is a necessary fail-safe mechanism in such situation. For example, the trehalose pathway in yeast consists of only a few metabolites that form a substrate cycle. It is governed by a surprisingly complex control system that is composed of several inhibiting or activating signaling mechanisms [8].

Proposition 8.1 [7]: Equivalently, the network robustness of the metabolic network in equation (8.9) can be rewritten as a more intuitive criterion for network robustness:

$$A_D^{-1}\Delta A_D \Delta A_D^T A_D^{-T} < \mathrm{I} \quad \text{or} \quad \Delta A_D \Delta A_D^T < A_D A_D^T \tag{8.14}$$

That is, $A_D A_D^T$ is the upper bound of $\Delta A_D \Delta A_D^T$ without violation of robust stability at steady state. If the network robustness criterion in equation (8.14) holds, then the steady state of the perturbative metabolic network still exists.

If a metabolic network cannot tolerate perturbations, i.e., the network robustness criterion in equation (8.14) is violated, then robust control via a biochemical circuit design is a suitable remedy. Based on robustness analysis, we develop a biochemical circuit design scheme for the robust control of metabolic networks. Consider the robust control system design of the metabolic network (in equation (8.1)) by specific biochemical feedback circuits in a more general metabolic network form [7]

$$\dot{X}_i = \alpha_i \prod_{j=1}^{n+m} X_j^{g_{ij}} \prod_{k=1}^{n} X_k^{f_{ik}} - \beta_i \prod_{j=1}^{n+m} X_j^{h_{ij}} \prod_{k=1}^{n} X_k^{l_{ik}}, \quad i = 1, \ldots, n \tag{8.15}$$

where $X_k^{f_{ik}}$ denotes a new biochemical control circuit with X_k regulating the production of X_i by the kinetic parameter f_{ik}. $X_k^{l_{ik}}$ denotes a new biochemical control circuit with X_k regulating the degradation of X_i by the kinetic parameter l_{ik}. The choice of regulating objects, X_k and X_i, and the specification of the kinetic parameters, f_{ik} and l_{ik}, are designed according to the feasibility of biochemical circuit linkage to achieve a desired robustness to tolerate ΔA_D within the prescribed range of kinetic parameter perturbations in a metabolic network. Since f_{ik} and l_{ik} represent the elasticities of the corresponding enzymes in the designed control circuits, the implementation of control circuits is heavily dependent on the elasticity specification of these enzymes.

Consider the robust control system of the metabolic network in equation (8.15). By using a similar procedure to equations (8.2)–(8.7) at steady state,

$$(A_D + F + \Delta A_D)(Y_D + \Delta Y_D) = (b + \Delta b) - (A_I + \Delta A_I)(Y_I + \Delta Y_I) \tag{8.16}$$

where f_{ij} and l_{il} are the kinetic parameters in F of the biochemical control circuit to be specified in equation (8.15).

Suppose we can find some F such that the inverse of $(A_D + F + \Delta A_D)$ exists. Equation (8.16) is equivalent then to

$$(A_D+F)\left(I+(A_D+F)^{-1}\Delta A_D\right)(Y_D+\Delta Y_D)=(b+\Delta b)-(A_I+\Delta A_I)(Y_I+\Delta Y_I) \quad (8.17)$$

Similar to equation (8.14), a robust design scheme for the controlled metabolic system in equation (8.17) is given by

$$\left\|(A_D+F)^{-1}\Delta A_D\right\|_2<1 \text{ or } \Delta A_D\Delta A_D^T<(A_D+F)(A_D+F)^T \quad (8.18)$$

In this case of robust circuit design, the design purpose is to specify feedback circuits such that the structural stability robustness of the metabolic network is improved, thus enabling the network robustness criterion in equation (8.18) to tolerate larger parameter perturbations ΔA_D. The phenotype (i.e., the steady state of the controlled metabolic network in equation (8.16)) is then given by

$$Y_D+\Delta Y_D=\left(I+(A_D+F)^{-1}\Delta A_D\right)^{-1}(A_D+F)^{-1}\left[(b+\Delta b)-(A_I+\Delta A_I)(Y_I+\Delta Y_I)\right] \quad (8.19)$$

This provides a simple illustration of network robustness analysis and circuit design.

Example 8.1:

Consider the cascaded network in Figure 8.1(a). Cascaded mechanisms are found in diverse areas of biochemistry and physiology, including hormonal control, gene regulation, immunology, blood clotting, and visual excitation [7, 8]. The S-system model is given as

$$\begin{aligned}
\dot{X}_1&=10X_2^{-0.1}X_3^{-0.05}X_4-5X_1^{0.5}, && X_1(0)=0.2\\
\dot{X}_2&=2X_1^{0.5}-1.44X_2^{0.5}, && X_2(0)=0.5\\
\dot{X}_3&=3X_2^{0.5}-7.2X_3^{0.5}, && X_3(0)=0.1,\\
\dot{X}_4&=0.75
\end{aligned} \quad (8.20)$$

$$A_D=\begin{bmatrix}-0.5 & -0.1 & -0.05\\ 0.5 & -0.5 & 0\\ 0 & 0.5 & -0.5\end{bmatrix},\ A_I=\begin{bmatrix}1\\0\\0\end{bmatrix}$$

The time responses of the cascaded network are shown in Figure 8.1(b). Suppose the kinetic parameters A_D suffer from parameter perturbations as follows:

$$\Delta A_D=\begin{bmatrix}0.04675 & 0.11756 & 0.1655\\ 0.04667 & 0.2826 & 0\\ 0 & 0.06494 & 0.0914\end{bmatrix} \quad (8.21)$$

System (8.19) is then perturbed as follows:

$$\begin{aligned} \dot{X}_1 &= 10X_2^{0.01756}X_3^{0.1155}X_4 - 5X_1^{0.45325}, & X_1(0) &= 0.2 \\ \dot{X}_2 &= 2X_1^{0.54667} - 1.44X_2^{0.2174}, & X_2(0) &= 0.5 \\ \dot{X}_3 &= 3X_2^{0.56494} - 7.2X_3^{0.4086}, & X_3(0) &= 0.1 \\ \dot{X}_4 &= 0.75 \end{aligned} \tag{8.22}$$

In this situation, robustness is violated and the steady state (phenotype) ceases to exist (Figure 8.1(c)). Hence, a robust circuit design is necessary to improve network robustness and tolerate this parameter perturbation. Suppose a biochemical control circuit can be designed (see Figure 8.1(d)) such that X_2 can self-regulate its production to achieve the desired robustness necessary to tolerate the parameter perturbations in equation (8.21). The second equation in (8.20) can then be modified as

$$X_2 = 2X_1^{0.5}X_2^{f_{22}} - 1.44X_2^{0.5} \tag{8.23}$$

The kinetic parameter f_{22} should be specified in Matlab such that the robust design criterion in equation (8.18) is satisfied. The range of f_{22} equired to tolerate ΔA_D in equation (8.21) is found to be [−1, −0.081].

$$\begin{bmatrix} 0.043396 & 0.035404 & 0.022761 \\ 0.035404 & 0.082041 & 0.018352 \\ 0.022761 & 0.018352 & 0.012571 \end{bmatrix} < \begin{bmatrix} -0.5 & -0.1 & -0.05 \\ 0.5 & -0.5+f_{22} & 0 \\ 0 & 0.5 & -0.5 \end{bmatrix} \begin{bmatrix} -0.5 & 0.5 & 0 \\ -0.1 & -0.5+f_{22} & 0.5 \\ -0.05 & 0 & -0.5 \end{bmatrix} \tag{8.24}$$

On the other hand, if enzyme activities can be adjusted via metabolite pathway engineering to change the kinetic parameters, an alternative design of enhancing an existing pathway by modulating its kinetic parameter value to tolerate ΔA_D can be considered. For instance, suppose a catalytic control circuit can be designed such that X_2 can regulate the production of X_1 (f_{12}) and X_2 can self-regulate its degradation (l_{22}; see Figure 8.1(f)) to satisfy the robust design scheme to tolerate ΔA_D. The differential equations of the cascaded metabolic network in equation (8.20) should then be modified as follows:

$$\begin{aligned} \dot{X}_1 &= 10X_2^{-0.1+f_{12}}X_3^{-0.05}X_4 - 5X_1^{0.5}, & X_1(0) &= 0.2 \\ \dot{X}_2 &= 2X_1^{0.5} - 1.44X_2^{0.5+l_{22}}, & X_2(0) &= 0.5 \\ \dot{X}_3 &= 3X_2^{0.5} - 7.2X_3^{0.5}, & X_3(0) &= 0.1,\ X_4 = 0.75 \end{aligned} \tag{8.25}$$

The biochemical circuit design work is reduced to the manner of specifying the ranges of f_{12} and l_{22} in equation (8.25) to simultaneously meet the robust design criterion in equation (8.18).

$$\begin{bmatrix} 0.043396 & 0.035404 & 0.022761 \\ 0.035404 & 0.082041 & 0.018352 \\ 0.022761 & 0.018352 & 0.012571 \end{bmatrix} < \begin{bmatrix} -0.5 & -0.1+f_{12} & -0.05 \\ 0.5 & -0.5-l_{22} & 0 \\ 0 & 0.5 & -0.5 \end{bmatrix} \begin{bmatrix} -0.5 & 0.5 & 0 \\ -0.1+f_{12} & -0.5-l_{22} & 0.5 \\ -0.05 & 0 & -0.5 \end{bmatrix} \quad (8.26)$$

The necessary ranges of f_{12} and l_{22} are found to be [−1, 0] and [0, 1], respectively. The simulation results of the robust circuit designs with $f_{12} = 0.08$ and $l_{22} = 0.31$ for the cascaded biochemical network are shown in Figure 8.1(g).

8.2. Multipurpose Circuit Control Design of Metabolic Networks

The robustness design above focuses on the tolerance of kinetic parameter perturbations ΔA_D. The effects of rate constant variations Δb and of environmental changes or upstream regulatory changes ΔY_I on output variations ΔY_D should also be considered in the circuit design of metabolic networks to guarantee robustness against both intrinsic parameter variations and extrinsic environmental perturbations. The sensitivity of ΔY_D to Δb in the designed metabolic network of equation (8.16) is given by [8]

$$\frac{\Delta Y_D}{\Delta b} = (A_D + F)^{-1} \quad (8.27)$$

The sensitivity of ΔY_D to ΔY_I is given by

$$\frac{\Delta Y_D}{\Delta Y_I} = -(A_D + F)^{-1} A_I \quad (8.28)$$

It is more appealing to design a robust metabolic network with desired sensitivities to variations in rate constants and environmental signals, i.e.,

$$\left\| \frac{\Delta Y_D}{\Delta b} \right\|_2 < s_1, \quad \left\| \frac{\Delta Y_D}{\Delta Y_I} \right\|_2 < s_2 \quad (8.29)$$

where the upper bounds s_1 and s_2 are prescribed in advance by the biochemical circuit designer. From equations (8.27)–(8.29), the equivalent sensitivity criteria for equation (8.29) are obtained as [7]

$$\begin{aligned} I - s_1^2 (A_D + F)(A_D + F)^T &< 0 \\ A_I A_I^T - s_2^2 (A_D + F)(A_D + F)^T &< 0 \end{aligned} \tag{8.30}$$

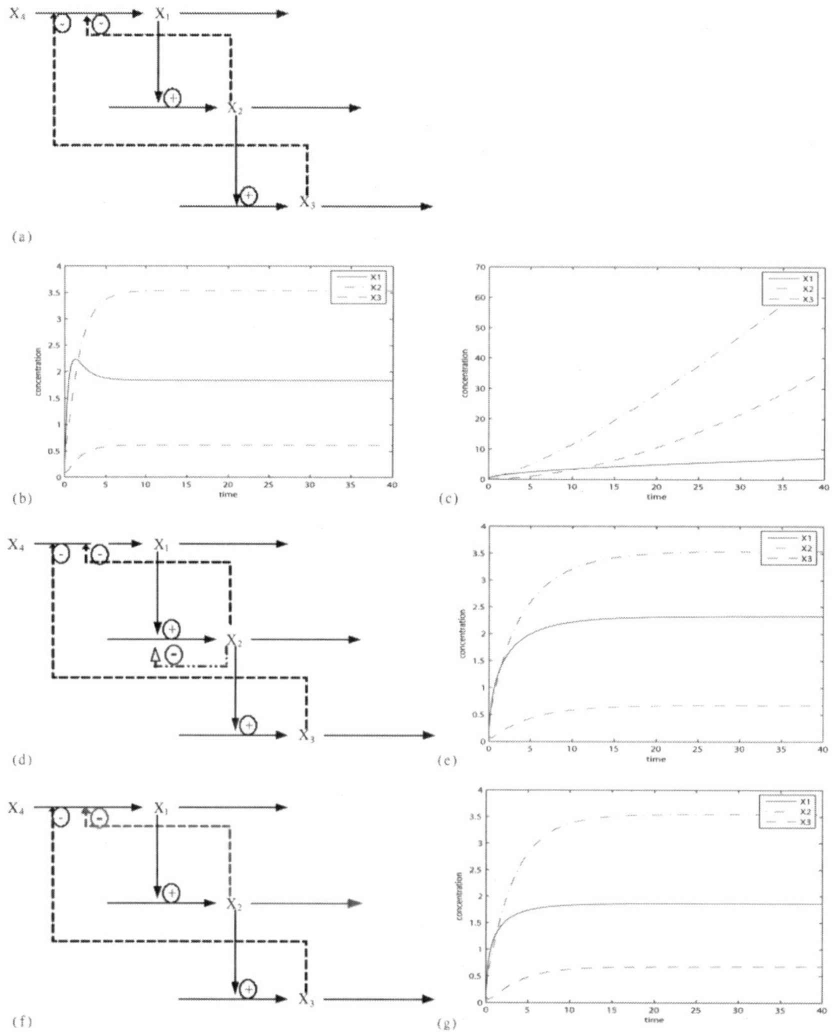

Figure 8.1. Robust circuit design of the cascaded biochemical network of example 8.1. (a) Cascaded biochemical network. (b) Time responses of (a) in the nominal parameter case. (c) Time responses of (a) under parameter perturbations in equation (8.21). (d) Designed cascaded biochemical network with $f_{22} = -0.407$ (blue dashed dotted line) following the multi-objective design in equations (8.24) and (8.31) in the perturbed biochemical network (8.22). (e) Time responses of the designed biochemical network in (d) under parameter perturbations in equation (8.21). (f) Designed cascaded biochemical network with $f_{12} = -0.08$ (green dashed line from X_2 to the production of X_1) and $l_{22} = 0.31$ (solid line) following the multi-objective design in equations (8.26) and (8.32) in the cascaded metabolic network (equation (8.25)). (g) Time responses of the designed biochemical networks in (f) under parameter perturbations in equation (8.21).

A multipurpose circuit design approach is therefore aimed to construct a biochemical circuit F to simultaneously satisfy the design requirements of network robustness in equation (8.18) and network sensitivity in equation (8.30). In example 8.1, suppose the goal is to design a robust biochemical circuit f_{22} in (8.23) to tolerate kinetic parameter variation ΔA_D in (8.21) and satisfy the network sensitivity in equation (8.29) or (8.30) with prescribed sensitivities of $\|\Delta Y_D/\Delta b\|_2 < \|\Delta Y_D/\Delta b\|_{2,\text{nominal}} \equiv s_1 = \|A_D^{-1}\|_2 = 3.42$ and $\|\Delta Y_D/\Delta Y_I\|_2 < \|\Delta Y_D/\Delta Y_I\|_{2,\text{nominal}} \equiv s_2 = \|A_D^{-1}A_I\|_2 = 2.66$. Then f_{22} should be specified to satisfy the robust circuit design in (8.24) and the following inequalities:

$$\begin{bmatrix}1&0&0\\0&1&0\\0&0&1\end{bmatrix}-3.42^2\begin{bmatrix}-0.5&-0.1&-0.05\\0.5&-0.5+f_{22}&0\\0&0.5&-0.5\end{bmatrix}\begin{bmatrix}-0.5&0.5&0\\-0.1&-0.5+f_{22}&0.5\\-0.05&0&-0.5\end{bmatrix}<0$$
$$\begin{bmatrix}1&0&0\\0&0&0\\0&0&0\end{bmatrix}-2.66^2\begin{bmatrix}-0.5&-0.1&-0.05\\0.5&-0.5+f_{22}&0\\0&0.5&-0.5\end{bmatrix}\begin{bmatrix}-0.5&0.5&0\\-0.1&-0.5+f_{22}&0.5\\-0.05&0&-0.5\end{bmatrix}<0 \quad (8.31)$$

simultaneously. With the help of Matlab, the range of f_{22} necessary to tolerate ΔA_D in (8.21) and satisfy the desired network sensitivity in equation (8.31) is found to be [−1, −0.081]. We choose $f_{22} = -0.407$ as a design example, which is a negative self-regulation. It has been found to efficiently eliminate the effect of parameter variations by negative compensation. About 10% of yeast genes encoding regulators are negatively self-regulating; thus, this mechanism seems to be important for maintaining robustness in yeast [7]. The metabolic network and time responses are shown in Figure 8.1(d) and (e), respectively.

Similarly, suppose the goal in the design case in equation (8.25) is to specify f_{12} and l_{22} to satisfy equations (8.26) and (8.31), i.e.,

$$\begin{bmatrix}1&0&0\\0&1&0\\0&0&1\end{bmatrix}-s_1^2\begin{bmatrix}-0.5&-0.1+f_{12}&-0.05\\0.5&-0.5-l_{22}&0\\0&0.5&-0.5\end{bmatrix}\begin{bmatrix}-0.5&0.5&0\\-0.1+f_{12}&-0.5-l_{22}&0.5\\-0.05&0&-0.5\end{bmatrix}<0$$
$$\begin{bmatrix}1&0&0\\0&0&0\\0&0&0\end{bmatrix}-s_2^2\begin{bmatrix}-0.5&-0.1+f_{12}&-0.05\\0.5&-0.5-l_{22}&0\\0&0.5&-0.5\end{bmatrix}\begin{bmatrix}-0.5&0.5&0\\-0.1+f_{12}&-0.5-l_{22}&0.5\\-0.05&0&-0.5\end{bmatrix}<0 \quad (8.32)$$

where s_1 and s_2 are the same as above. Similarly, the ranges of f_{12} and l_{22} necessary to tolerate ΔA_D in equation (8.21) and satisfy the desired sensitivity criteria in equation (8.32) are found to be [−1, 0] and [0, 1], respectively. $f_{12} = -0.08$ and $l_{22} = 0.31$ are chosen as a design example. The metabolic network and time responses are shown in Figure 8.1(f) and (g), respectively.

Example 8.2

Consider the tricarboxylic acid (TCA) cycle metabolic network in *Dictyostelium discoideum* [8]. The TCA cycle, a cyclic reaction, can produce ATP very efficiently while decomposing pyruvate to water and CO_2 via acetyl-CoA and serve as the core of the metabolic network in most living cells. The condensation of acetyl coenzyme A (acetyl CoA) and oxaloacetic acid (OAA) results in the products citric acid and acetyl CoA. In succeeding reactions, the products cooperate with the electron-delivering mechanism and oxidative phosphorylation (ADP→ATP) at the cell membrane of prokaryotes or at the intima of eukaryotic mitochondria to oxidize an oxaloacetic acid molecule to equivalent water, CO_2 and 12 ATP molecules. In this example, the TCA cycle mode (Figure 8.2(a)) is reasonably simplified to involve the following 13 dependent metabolites X_1 through X_{13}, 35 independent metabolites, and 26 enzyme-catalyzed processes coded as X_{14} through X_{39} [7, 8]. In addition, three cofactors are coded as X_{46}, X_{47}, and X_{48}. The cofactors are modeled as independent variables, because they are assumed to be constant. Finally, the model contains two fixed reservoirs for CO_2 and protein (X_{49} and X_{50}).

X_1	Oxaloacetate 1 (OAA 1)	X_{25}	Aconitase
X_2	Oxaloacetate 2 (OAA 2)	X_{26}	Isocitrate dehydrogenase
X_3	Acetyl-CoA (ACO)	X_{27}	Glu → Suc
X_4	Isocitrate (ISOC)	X_{28}	Aspartate transaminase
X_5	Pyruvate (PYR)	X_{29}	Alanine transaminase
X_6	Glutamate (GLU)	X_{30}	Oaa1 → Oaa 2
X_7	Aspartate (ASP)	X_{31}	Asp → Oaa 1
X_8	Alanine (ALA)	X_{32}	Suc → Glu
X_9	Citrate 1 (CIT 1)	X_{33}	Oaa1 → Asp
X_{10}	α-Ketoglutarate (KG1)	X_{34}	Protein → Asp
X_{11}	Succinate (SUC)	X_{35}	Protein → AcCoA
X_{12}	Fumarate (FUM)	X_{36}	Protein → Suc
X_{13}	Malate (MAL 1)	X_{37}	Protein → Fum
X_{14}	Glutamate dehydrogenase	X_{38}	Protein → Ala
X_{15}	α-Ketoglutarate dehydrogenase complex	X_{39}	Protein → Glu
X_{16}	Succinate dehydrogenase	X_{40}	Asp → Protein
X_{17}	Fumarase	X_{41}	Acetyl-CoA → Protein
X_{18}	Malate dehydrogenase	X_{42}	Suc → Protein
X_{19}	Malic enzyme	X_{43}	Fum → Protein
X_{20}	Ala → Pyr	X_{44}	Ala → Pro tein
X_{21}	Pyruvate dehydrogenase complex	X_{45}	Glu → Protein
X_{22}	Oaa 2 → Asp	X_{46}	NAD
X_{23}	Asp → Oaa 2	X_{47}	CoA
X_{24}	Citrate synthetase	X_{48}	NADH

The S-system model of the TCA cycle network in *D. discoideum* is written as follows [8]:

$$
\begin{aligned}
\dot{X}_1 &= 0.8282 X_1^{-0.038} X_6^{-0.0204} X_7^{0.106} X_{10}^{0.114} X_{13}^{0.7} X_{18}^{0.807} X_{28}^{0.108} X_{31}^{0.0848} X_{46}^{0.599} X_{48}^{-0.181} \\
&\quad -1.3423 X_1 X_{30}^{0.915} X_{33}^{0.0847} \\
\dot{X}_2 &= 1.3401 X_1^{0.915} X_7^{0.0848} X_{23}^{0.0848} X_{30}^{0.915} - 17.166 X_2^{0.706} X_3^{0.0716} X_{22}^{0.0848} X_{24}^{0.915} X_{47}^{-0.0341} \\
\dot{X}_3 &= 0.3231 X_3^{-0.405} X_5^{0.156} X_{21}^{0.427} X_{35}^{0.573} X_{46}^{0.422} X_{47}^{0.405} X_{48}^{-0.418} \\
&\quad -9.6952 X_2^{0.376} X_3^{0.489} X_{24}^{0.554} X_{41}^{0.446} X_{47}^{-0.00206} \\
\dot{X}_4 &= X_9 X_{25} - 0.152 X_4^{0.958} X_{26} X_{46}^{0.0348} X_{48}^{-0.862} \\
\dot{X}_5 &= 1.875 X_7^{0.0274} X_8^{0.465} X_{13}^{0.336} X_{19}^{0.535} X_{20}^{0.465} - 0.01923 X_3^{-0.717} X_5^{0.413} X_6^{0.306} X_8^{-0.29} X_{10}^{-0.0883} \\
&\quad \times X_{21}^{0.756} X_{29}^{0.244} X_{46}^{0.748} X_{47}^{0.718} X_{48}^{-0.741} \\
\dot{X}_6 &= 2.459 X_1^{-0.00921} X_6^{-0.0154} X_7^{0.0162} X_{10}^{0.086} X_{11}^{0.276} X_{28}^{0.813} X_{32}^{0.276} X_{39}^{0.6413} \\
&\quad -1.1528 X_5^{0.0963} X_6^{1.01} X_8^{-0.204} X_{10}^{-0.062} X_{14}^{0.0518} X_{27}^{0.277} X_{29}^{0.171} X_{45}^{0.5} X_{46}^{0.0222} X_{48}^{-0.0191} \\
\dot{X}_7 &= 2.1167 X_1^{0.129} X_2^{0.129} X_{22}^{0.129} X_{33}^{0.129} X_{34}^{0.741} - 3.4893 X_1^{-0.0187} X_6^{-0.0311} X_7^{0.868} X_{10}^{0.174} \\
&\quad \times X_{23}^{0.129} X_{28}^{0.165} X_{31}^{0.129} X_{40}^{0.577} \\
\dot{X}_8 &= 0.5724 X_5^{0.111} X_6^{0.247} X_8^{0.234} X_{10}^{0.0713} X_{29}^{0.197} X_{38}^{0.803} - 1.9369 X_8 X_{20}^{0.375} X_{44}^{0.625} \\
\dot{X}_9 &= 16.242 X_2^{0.679} X_3^{0.0782} X_{24} X_{47}^{-0.0372} - X_9 X_{25} \\
\dot{X}_{10} &= 0.156 X_4^{0.724} X_5^{0.106} X_6^{0.259} X_8^{-0.223} X_{10}^{-0.0679} X_{14}^{0.0568} X_{26}^{0.756} X_{29}^{0.188} X_{46}^{0.0506} X_{48}^{-0.672} \\
&\quad -0.8063 X_1^{-0.0101} X_6^{-0.0168} X_7^{0.0177} X_{10}^{0.99} \times X_{11}^{-0.879} X_{15}^{0.911} X_{28}^{0.0891} X_{46}^{0.882} X_{47}^{0.879} X_{48}^{-0.881} \\
\dot{X}_{11} &= 2.0031 X_6^{0.166} X_{10}^{0.491} X_{11}^{-0.481} X_{15}^{0.499} X_{27}^{0.166} X_{36}^{0.335} X_{46}^{0.483} X_{47}^{0.481} X_{48}^{-0.483} \\
&\quad -2.4373 X_{11}^{0.495} X_{12}^{-0.00542} X_{16}^{0.574} X_{32}^{0.166} X_{42}^{0.261} \\
\dot{X}_{12} &= 1.271 X_{11}^{0.106} X_{12}^{-0.00836} X_{16}^{0.885} X_{37}^{0.115} - 9.1694 X_{12}^{1.89} X_{13}^{-1.24} X_{17}^{0.911} X_{43}^{0.0893} \\
\dot{X}_{13} &= 8.289 X_{12}^{1.98} X_{13}^{-1.36} X_{17} - 0.9387 X_1^{-0.0197} X_7^{0.0196} X_{13}^{0.775} X_{18}^{0.618} X_{19}^{0.382} X_{46}^{0.458} X_{48}^{-0.139}
\end{aligned}
\tag{8.33}
$$

where

$$
\begin{aligned}
&X_1(0) = 0.003, X_2(0) = 0.003, X_3(0) = 0.065, X_4(0) = 0.01, X_5(0) = 0.32, X_6(0) = 6.63 \\
&X_7(0) = 2.035, X_8(0) = 5.313, X_9(0) = 0.0275, X_{10}(0) = 0.01, X_{11}(0) = 0.9, X_{12}(0) = 0.04 \\
&X_{13}(0) = 0.24, X_{14} = 0.977, X_{15} = 7610, X_{16} = 3.15, X_{17} = 25.7, X_{18} = 77.8, X_{19} = 3.08 \\
&X_{20} = 0.196, X_{21} = 258, X_{22} = 74, X_{23} = 0.1, X_{24} = 8.24, X_{25} = 80, X_{26} = 271, X_{26} = 271 \\
&X_{27} = 0.133, X_{28} = 9.95, X_{29} = 2.67, X_{30} = 800, X_{31} = 0.1, X_{32} = 1, X_{33} = 74, X_{34} = 1.06 \\
&X_{35} = 2.07, X_{36} = 1.62, X_{37} = 0.36, X_{38} = 2.03, X_{39} = 1.86, X_{40} = 0.446, X_{41} = 27.2, \\
&X_{42} = 1.57, X_{43} = 7, X_{44} = 0.326, X_{45} = 0.24, X_{46} = 0.072, X_{47} = 0.1, X_{48} = 0.18
\end{aligned}
$$

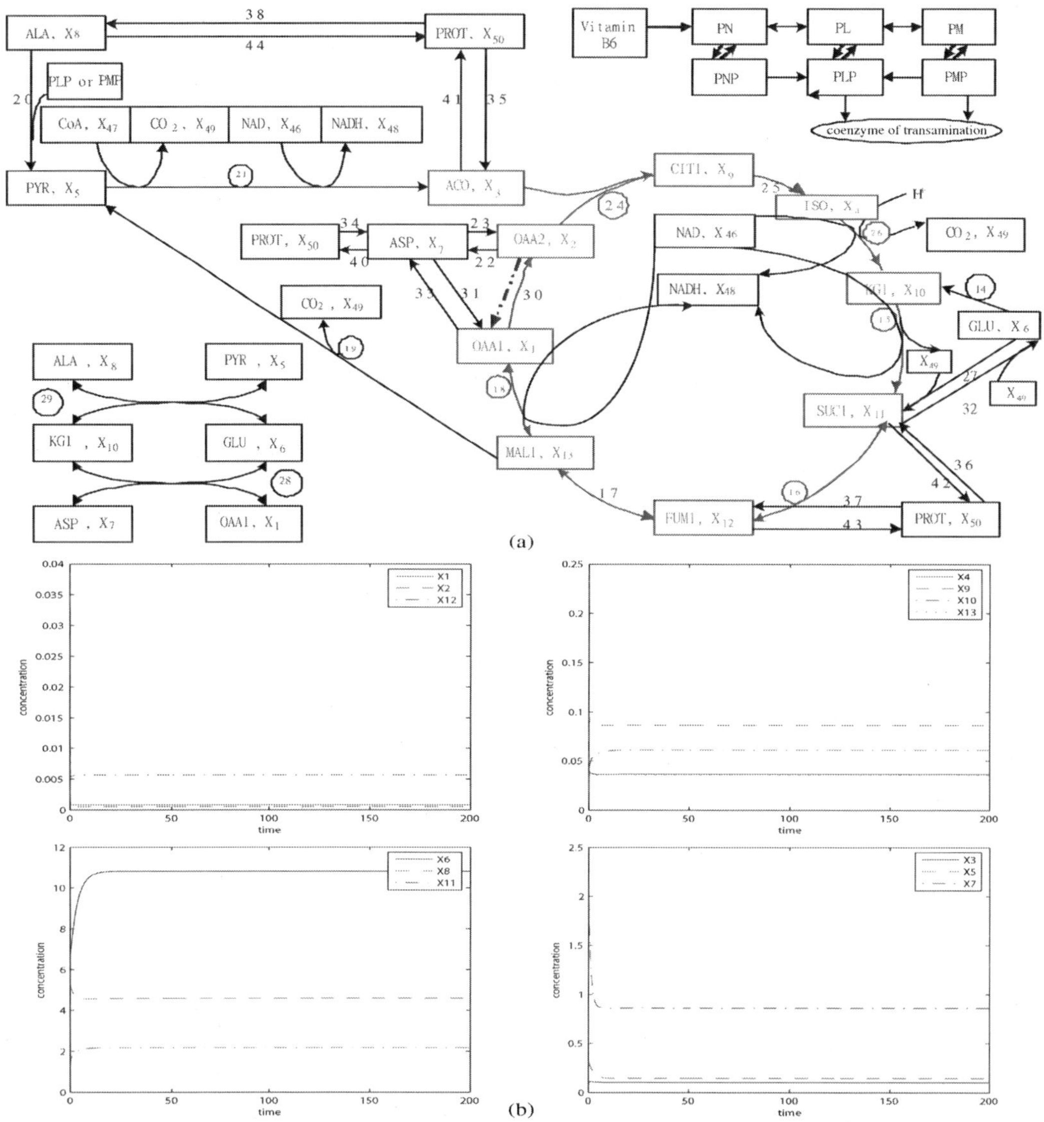

Figure 8.2. (a) TCA cycle metabolic network of example 8.2 in *D. discoideum* redrawn from the KEGG database [7, 8] The S-system model of the TCA cycle metabolic network is given in equation (8.33) (b) Time responses of the TCA cycle metabolic network in the nominal parameter case.

The time responses of the TCA cycle metabolic network in equation (8.33) are shown in Figure 8.2(b). Suppose the metabolic network suffers an intrinsic parameter perturbation ΔA_D in equation (8.34), which violates the upper bound of the robustness condition in equation (8.14), so that the steady state of the TCA cycle network ceases to exist. The corresponding time responses are shown in Figure 8.2(c).

$$\Delta A_D = \left[\begin{array}{ccccccccccccc} 0.0209 & 0 & 0 & 0 & 0 & 0.0267 & 0.000292 & 0 & 0 & 0.0565 & 0 & 0 & 0.0142 \\ 0.0219 & 0.0412 & 0 & 0 & 0 & 0 & 0.000305 & 0 & 0 & 0 & 0 & 0 & 0 \\ 0 & -0.0278 & 0 & 0 & 0.058 & 0 & 0 & 0 & 0 & 0 & 0 & 0 & 0 \\ 0 & 0 & 0.0157 & 0.0157 & 0 & 0 & 0 & 0 & 0.0135 & 0 & 0 & 0 & 0 \\ 0 & 0 & 0 & 0 & 0.0611 & -0.0198 & -0.0002 & 0.00571 & 0 & -0.0418 & 0 & 0 & -0.0105 \\ 0.0087 & 0 & 0 & 0 & -0.0344 & 0.0111 & 0.0001 & -0.0032 & 0 & 0.0235 & 0.0157 & 0 & 0 \\ 0.00039 & 0.00726 & 0 & 0 & 0 & 0.00492 & 0.000054 & 0 & 0 & 0.0104 & 0 & 0 & 0 \\ 0 & 0 & 0 & 0 & 0.0413 & -0.0133 & 0 & 0.0039 & 0 & -0.0282 & 0 & 0 & 0 \\ 0 & 0.0216 & 0 & 0 & 0 & 0 & 0 & 0 & 0.0149 & 0 & 0 & 0 & 0 \\ 0.0119 & 0 & 0.0179 & 0.0179 & -0.0472 & 0.0153 & 0.00017 & -0.0044 & 0 & 0.0323 & 0.0215 & 0 & 0 \\ 0 & 0 & 0 & 0 & 0 & 0.0207 & 0 & 0 & 0 & 0.0438 & 0.0293 & 0.0103 & 0 \\ 0 & 0 & 0 & 0 & 0 & 0 & 0 & 0 & 0 & 0 & 0.0239 & 0.00843 & 0.009 \\ 0.0124 & 0 & 0 & 0 & 0 & 0 & 0.00017 & 0 & 0 & 0 & 0 & 0.0079 & 0.0084 \end{array}\right] \tag{8.34}$$

Given that the goal is to design a robust metabolic circuit to tolerate the perturbation ΔA_D and to achieve the desired sensitivities s_1 and s_2 in equation (8.30), the robust control circuit design problem can be reduced to specifying F to satisfy the following multipurpose control circuit design derived from equations (8.18) and (8.30):

$$\begin{aligned} &\Delta A_D \Delta A_D^T < (A_D + F)(A_D + F)^T \\ &I - s_1^2 (A_D + F)(A_D + F)^T < 0 \\ &A_I A_I^T - s_2^2 (A_D + F)(A_D + F)^T < 0 \end{aligned} \tag{8.35}$$

with nominal sensitivities $s_1 = \|A_D^{-1}\|_2 = 8.3685$ and $s_2 = \|A_D^{-1} A_I\|_2 = 7.5464$.

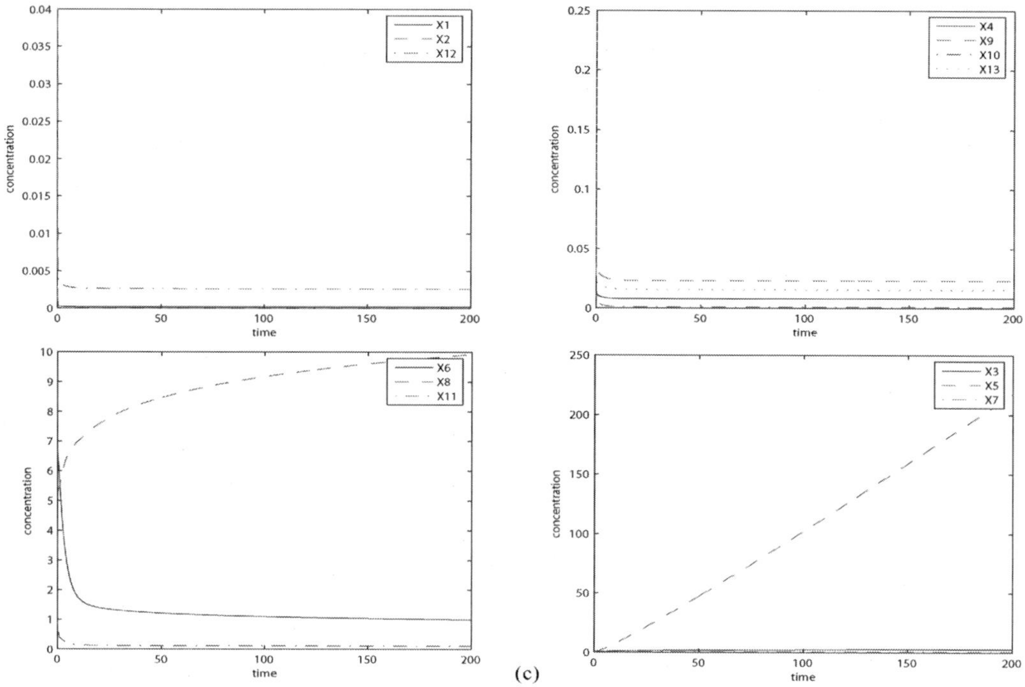

(c)

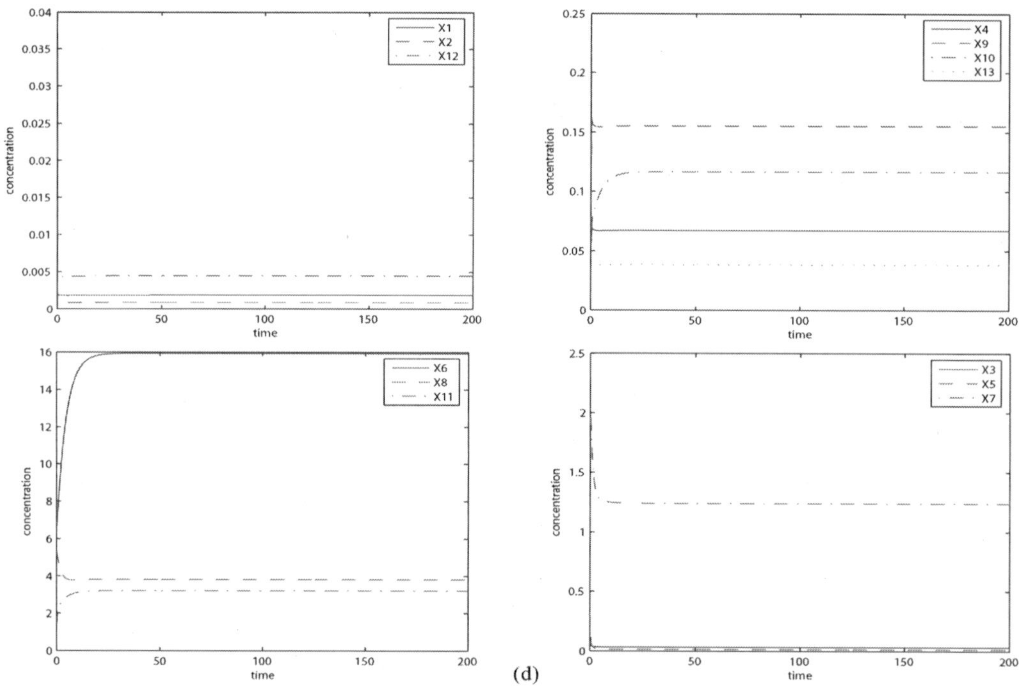

(d)

Figure 8.2. (Continued) (c) Time responses of the TCA cycle metabolic network in example 8.2 under parameter perturbations ΔA_D in equation (8.34). (d) Time responses of the designed TCA cycle metabolic network with $f_{12} = -0.2$ (the dashed dotted line from X_2 to X_1) under parameter perturbations in equation (8.34).

If one biochemical control pathway with kinetic parameter f_{12} is designed to satisfy the multi-objective design criteria in equation (8.35), then the range of f_{12} is found to be within $[-0.8, -0.1]$. If f_{12} is chosen as -0.2 (Figure 8.2(a), blue line), then the time responses of the designed TCA cycle network shown in Figure 8.2(d), which match the desired properties of the proposed design method, are obtained. That is, the robust controlled biochemical network not only can tolerate ΔA_D (to preserve its phenotype under parameter perturbations) but also retains sensitivity to environmental molecules in the nominal case. If dynamic circuit design is employed to implement the biochemical circuit f_{12}, then an enzyme capable of catalyzing the reaction $X_2 \rightarrow X_1$ is required. Additionally, a TF (Z) has to be found such that oxaloacetate2 (X_2) can bind to the promoter of the enzyme's inhibitor gene, as f_{12} is negative. The concentration of X_2 could therefore regulate X_1 through the kinetic parameter f_{12}. The elasticity of the enzyme inhibitor's gene sequence has to be modulated then to the specified performance by rational design or directed evolution. This allows the construction of the biochemical control circuit f_{12}.

The previous section discusses robust circuit design of metabolic networks based on the steady state of an S-system model. In the following section, a stochastic dynamic model approach is taken based on the systems biology approach on stochastic dynamic model outlined in the next section.

8.3. Robust Circuit Design of Stochastic Dynamic Metabolic Networks

Based on sampled data from biochemical experiments and the delayed effect of molecular diffusion and transport in cells, the linear metabolic regulatory network can be suitably modeled by the following discrete-time dynamic system:

$$\begin{aligned} X(t+1) &= AX(t), \quad X(0) = X_0 \\ y(t) &= CX(t) \end{aligned} \tag{8.36}$$

where the state vector $X(t)$ denotes the discrete-time expression vector of the molecules (mRNAs, proteins, or other chemical complexes in the biochemical regulatory network) at time t, A denotes the stoichiometric (interactive) matrix among these molecules (see Figure 8.3), and $y(t)$ denotes the metabolic output. Thus,

$$X(t) = \begin{bmatrix} x_1(t) \\ x_2(t) \\ \vdots \\ x_i(t) \\ \vdots \\ x_n(t) \end{bmatrix}, \quad A = \begin{bmatrix} a_{11} & a_{12} & \cdots & \cdots & a_{1n} \\ a_{21} & \ddots & & & \vdots \\ \vdots & & a_{ij} & & \vdots \\ \vdots & & & \ddots & \vdots \\ a_{n1} & \cdots & \cdots & \cdots & a_{nn} \end{bmatrix}$$

where $x_i(t)$, $i = 1,\cdots, n$ are molecular concentrations of mRNAs, proteins, and other complexes in the biochemical network.

Suppose the kinetic parameters of a biochemical regulatory network of metabolic processes are affected by the following intrinsic perturbation and environmental disturbance:

$$
\begin{aligned}
X(t+1) &= \left(A + \sum_{i=1}^{L} A_i n_i(t) \right) X(t) + B_v v(t) \\
y(t) &= CX(t)
\end{aligned}
\tag{8.37}
$$

where $\sum_{i=1}^{L} A_i n_i(t)$ denotes the intrinsic parameter fluctuations due to an L random fluctuation source (e.g., thermal fluctuation, alternative splicing, molecular diffusion, etc.), and $n_i(t)$ denotes the i-th random noise with the statistics $E[n_i(t)] = 0$ and $E[n_i^2(t)] = \sigma_i^2$, $i = 1,\cdots, L$. $v(t)$ denotes the environmental disturbance.

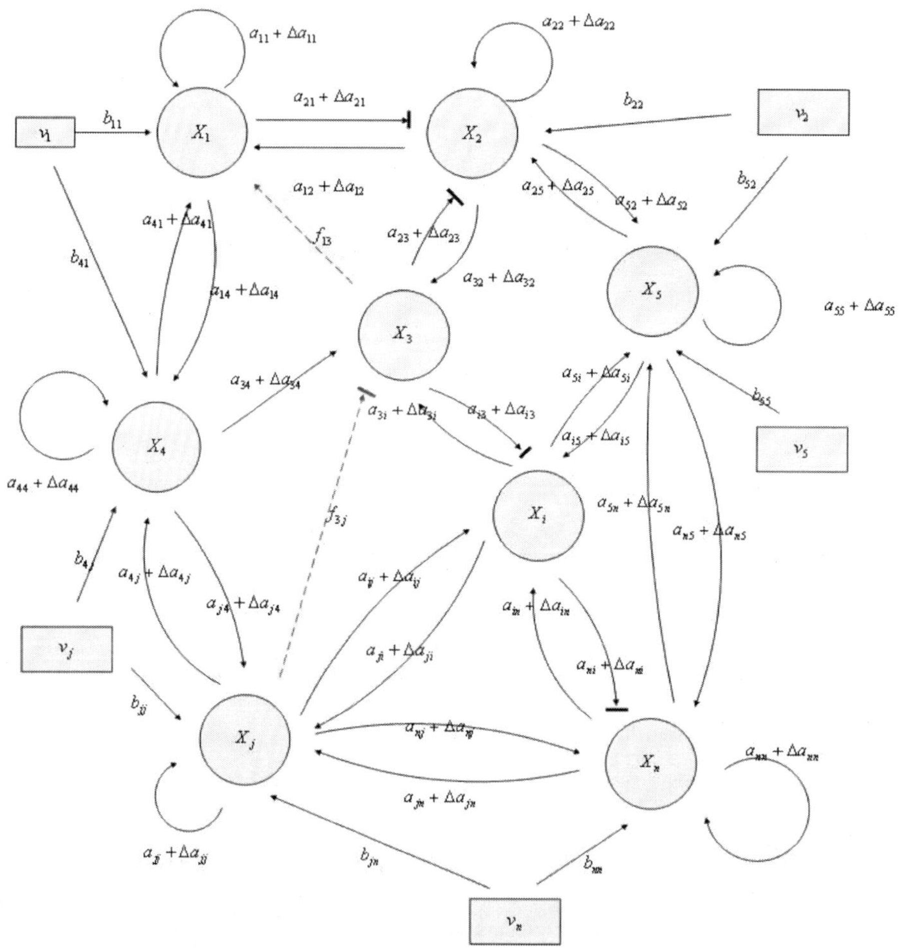

Figure 8.3. Linear metabolic network of n molecules with intrinsic parameter fluctuation Δa_{ij} and extrinsic noise v_i. f_{ij} denotes the biochemical circuit design from x_j to x_i to improve network robustness stability and noise-filtering ability.

Based on the robust stabilization and disturbance filtering described in section 8.2, the environmental disturbance attenuation level ρ of the metabolic network is denoted as

$$\frac{\|y(t)\|_2}{\|v(t)\|_2} \leq \rho \tag{8.38}$$

Following the systems biology approach in section 8.2, the following robust result for a linear stochastic metabolic network is derived.

Proposition 8.2 [10]: For a linear stochastic metabolic network with intrinsic parameter fluctuation and environmental disturbance in (8.37), if there exists a positive definite matrix $P=P^T > 0$ such that the following matrix inequality holds for a desired disturbance attenuation level ρ

$$A^T PA - P + \sum_{i=1}^{L} \sigma_i^2 A_i^T PA_i + C^T C + (A^T PB_v)^T (B_v^T PB_v - \rho^2 I)^{-1} A^T PB_v < 0 \tag{8.39}$$

then the intrinsic parameter perturbation can be tolerated and environmental disturbance can be attenuated to a level ρ in the stochastic metabolic network equation (8.37).

Proof: see 8.6.1

The phenotype robustness criterion in equation (8.39) could be rewritten as

$$\underbrace{\sum_{i=1}^{L} \sigma_i^2 A_i^T PA_i}_{\text{intrinsic robustness}} + \underbrace{C^T C + (A^T PB_v)^T (B_v^T PB_v - \rho^2 I)^{-1} A^T PB_v}_{\text{environmental robustness}} \leq \underbrace{P - A^T PA}_{\text{network robustness}} \tag{8.40}$$

Remark 8.2:

(i) The physical interpretation of the phenotype robustness criterion of the metabolic network in equation (8.40) is that if intrinsic robustness allowing tolerance of intrinsic parameter fluctuation and environmental robustness, as well as attenuation of environmental disturbance are simultaneously conferred by the network robustness, then the phenotype of the metabolic network is maintained. It can be shown that if the eigenvalues of A are closer to the origin, then the network robustness is commensurately larger.

(ii) To solve the phenotype robustness criterion in equation (8.39), it can be transformed into the following equivalent LMI:

$$\begin{bmatrix} A^T PA - P + C^T C + \sum_{i=1}^{L} \sigma_i^2 A_i^T PA_i & A^T PB_v \\ B_v^T PA & B_v^T PB_v - \rho^2 I \end{bmatrix} \leq 0 \tag{8.41}$$

According to its definition, the disturbance attenuation (filtering) ability ρ_0 (i.e., the minimum ρ) can be obtained by solving the following constrained optimization problem [10]:

$$\rho_0 = \min_P \rho \quad \text{subject to } P > 0, \text{ and LMI in (8.41)} \tag{8.42}$$

If the metabolic network in equation (8.37) cannot achieve a prescribed disturbance attenuation ability ρ_P (i.e., $\rho_P < \rho_0$) specified for therapeutic or biotechnological purposes, then a biochemical circuit design using state feedback is necessary:

$$\begin{aligned} X(t+1) &= \left(A + F + \sum_{i=1}^{L} A_i n_i(t) \right) X(t) + B_v v(t) \\ y(t) &= CX(t) \end{aligned} \tag{8.43}$$

The phenotype robustness criterion in equation (8.40) is then modified as

$$\underbrace{\sum_{i=1}^{L} \sigma_i^2 A_i^T P A_i}_{\text{intrinsic robustness}} + \underbrace{C^T C + (A^T P B_v)^T (B_v^T P B_v - \rho_P^2 I)^{-1} A^T P B_v}_{\text{environmental robustness}} \le \underbrace{P - (A+F)^T P (A+F)}_{\text{network robustness}} \tag{8.44}$$

The negative feedback circuit F could shift the eigenvalues of A to the origin, thereby improving the network robustness of the metabolic network on the right-hand side of equation (8.44) and achieving the prescribed attenuation level $\rho_P < \rho_0$. In general, the interactions of a biochemical regulatory network in metabolic processes are nonlinear in real biosystems. In this situation, a nonlinear biochemical regulatory network of metabolic pathways under intrinsic stochastic parameter perturbation and environmental disturbance can be represented based on the stochastic dynamic model of systems biology in section 2:

$$\begin{aligned} X(t+1) &= f(x) + \sum_{i=1}^{L} f_i(x) n_i(t) + B_v v(t) \\ y(t) &= CX(t) \end{aligned} \tag{8.45}$$

In the following, the robust stability and filtering ability ρ_0 on $v(k)$ at $y(k)$ in the stochastic nonlinear metabolic network in equation (8.45) is discussed.

Proposition 8.3 [10]: If the following matrix inequality holds for a Lyapunov function $V(x(t)) > 0$ and a noise attenuation level ρ

$$\begin{aligned} &\left(\frac{\partial V(X(t))}{\partial X(t)} \right)^T \left(f(X(t)) - X(t) \right) + \frac{1}{2} \left(\frac{\partial V(X(t))}{\partial X(t)} \right)^T \left(\frac{\partial V(X(t))}{\partial X(t)} \right) + \frac{1}{2} \sum_{i=1}^{L} f_i^T(X(t)) \\ &\times \frac{\partial^2 V(X(t))}{\partial X^2(t)} f_i(X(t)) + \frac{1}{2\rho^2} \left(\frac{\partial V(X(t))}{\partial X(t)} \right)^T B_v B_v^T \left(\frac{\partial V(X(t))}{\partial X(t)} \right) + \frac{1}{2} X^T(t) C^T C X(t) \le 0 \end{aligned} \tag{8.46}$$

then the effect of environmental disturbance $v(k)$ on $y(k)$ is less than ρ, i.e., the robust disturbance attenuation level ρ is achieved for the nonlinear metabolic network in equation (8.45).

Proof: see 8.6.2

The filtering ability ρ_0 of the nonlinear stochastic metabolic network in equation (8.45) that attenuates environmental disturbance can be obtained by solving the following constrained optimization problem:

$$\rho_0 = \min_{V(X)>0} \rho$$
$$\text{subject to HJI in (8.46)} \tag{8.47}$$

Following the robustness result discussed in section 8.2, the phenotype robustness criterion for the nonlinear stochastic metabolic network in equation (8.45) is given as

$$\underbrace{\frac{1}{2}\sum_{i=1}^{L} f_i^T(X)\frac{\partial^2 V(X(t))}{\partial X^2(t)} f_i(X)}_{\text{intrinsic robustness}}$$
$$\underbrace{+\frac{1}{2}\left(\frac{\partial V(X(t))}{\partial X(t)}\right)^T\left(\frac{\partial V(X(t))}{\partial X(t)}\right)+\frac{1}{2\rho^2}\left(\frac{\partial V(X(t))}{\partial X(t)}\right)^T B_v B_v^T\left(\frac{\partial V(X(t))}{\partial X(t)}\right)+\frac{1}{2}X^T(t)C^T CX(t)}_{\text{environmental robustness}}$$
$$\underbrace{\le -\left(\frac{\partial V(X(t))}{\partial X(t)}\right)^T\left(f(X(t))-X(t)\right)}_{\text{network robustness}} \tag{8.48}$$

Suppose the network robustness of the metabolic network cannot confer enough intrinsic robustness and environmental robustness to maintain the phenotype of the metabolic network. In this situation, some negative feedback loops $Fg(X(t))$ should be implemented to improve network robustness as follows:

$$X(t+1) = f(X(t)) + Fg(X(t)) + \sum_{i=1}^{L} f_i(X(t))n_i(t) + B_v v(t)$$
$$y(t) = CX(t) \tag{8.49}$$

In this case, the phenotype robustness criterion in equation (8.48) should be modified as follows:

$$\underbrace{\frac{1}{2}\sum_{i=1}^{L} f_i^T(X)\frac{\partial^2 V(X(t))}{\partial X^2(t)} f_i(X)}_{\text{intrinsic robustness}} \tag{8.50}$$
$$\underbrace{+\frac{1}{2}\left(\frac{\partial V(X(t))}{\partial X(t)}\right)^T\left(\frac{\partial V(X(t))}{\partial X(t)}\right)+\frac{1}{2\rho^2}\left(\frac{\partial V(X(t))}{\partial X(t)}\right)^T B_v B_v^T\left(\frac{\partial V(X(t))}{\partial X(t)}\right)+\frac{1}{2}X^T(t)C^T CX(t)}_{\text{environmental robustness}}$$

$$\le \underbrace{-\left(\frac{\partial V(X(t))}{\partial X(t)}\right)^T \big(f(X(t))+Fg(X(t))-X(t)\big)}_{\text{network robustness}}$$

That is, the negative feedback loops $Fg(X)$ are employed to improve network robustness to tolerate more fluctuations of the intrinsic parameter and to filter more environmental disturbances.

It is generally very difficult to solve the HJI in equations (8.46), (8.48), and (8.50) for robust chemical circuit design of metabolic networks. Based on the global linearization method in equation (7.12) [11, 12] or fuzzy interpolation methods [13, 14], the nonlinear stochastic metabolic network can be approximated by interpolating M local metabolic networks as follows:

$$\begin{aligned} X(t+1) &= \sum_{j=1}^{M} \alpha_j\big(X(t)\big)\left(A_j X(t) + FG_j X(t) + \sum_{i=1}^{L} A_{ij} X(t) n_i(t)\right) + B_v v(t) \\ y(t) &= CX(t) \end{aligned} \tag{8.51}$$

where $f(X)$, $g(X)$, and $f_i(X)$ are approximated by $\sum_{i=1}^{M} \alpha_j(X) A_j X(t)$, $\sum_{i=1}^{M} G_j X(t)$, and $\sum_{i=1}^{M} A_{ij} X(t)$, respectively. In this situation, the following robust chemical circuit design for a metabolic network with a prescribed noise attenuation level ρ_P is derived.

Proposition 8.4 [10]:

Suppose the negative feedback loop F is designed for the metabolic network in equation (8.49) or (8.51), such that the following LMIs have a positive solution $P > 0$:

$$\begin{bmatrix} -P + \sum_{i=1}^{L} A_{ij}^T P A_{ij} + C^T C & \left(A_j + FG_j\right)^T P & \left(A_j + FG_j\right)^T P B_v \\ P\left(A_j + FG_j\right) & -P & 0 \\ B_v^T P \left(A_j + FG_j\right) & 0 & B_v^T P B_v - \rho_P^2 I \end{bmatrix}, \text{for } j = 1, 2, \ldots, M \tag{8.52}$$

Thus, the prescribed disturbance attenuation level ρ_P is achieved by the biochemical circuit design.

If the optimal robust filtering design is employed for the nonlinear stochastic metabolic network in equation (8.49) or (8.51), then the control matrix F is specified to solve the following constrained optimization problem:

$$\begin{gathered} \rho_0 = \min_F \rho_P \\ \text{subject to } P > 0 \text{ and LMIs in (8.52)} \end{gathered} \tag{8.53}$$

where ρ_0 in equation (8.53) is the disturbance (noise)-filtering ability of the optimally controlled metabolic network in equation (8.49) or (8.51).

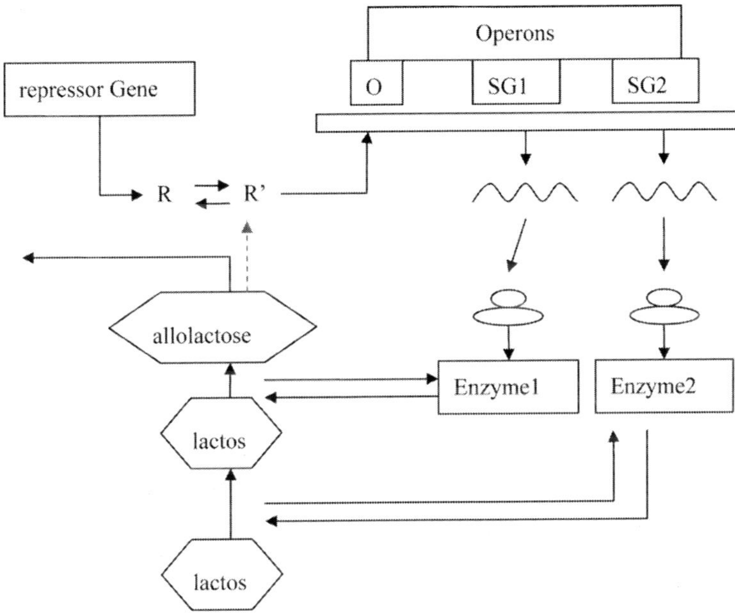

Figure 8.4. Benchmark design example of *E. coli* in a metabolic network of the operons in equation (8.54). The metabolic network suffers from intrinsic parameter fluctuation and environmental disturbance, as shown in equation (8.55).

Example 8.3

This example treats a real biochemical regulatory network of operons in *E. coli*, which prefers glucose as its energy source. When glucose is in short supply, it starts metabolizing lactose (Figure 8.4). Since the metabolite pathway is very complex, some assumptions are adopted to simplify the corresponding reaction: (1) The quasi-steady-state approximation applies to the concentration of mRNA. (2) Concentrations of enzymes are equal. (3) There is a sustained lactose source outside the cell. (4) There is one delay in the conversion of lactose into *E. coli*. If the above conditions hold, then this operon model can be considered as the following discrete-time dynamic system [8] for the enzyme e, lactose lac, and allolactose a:

$$\begin{aligned} e[t+1] &= 0.15 + \frac{0.1a[t]}{1+a[t]} + 0.85e[t] \\ a[t+1] &= \frac{0.1e[t]lac[t]}{1+lac[t]} - \frac{0.1e[t]a[t]}{1+a[t]} + a[t] \\ lac[t+1] &= 0.05 - \frac{0.1e[t]lac[t]}{1+lac[t]} + lac[t] \end{aligned} \tag{8.54}$$

The dynamic time response in this case is shown in Figure 8.5(a). Suppose the operon regulatory network is also affected by intrinsic parameter fluctuations and environmental disturbance $v(k)$ as follows:

$$e[t+1]=0.15+\frac{0.1a[t]}{1+a[t]}+0.85e[t]+n_1[t]\left(0.15+\frac{0.1a[t]}{1+a[t]}+0.85e[t]\right)$$
$$a[t+1]=\frac{0.1e[t]lac[t]}{1+lac[t]}-\frac{0.1e[t]a[t]}{1+a[t]}+a[t]+n_2[t]\left(\frac{0.1e[t]lac[t]}{1+lac[t]}-\frac{0.1e[t]a[t]}{1+a[t]}+a[t]\right) \quad (8.55)$$
$$lac[t+1]=0.05-\frac{0.1e[t]lac[t]}{1+lac[t]}+lac[t]+n_2[t]\left(0.05-\frac{0.1e[t]lac[t]}{1+lac[t]}+lac[t]\right)+v[t]$$

The stochastic intrinsic parameter fluctuations are $n_1[t]$, which includes transcriptional and translational noise and $n_2[t]$, which includes transport noise. Both are zero mean white noises with variances of $\sigma_1 = 0.1$ and $\sigma_2 = 0.04$. External disturbance is $v[t] = 5e^{-0.05t}(\cos(0.2\pi t) + 1)$.

Again, the goal is to design a pathway from the final product allolactose to the regulatory gene. This design produces the corresponding enzyme to robustly stabilize allolactose production $a(t)$. In this case of robust feedback circuit design, the design objective is based on the transfection technique. This method involves modifying the binding site of the promoter region of the corresponding gene of the control enzyme to change transcriptional ability and basal production rate. The engineered single control pathway circuit has two kinetic parameters F_1 and F_2 to robustly stabilize the biochemical network. The following dynamic equation has to be modified with the control terms:

$$e[t+1]=0.15F_1+\frac{0.1a^{F_2}[t]}{1+a^{F_2}[t]}+0.85e[t]+n_1[t]\left(0.15+\frac{0.1a[t]}{1+a[t]}+0.85e[t]\right) \quad (8.56)$$

The design parameters F_1 and F_2 are Michaelis constants. Since the designed steady states are kept close to the nominal ones, the relationship between F_1 with F_2 is $F_1 = 1.93014 - (0.6546^{F2-1}/(1 + 0.6546^{F2}))$ (equation (8.56)). Because Michaelis constants are positive, we choose $F_2 > 0$. Based on the global linearization scheme and the proposed robust filter design with a desired disturbance attenuation level $\rho_P = 1$ in equation (8.52), we choose four convex hull vertices in the form of four globally linearized systems, and set $\alpha_j(X) = 0.25$. The independent control parameter F_2 is then chosen such that the following designed metabolic network with desired disturbance attenuation level $\rho_P = 1$ is guaranteed:

$$X[t]=\begin{bmatrix} e[t] & a[t] & lac[t]\end{bmatrix}^T$$
$$f(X,F)=\begin{bmatrix} 0.15F_1+\dfrac{0.1a^{F_2}[t]}{1+a^{F_2}[t]}+0.85e[t] \\ \dfrac{0.1e[t]lac[t]}{1+lac[t]}-\dfrac{0.1e[t]a[t]}{1+a[t]}+a[t] \\ 0.05-\dfrac{0.1e[t]lac[t]}{1+lac[t]}+lac[t] \end{bmatrix} \quad (8.57)$$

$$f_1(X) = \begin{bmatrix} 0.15 + \dfrac{0.1a[t]}{1+a[t]} + 0.85[t] & 0 & 0 \end{bmatrix}^R$$

$$f_2(X) = \begin{bmatrix} 0 & \dfrac{0.1e[t]lac[t]}{1+lac[t]} + a[t] - \dfrac{0.1e[t]a[t]}{1+a[t]} & 0.05 + lac[t] - \dfrac{0.1e[t]lac[t]}{1+lac[t]} \end{bmatrix}^T$$

$$B_v = [0 \quad 0 \quad 0.1]^T, \qquad C_Z = [0 \quad 1 \quad 0]$$

With the help of the LMI toolbox, we found that $F_2 \in (1,\infty)$ could satisfy the LMIs in equation (8.52) to guarantee $\rho_P = 1$. For convenience of design, let $F_2 = 2$. The corresponding control kinetic parameters are $[F_1 \; F_2] = [2.0968 \; 2]$. The solution P for the robust biochemical circuit design in equation (8.52) for the prescribed disturbance attenuation level $\rho_P = 1$ is given by

$$P = \begin{bmatrix} 0.5397 & -1.1067 & 0 \\ -1.1067 & 1504.3942 & 0 \\ 0 & 0 & 0.01 \end{bmatrix} > 0 \tag{8.58}$$

Figure 8.5(b) shows the simulation results. The network filtering ability can be calculated as $\rho \approx 0.6814 < 1 = \rho_P$. The result of the theoretical attenuation level is clearly more conservative. However, using the proposed robust method of biochemical circuit design, the prescribed disturbance attenuation level $\rho_P \leq 1$ can be guaranteed for the metabolic network. The conservative result is mainly due to the conservativeness of both Lyapunov stability and LMIs in the design procedure.

This robust biochemical circuit design example in a metabolic network illustrates that the proposed systematic method of biochemical circuit design produces metabolic networks that not only tolerate more fluctuations of the intrinsic random parameter but also filter more environmental disturbances. Systems metabolic engineering design is thus capable of improving the network robustness of metabolic networks and efficiently attenuating the effect of environmental noises. This approach may serve as a basis for drug designs against genetic perturbations, pathological environmental cues (such as infectious agents or chemical carcinogens), or both.

Network robustness is a systematic property that allows a metabolic network to maintain its biochemical function or to generate biochemical products despite external disturbance and intrinsic parametric fluctuation. It is from a class of fundamental and ubiquitously observed systems-level phenomena that cannot be understood by observing individual components only. Thus, a study of network robustness at one level is simply a study of why, when, and how metabolic networks function properly or otherwise [15-17]. If network robustness is not large enough to obtain sufficient intrinsic robustness to tolerate random parameter fluctuations and environmental robustness to filter environmental molecular noises, then a robust biochemical circuit design is necessary to improve network robustness such that the network can maintain its function or phenotype. The proposed robust circuit design principles could potentially be used for robust biosynthetic network design with applications in drug design, gene therapy, and metabolic engineering. Future focus may be on robust chemical circuit design for such applications or construction of other pathways by nanotechnology and metabolic engineering [10]. If network robustness, intrinsic robustness, and environmental robustness are taken into consideration in metabolic network engineering, such networks would function more reliably and efficiently.

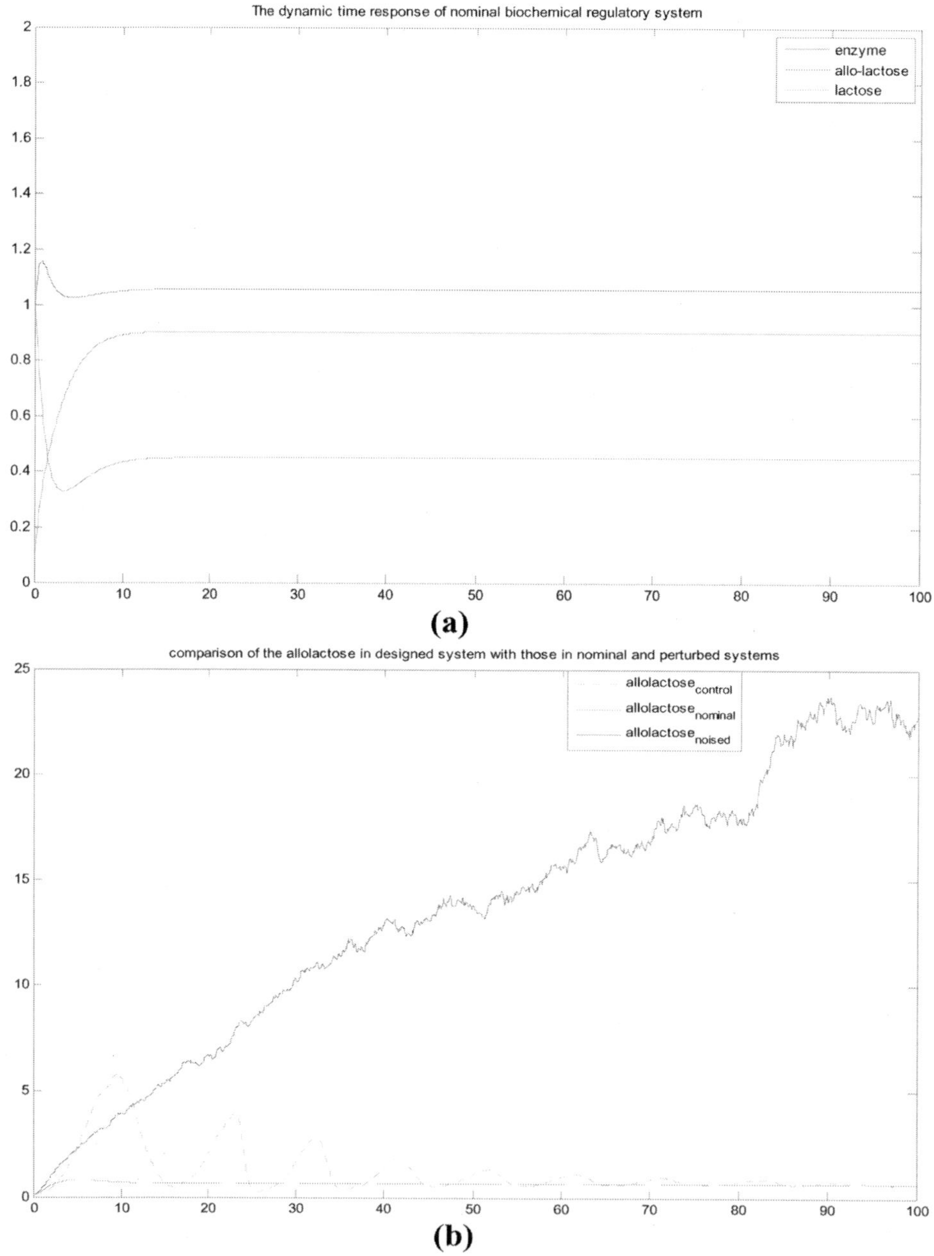

Figure 8.5. (a) The dynamic time response of each molecule of the nominal biochemical regulatory system od example 8.3 in equation (8.54). (b) Comparison of the time response of allolactose ($a(t)$) of the designed metabolic system with the time responses of the nominal metabolic system and perturbed metabolic system. The proposed robust circuit design tolerates intrinsic fluctuation and significantly improves the filtering of extrinsic noise.

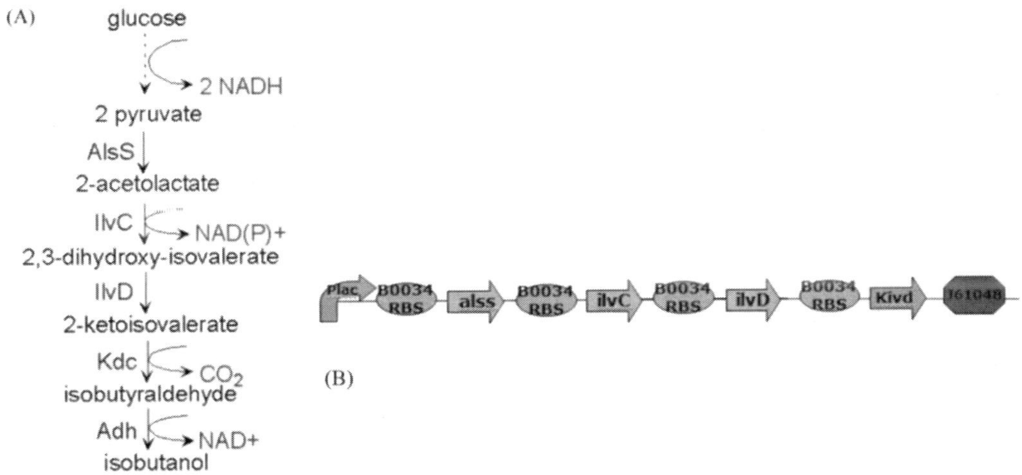

Figure 8.6. Engineered synthetic metabolic pathway for isobutanol production in *E. coli*. (A) Schematic representation of engineered isobutanol production pathway. (B) Engineered synthetic genetic circuit to generate the enzymes necessary for pathway in (A) for isobutanol production in *E. coli*.

8.4. Isobutanol Biosynthetic Pathway in *E. coli* for Biofuel via Systems Synthetic Biology

With the exception of implementing feedback biochemical circuits in metabolic networks, systems synthetic biology techniques detailed in section 8.3 could be combined with systems metabolic engineering approaches to engineer complete metabolic pathways or networks to produce biochemical products that could not be produced in the host cell. As an example, *E. coli* does not have non-fermentative pathways for isobutanol (see Figure 8.6(A)). Suppose a synthetic pathway in *E. coli* is required for the production of butanol as biofuel. In this application, it is necessary to engineer transcription and translation genes to produce the enzymes AlsS, IlvC, IlvD, and Kdc (KIVD), which are necessary to catalyze the isobutanol metabolic pathway (Figure 8.6(B)). The last enzyme of the metabolic pathway (Adh) is already present in *E. coli*. Promoters and RBSs with different regulatory abilities can be selected from the corresponding libraries to engineer the genetic circuit in Figure 8.6(B) to produce corresponding enzymes [18-20]. However, in order to improve the yield of isobutanol with adequate computation and simulation to estimate adequate concentrations of enzyme following the systems biology approach (see section 8.2), a temperature-controllable synthetic metabolic pathway in Figure 8.7 could be designed and engineered (in *E. coli*) to achieve high-yield and high-specificity metabolic production of isobutanol in the following section.

8.4.1. Integration of a Temperature-Controllable Enzyme Expression Circuit into an Isobutanol Biosynthetic Pathway to Improve Isobutanol Yield

Synthetic genetic circuits have recently been applied in a wide range of metabolic engineering problems as tools to construct or modify metabolic pathways to generate many biochemical products that do not exist in the host cell [21-26]. The standard method previously used to induce bacterial protein production is to use the linkage of the genes of interest to an inducible promoter and the corresponding inducers, such as IPTG, to activate the protein expression. However, the overexpression of foreign proteins may affect the bacterial physiology differently, and bacterial hosts sometimes produce inactive proteins.

The bacterial physiology varies in order to accommodate environmental stresses, such as sudden changes in osmotic pressure, pH level, and temperature [27]. It has been found that different regulatory proteins are selectively activated on the translational level in order to respond to such environmental changes [28]. In addition, the structures of some mRNAs were found to vary in light of changes in the environmental temperature, and these are used as RNA thermometers or RNA thermo-sensors [29-32]. The mechanism of an RNA thermometer functions by unwinding its secondary structure to allow ribosome binding when exposed to relatively high temperatures. The discovery of these RNA thermometers allows us to control protein expressions through temperature variation without the introduction of external inducers.

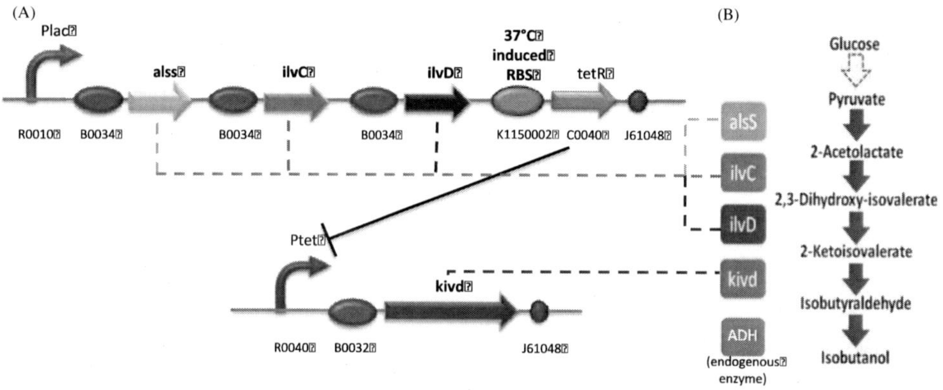

Figure 8.7. The temperature-controllabe enzyme expression circuit and the isobutanol synthetic pathway. (A) Genetic circuit to generate the enzymes for isobutanol synthesis. Plac is the LacI-regulated promoter used to initiate the genetic circuit to produce the relative enzymes during isobutanol synthesis. Genes alss, ilvC, and ilvD will be translated into enzymes and catalyze pyruvate into 2-ketoisovalerate. Ptet is the TetR-regulated promoter. The gene kivd encodes 2-keto-acid decarboxylase, which can convert 2-ketoisovalerate into isobutyraldehyde, which is the substrate of the endogenous alcohol dehydrogenase (ADH) in E. coli and is catalyzed to isobutanol. A temperature-sensitive RBS is inserted into the genetic circuit to regulate the KivD enzyme expression to further control the yield of isobutanol. (B) Isobutanol synthetic pathway induced by the temperature-controllable enzyme expression circuit. After glycolysis, two molecules of pyruvate are condensed into 2-acetolactate by acetolactate synthase (alsS; B. subtilis), which is then converted into 2-ketoisovalerate by acetohydroxy acid isomeroreductase and dihydroxy acid dehydratase (ilvC and ilvD; *E. coli*). The 2-ketoisovalerate is then decarboxylated to form isobutyraldehyde by keto-acid decarboxylase (kivd; L. lactis). The target biofuel, isobutanol, can be converted from isobutyraldehyde by various aldehyde reductases in *E. coli*.

RNA thermometers can be classified into three major families by their distinct RNA secondary structures: (1) Repressor Of heat-Shock gene Expression, (ROSE), (2) PrfA, and (3) FourU. All families have been extensively studied in the literature with respect to their characteristics and potential applications.[32-36] In our experiment, FourU, a family of artificially modified RNA thermometers originating from the *Salmonella* genome, was used to control the protein production. It has four closed uridine nucleotides, which can interact with the corresponding Shine-Dalgarno (SD) sequence (5′-AGGAG-3′) at low temperatures. As temperature increases, the FourU's secondary structures lose stability and unwind, allowing the expression of the downstream coding protein.[36-38] Previous studies have also shown that RNA thermometers are ideal modification targets, whose changes can specifically alter the bacterial protein expression.[35, 39] Based on the concept of temperature regulation, we introduced temperature-sensitive ribosome binding sites (RBS) into genetic circuits to create a genetic circuit controllable by a temperature switch for adequate protein expression. This design for a temperature-controllable enzyme expression system is more economically viable, since it does not depend on the addition of external inducers, such as IPTG, to turn on the protein production. This is an ideal candidate for a low-cost solution for the fermentation industry.

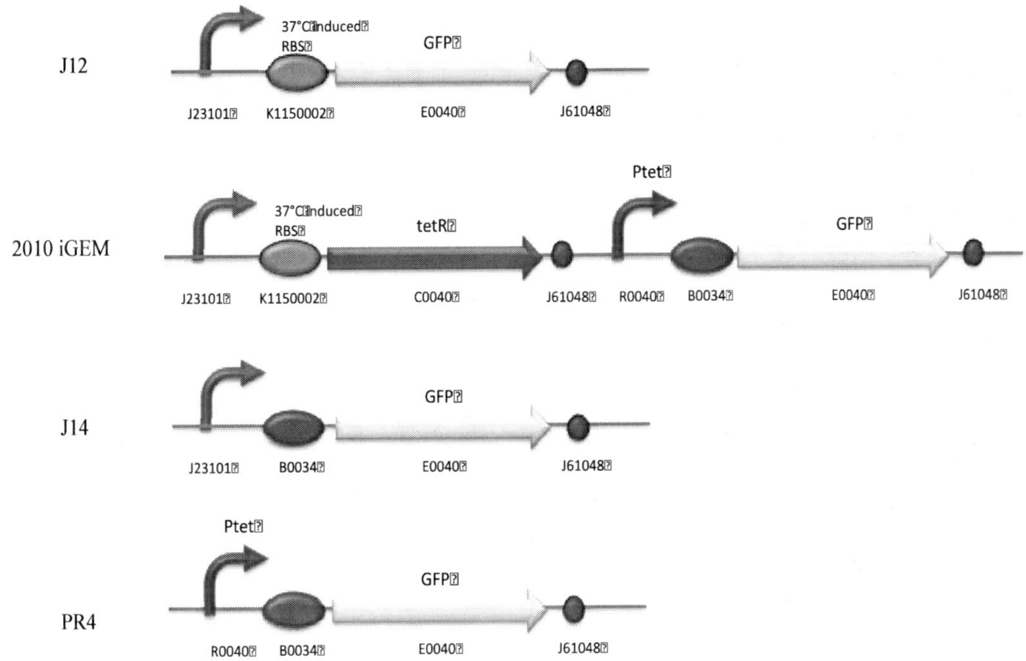

Figure 8.8. Diagram of the temperature-controllable genetic circuits. These genetic circuits were applied in the experiment to detect the fluorescence data, which were further used in the simulation. J12 represents the genetic circuit of the high-temperature protein expression system; 2010 iGEM represents the low-temperature expression circuit; J14 is similar to J12 but with the temperature-sensitive RBS being replaced by a temperature-insensitive RBS; and PR4 represents the second part of the 2010 iGEM circuit.

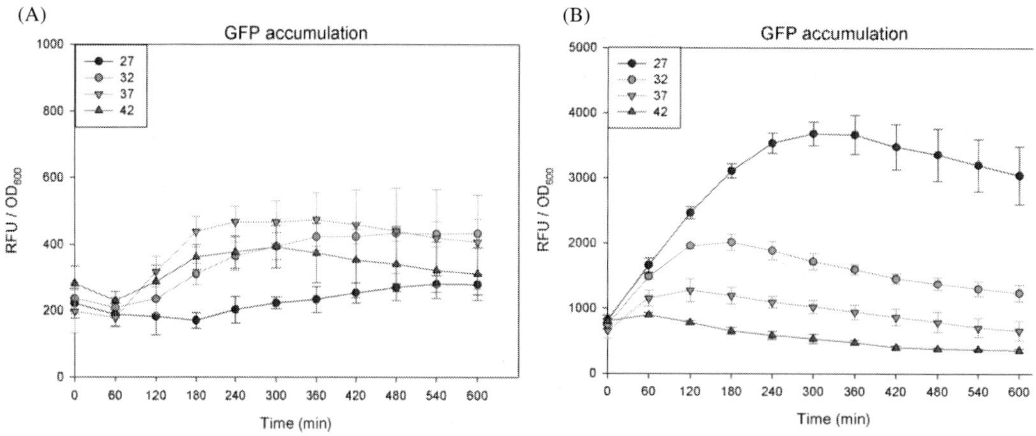

Figure 8.9. The fluorescence exhibition of the temperature-controllable protein expression system. The GFP fluorescence exhibition of the (A) J12 and (B) 2010 iGEM circuits. Time profiles were GFP accumulations of the E. coli host **strain DH5α containing the respective plasmid** at different temperatures (27, 32, 37, and 42 °C). The data was deducted from the background fluorescence and normalized with the absorbance at 600 nm.

Isobutanol is one of the most promising clean alternative biofuels and has a similar energy density to gasoline [40]. In addition, it offers 25% more energy per molecule than ethanol, and its solubility in water and its vapor pressure are also lower than those of ethanol.[18-20] However, due to a lack of corresponding enzymes, there exists no isobutanol pathway in *E. coli.* Therefore, an enzyme synthetic circuit is engineered with temperature control to generate the corresponding enzymes for an isobutanol synthetic pathway (Figure 8.7). We applied the temperature-controllable enzyme expression system and used GFP as the reporter in order to characterize the system response to different temperatures for adequate enzyme expression to improve the isobutanol yield in bacteria. In this work, we used the temperature-switching strategy to avoid the complex manipulation of the genomic insertion or knockout, [41] and sequentially regulate a number of the related protein expressions of enzymes for the isobutanol pathway without the chemical inducers. Therefore, the yield of isobutanol by the proposed temperature-contrllable circuit is much better than the control circuit without temperature-sensitive RBS. Moreover, we established a mathematical model to confirm the temperature-switching scheme for the control of enzyme production in our synthetic genetic system to achieve a high yield of isobutanol in *E. coli* based on the prediction of this model system. This temperature-switching control circuit could provide a potential application for industrial fermentation.

Establishment of the Temperature-Controllable Protein Expression System

In order to avoid the expenditure of chemical inducers, we used an RNA thermometer to create the temperature-controllable protein expression system and introduced the identified kinetic strengths of promoter-RBS library in the previous study [42]. The iGEM Biobrick™ construction principles were used, along with methods to design our genetic circuit for the protein expression system to generate enzymes for isobutanol biosynthesis. The key components we used included temperature-sensitive RBS K1150002 (iGEM classification), and GFP as the reporter protein to identify the RBS behavior. The diagram of the circuit is shown as J12 in Figure 8.8. As the *E. coli* grew, the green fluorescence levels were measured at four different temperatures: 27, 32, 37, and 42 °C. The fluorescence readings reflect an upward increasing trend with increasing temperature. The genetic circuit functioned as expected and exhibited higher fluorescence at higher temperatures, acting as a protein expression system by increasing temperature (Figure 8.9A). However, the fluorescence level at 42 °C was only twice that at 27 °C.

In order to obtain different levels of the protein expression at different temperatures, we utilized tetR and a tetR-regulated promoter to reverse the protein expression patent. The circuit was assembled as in the circuit diagram, 2010 iGEM, shown in Figure 8.8. Therefore, as the temperature decreases, the protein accumulation will increase. As expected, the fluorescence increased as the temperature decreased and the fluorescence level at 27 °C was five times that at 42 °C (Figure 8.9B). Furthermore, it can be seen that the fluorescence level was distinguishable even after 10 hours. Comparison of the two circuits suggests that the 2010 iGEM circuit has a sharper distinction in the protein expression between the different temperatures.

Temperature-Controllable Isobutanol Biosynthesis Circuit

In order to demonstrate the effectiveness of the 2010 iGEM circuit, we employed it to construct a temperature-controllable enzyme synthetic circuit to produce the enzymes required to conduct an isobutanol biosynthesis pathway in *E. coli*, which can yield isobutanol and whose product can be controlled by the adjustment of temperature. As in previous studies, glucose is used as a raw material, and after several distinctive enzymatic steps in the isobutanol metabolic pathway, it will be converted into isobutanol (Figure 8.6), [43]. Since several important enzymes are lacking in the isobutanol biosynthesis pathway, which includes alss (from *Bacillus subtilis*), ilvC, ilvD (from *Escherichia coli* MG1655), and KivD (from *Lactococcus lactis*)[44] (Table 8.1), we cloned these genes of the enzymes to assemble the temperature-regulated synthetic genetic circuit to yield isobutanol in *E. coli*.

Table 8.1. The profiles of the enzymes used for isobutanol yield and its modified nucleotides for the standard iGEM assembly

Enzyme	Strain	Peptide length (aa)	Modified nucleotides
Ketol-acid reductoisomerase, NAD(P)-binding, [ilvC]	*Escherichia coli str. K-12 substr. MG1655*	491	C84T, C936T
Dihydroxyacid dehydratase, [ilvD]	*Escherichia coli str. K-12 substr. MG1655*	616	T1146A
Alpha-acetolactate synthase protein, [AlsS]	*Bacillus subtilis*	571	T285A, T759C, T1101A, C1686T
Alpha-ketoisovalerate decarboxylase, [kivd]	*Lactococcus lactis subsp. Lactis KF147*	548	A804G

After cloning the aforementioned genes, we confirmed the expression and activity of the enzymes in different bacterial strains using sodium dodecyl sulfate polyacrylamide gel electrophoresis (SDS-PAGE) after IPTG induction (Figure 8.10) and GC-FID to detect the isobutanol (Figure 8.11). The results show that these enzymes are functional in *E. coli*, as found in a previous study. [45] In order to accommodate the standard iGEM assembly, we employed strategic point mutations (Table 8.2) to eliminate internal restriction digestion sites in the sequences before the key enzymes for the isobutanol biosynthetic pathway are assembled together in the circuit. The final genetic circuit to produce the enzymes for the isobutanol yield is shown in Figure 8.7(A). Briefly, to avoid the expression of these enzymes becoming a burden on the *E. coli*, the first promoter was selected to be the LacI-regulated promoter, which is a weak promoter without its inducers, and subsequently conjugated with three sets of RBSs and the enzyme of the isobutanol biosynthetic pathway. The expression of the exogenous enzyme, KivD, was regulated using the temperature-controllable expression system to form the most important role of controlling the final product yield, isobutanol. Through the distinct enzyme expression levels of the temperature-regulated synthetic circuit by different temperature regulation schemes, the genetic circuit may produce different yields of isobutanol.

Optimization of the Isobutanol Yield Using the Temperature-Controllable Circuit

In order to optimize the incubating conditions to achieve the maximum yield of isobutanol and avoid the overexpression of exogenous proteins that affect the growth of bacteria, we attempted to switch the incubation temperature to control enzyme production for the isobutanol biosynthetic pathway using different switching schemes. The bacteria were cultured at 37 °C to 0.2 at OD_{600}, after which the culturing temperature was changed to 27, 32, 37, and 42 °C, respectively, by different switching schemes (Figure 8.12). A number of samples were collected after 24 h and the concentration of isobutanol produced was analyzed by GC-FID.

In the 27-°C switching strategy, the best performance occurred with the following switching conditions: transferring the culture into 27 °C for 8 h after 16 h in 37 °C (Figure 8.12B). Similarly, the switching strategy for 32 °C appeared effective when switched at 16 h, and had a better yield when switched at 20 h (Figure 8.12C). According to the results, the temperature-switching strategy improved the yield of isobutanol compared to its control circuit, which lacked the temperature-regulated RBS (Figure 8.13).

The control results without temperature switching at 37 °C are shown in Figure 8.14. For the switching strategy from 37 °C to 42 °C, the yield increased dramatically to the maximal level at the first switching scheme, but decreased for the later switching scheme (Figure 8.12D). Taken together, the maximal isobutanol concentration for switching schemes at 27, 32, and 37 °C reached approximately 0.3–0.6%, and peaked for the switching scheme at 42 °C with the best concentration of 0.75% isobutanol (Figure 8.13). Obviously, the yield of isobutanol by the proposed temperature-controllable circuit is much better than the control circuit without temperature-sensitive RBS at different temperatures. Consequently, the temperature-controllable circuit could enhance the isobutanol yield. The yield with the switching scheme at 42 °C demonstrated that the maximal production of isobutanol may have an inverse relation with the KivD expression.

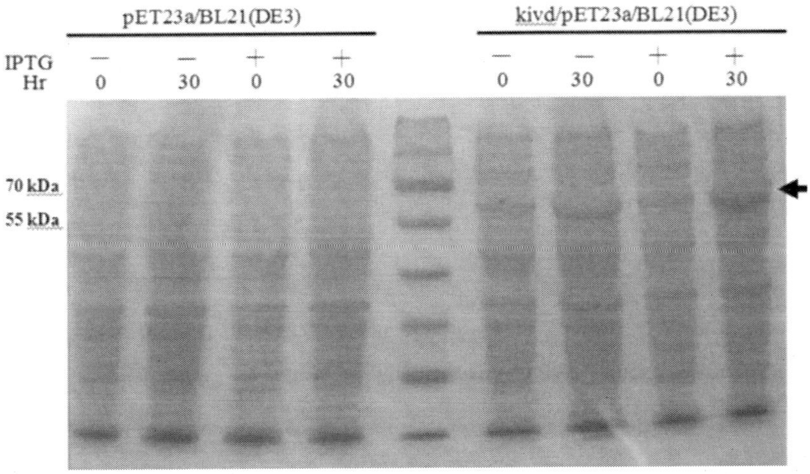

Figure 8.10. Expression of recombinant KivD in E. coli. The protein extract of E. coli BL21 (DE3) containing the kivd/pET-23b plasmid was analyzed using 10% Tricine SDS-PAGE. The left four lanes are the protein extracts of the cell with vector (pET-23b) only, while the right four lanes are the protein extracts containing the KivD expression. The middle lane is the protein marker. After 1 mM IPTG induction for 30 h, the KivD had been over-expressed and showed an obvious band on the rightmost lane. The approximate size of the KivD protein is approximately 62 kDa as the arrow indicates.

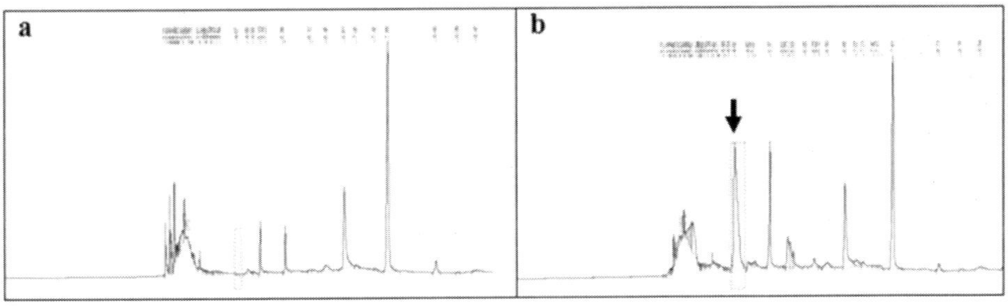

Figure 8.11. Determination of the isobutanol yielded by the GC-FID. The cell cultures were induced with 1 mM IPTG for 30 h, and the culture media were collected by centrifugation and filtered for GC-FID analysis to detect the isobutanol content. The chromatograms are: (A) the culture medium of the cells containing the vector only, and (B) the culture medium of the cells containing the KivD-expressing plasmid. The arrow indicates the retention time of isobutanol is 4.6 minutes.

Table 8.2. Primers used in this chapter

Primer name	Seqence 5'→3'
lssPF	GTTTCTTCGAAT-TCGCGG-CCGCTTCTAGATGTTGACAAAAGCAACAAAAG
alssPR	GTTTCTTCCTGCAGCGGCCGCTACTAGTACTAGAGAGCTTTCGTTTTC
ilvCPF	GTTTCTTCGAATTCGCGGCCGCTTCTAGATGGCTAACTACTTCAATACAC
ilvCPR	GTTTCTTCCTGCAGCGGCCGCTACTAGTATTAACCCGCAACAGCAATA-CG
ilvDPF	GTTTCTTCGAATTCGCGGCCGCTTCTAGATGCCTAAGTACCGTTCCG
ilvDPR	GTTTCTTCCTGCAGCGGCCGCTACTAGTATTAACCCCCCAGTTTCGATTTATC
kivdPF	GTTTCTTCGAATTCGCGGCCGCTTCTAGATGTATACAGTAGGAGATTACCTAT
kivdPR	GTTTCTTCCTGCAGCGGCCGCTACTAGTTTATGATTTATTTTGTTCAGCAAATAG

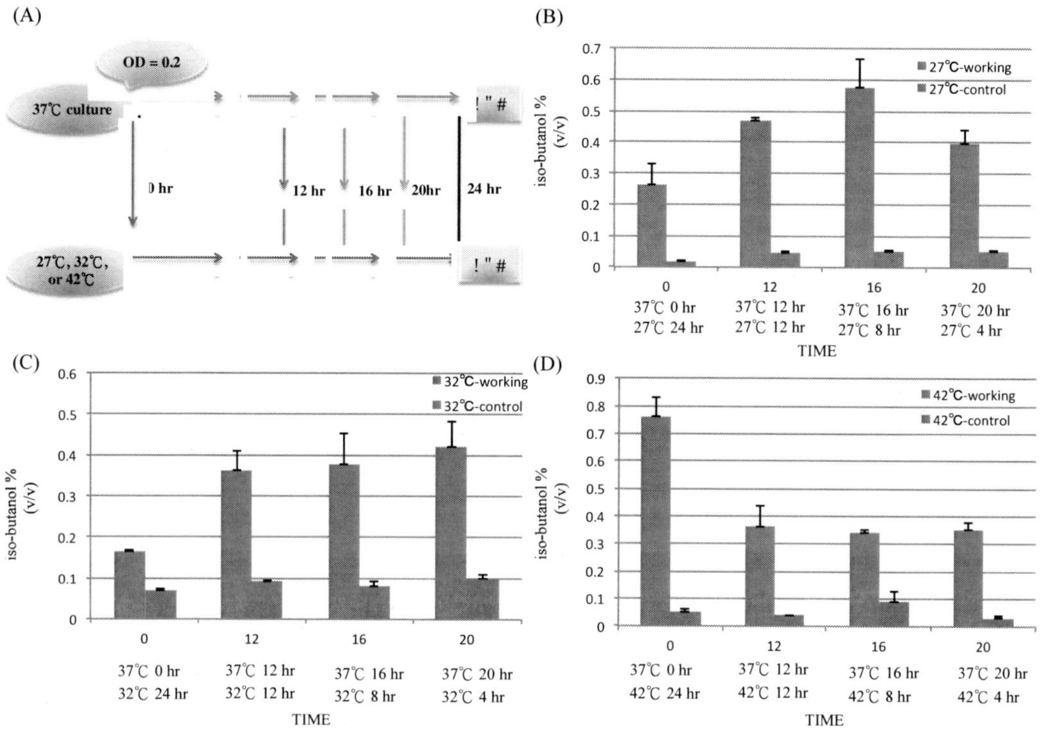

Figure 8.12. Temperature switching schemes and their isobutanol yields. *E. coli* was cultured at 37 °C until OD_{600} = 0.2 and then transferred to various culturing temperatures. (A) The switching schemes at different switching time points are indicated. The isobutanol yields of the different temperature-switching schemes when transferred from 37 °C into (B) 27 °C, (C) 32 °C, and (D) 42 °C at different switching time points are shown in the bar charts. The blue column represents the yield of different temperature-switching circuits, while the red column is the yield of the control circuit without temperature-sensitive RBSs.

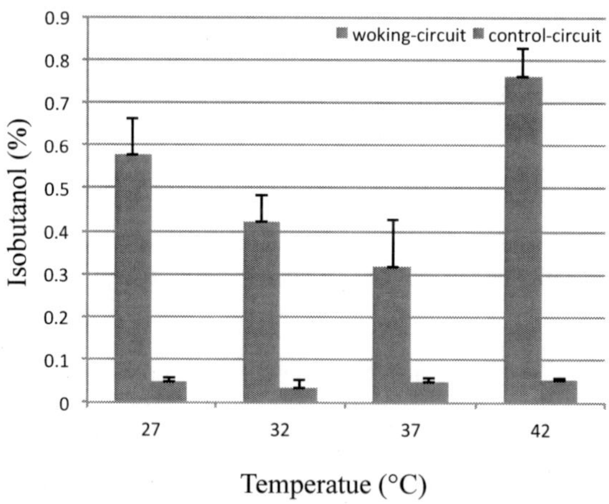

Figure 8.13. Maximal yield of isobutanol at different temperatures. The maximal isobutanol production by the DH5α strain with the genetic circuit in the different temperature-switching experiments. The blue column represents the yield of the different temperature-regulated schemes, while the red one is the yield of the control circuit without temperature-sensitive RBS.

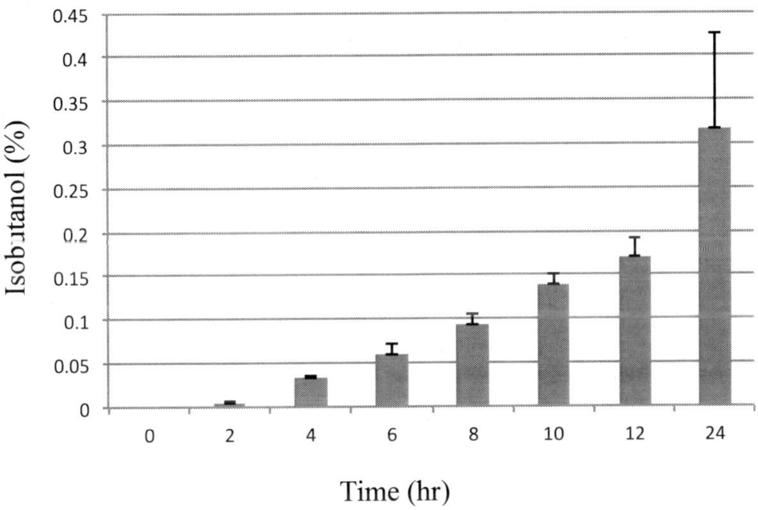

Figure 8.14. Time profile of isobutanol production at 37 °C without the Switching Temperature Scheme. *E. coli* DH5α containing the isobutanol circuit was incubated at 37 °C. The isobutanol production was quantified by the GC-FID at a different time point.

Simulating the Time Profile of Isobutanol Production of the Genetic Circuit in E. coli in silico

Through simulations of the collected experimental data, we can plot the time profile of the isobutanol yield at different temperatures (Figure 8.15), and establish a dynamic model of the enzymatic transformation for prediction in the following. The process of enzymatic transformation is for the reaction catalyzed by an enzyme, which may help convert substrates into productions. Here we suppose that 2-ketoisovalerate combining with KivD can form a complex, which subsequently dissociates into KivD and isobutyraldehyde (Figure 8.7). Similarly, isobutyraldehyde is further catalyzed by the endogenous alcohol dehydrogenase (ADH) of *E. coli* to yield isobutanol. In these enzymatic transformation processes, the backward reactions are also considered for the complex molecules. Thus, the kinetic scheme can be written as:

$$S_1 + E_1 \underset{k_{-1}}{\overset{k_1}{\rightleftharpoons}} E_1S_1 \xrightarrow{k_2} E_1 + P_1$$

$$P_1 + E_2 \underset{k_{-3}}{\overset{k_3}{\rightleftharpoons}} E_2P_1 \xrightarrow{k_4} E_2 + P_2$$

where S_1, P_1, P_2, E_1, and E_2 denote 2-ketoisovalerate, isobutyraldehyde, isobutanol, KivD, and ADH, respectively. The parameters k_1 and k_3 are the forward reaction rate constants, and k_{-1} and k_{-3} are the backward ones. The production formation rate constants are k_2 and k_4. The changes in concentration of the substrates and enzymes in the isobutanol synthesis pathway with respect to time are given by the dynamic equations:

$$
\begin{aligned}
\dot{x}_{S_1} &= -k_1 x_{S_1} x_{E_1} + k_{-1} x_{E_1 S_1} \\
\dot{x}_{E_1} &= -k_1 x_{S_1} x_{E_1} + \left(k_{-1} + k_2\right) x_{E_1 S_1} \\
\dot{x}_{E_1 S_1} &= k_1 x_{S_1} x_{E_1} - \left(k_{-1} + k_2\right) x_{E_1 S_1} \\
\dot{x}_{P_1} &= k_2 x_{E_1 S_1} - k_3 x_{P_1} x_{E_2} + k_{-3} x_{E_2 P_1} \\
\dot{x}_{E_2} &= -k_3 x_{P_1} x_{E_2} + \left(k_{-3} + k_4\right) x_{E_2 P_1} \\
\dot{x}_{E_2 P_1} &= k_3 x_{P_1} x_{E_2} - \left(k_{-3} + k_4\right) x_{E_2 P_1} \\
\dot{x}_{P_2} &= k_4 x_{E_2 P_1}
\end{aligned}
\qquad (8.59)
$$

From the data of the time profile of the concentration of 2-ketoisovalerate, isobutyraldehyde, and isobutanol, we can calculate the K_m and k_{cat} of KivD and ADH at different temperatures by definition [45, 46]. However, the isobutyraldehyde and 2-ketoisovalerate were rarely detectable. Therefore, we suggest that the conversions of the two intermediates were fast and barely accumulated. We further used the circuit without KivD to detect the maximum accumulation of 2-ketoisovalerate, the substrate for KivD by HPLC. We found that the accumulation was still low. Therefore, we suggest that the intermediate would be converted to another product, such as L-valine or L-leucine [47]. Through the simulation data, the higher K_m may be caused by the higher k_2 and k_4 in the reaction, yet the values increased with increasing temperature and were consistent with the isobutanol yield. Since the production rate of isobutanol is quicker with rising temperature, the k_{cat} values of KivD and ADH were also increased (Figure 8.15). The K_m and k_{cat} values were different from those of previous studies, which used the purified enzymes to calculate the parameters from the enzymatic assay [45, 46]. In this chapter, however, we used a systematic viewpoint to evaluate the parameters of the isobutanol biosynthetic pathway in equation (1) at different temperatures for the prediction of isobutanol yield (Table 8.3). In conclusion, by using the temperature-controllable system, we can regulate the expression amount of the enzymes to achieve the highest yield without additive chemical inducers.

8.4.2. Discussion

The use of prokaryotes such as *E. coli* to express exogenous proteins has become a very important tool in biotechnology, owing to its high protein expression, fast growth, easy gene manipulation, and low culturing cost. It has been widely applied to the production of proteins, medicine, and biofuels [43, 48-51]. However, the proteins expressed in *E. coli* may often be inclusion bodies or in an inactive form. To solve the inclusion body problem, researchers must spend a considerable amount of time refolding the inclusion body or reducing the formation of the inclusion body by decreasing the culture temperature. In this chapter, we bypassed the use of inducers entirely and constructed a protein expression system that can be induced only by temperature switching in order to regulate the enzymes of the isobutanol synthetic pathway and control the yield of isobutanol.

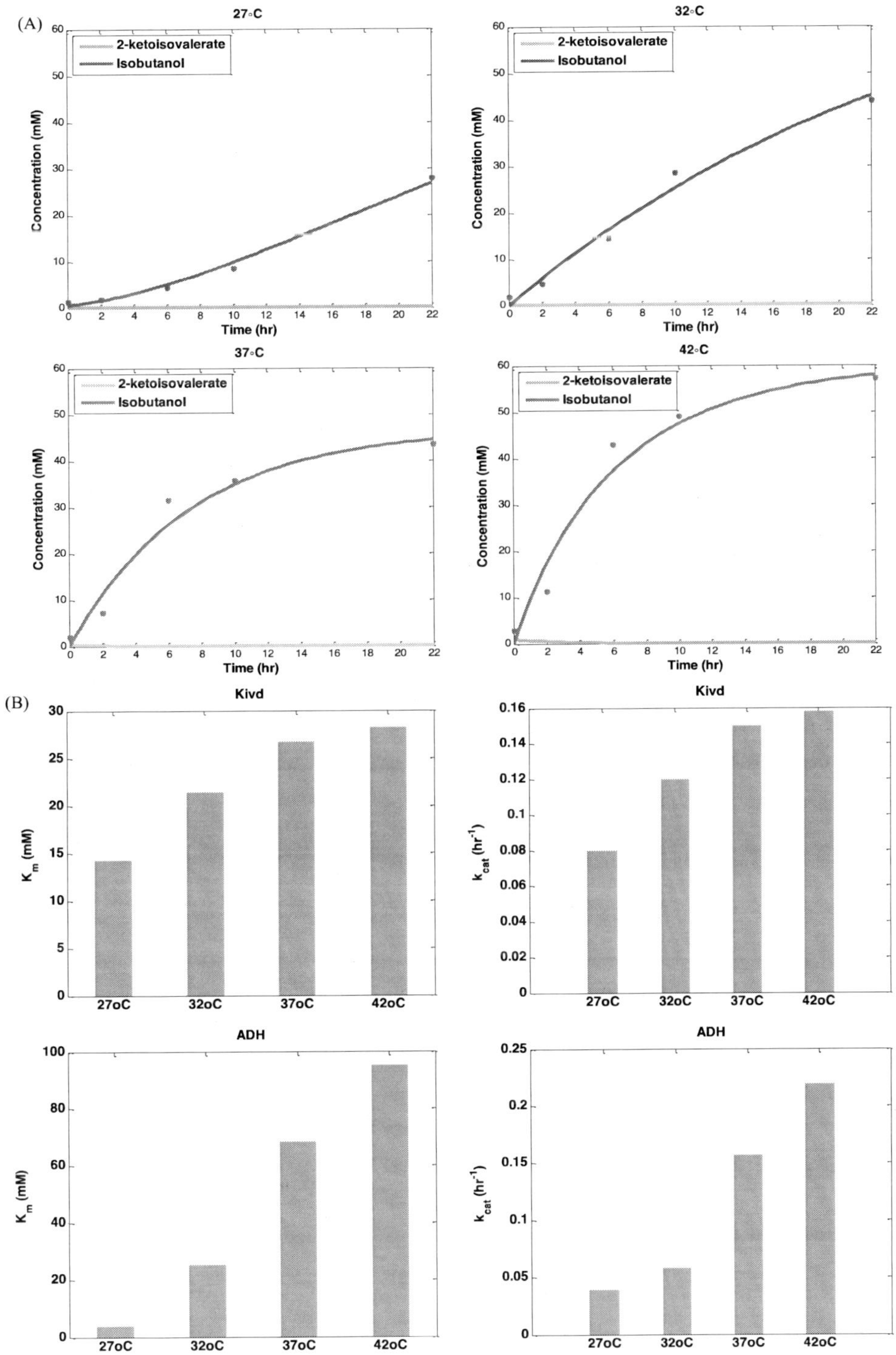

Figure 8.15. Time profiles of isobutanol production at different temperatures. (A) The concentrations of isobutanol and 2-ketoisovalerate in different time points and at different temperatures. The solid line is the trend line simulated by dynamic equation (1), while the dots represent the experimental data. (B) The enzymatic parameters of KivD and ADH in the isobutanol synthesis system in equation (1) are estimated from the experimental and simulated data in (A).

We have achieved the following goals: (1) Every gene expression for enzymes in the biosynthetic metabolic pathway is controlled such that the gene group can be properly expressed into enzymes with a lower chance of over-expression, which may compete with the resources required for bacterial growth and increase the risk of the inactive enzymes, or inclusion bodies occurring. (2) We created a temperature-inducible multi-gene biosynthetic metabolic pathway such that the metabolic pathway of isobutanol can be regulated through temperature. In the experiment of the J12 circuit, we utilized the reporter protein, GFP, to characterize our temperature-sensitive RBS, and created a dynamical mathematical model to describe how the target enzyme production is induced and increased in correlation with an increasing temperature.

Our experimental results show that at different temperatures *i.e.*, 27 °C, 32 °C, and 37 °C, the fluorescence level increased significantly with temperature. However, the protein accumulation at 42 °C is not increased (Figure 8.9). To explore this phenomenon further, we replaced the temperature-sensitive RBS with a conventional RBS to construct the J14 and PR4 circuits (Figure 8.8), and tested the protein expression at different temperatures. The results show that the GFP expression at 42 °C is less than other temperatures at 27, 32, and 37 °C (Figure 8.16 and Figure 8.17). The *E. coli* may not accommodate itself to normal growth and proper protein folding at 42 °C [52], and could subsequently cause the J12 circuit to have a lower level of protein expression than expected as the temperature reached 42 °C. It is worth noting that the OD_{600} values were increased, but that the values of the related fluorescence units (RFU) reached the steady state or slightly decreased after 5 h, suggesting that it was the maximal tolerance to the exogenous proteins for *E. coli* when the temperature was 37 °C and over (Figure 8.16 and Figure 8.17).

We subsequently used a negative regulatory system, a TetR-repressor promoter, to create a low-temperature protein expression ability for the 2010 iGEM circuit. A comparison of the fluorescence intensities of the J12 and 2010 iGEM circuits, shows that the latter system yielded significantly higher fluorescence, and was also more sensitive (with a larger dynamic range) to the changes in culturing temperatures. This suggests that the 2010 iGEM circuit is more suitable to our goal of creating temperature-controllable genetic circuits. From the time-fluorescence data in Figure 8.9B, it was seen that the fluorescence expression at each temperature approaches a steady state within 3–5 hours. Furthermore, this system could be applied to proteins that are sensitive to temperature, *e.g.*, carotenoid [53, 54]; DegP, a heat shock protein, which can switch the chaperone or proteolytic activities by temperature [55]; and TlpA, which is a thermometer for gene regulators [37].

Table 8.3. The parameters estimated in this chapter

Kivd	27°C	32°C	37°C	42°C
K_m (mM)	14.285	21.428	26.753	28.214
k_{cat} (hr^{-1})	0.08	0.12	0.15	0.158
ADH	27°C	32°C	37°C	42°C
K_m (mM)	3.861	25.387	68.233	95.03
k_{cat} (hr^{-1})	0.039	0.058	0.156	0.218

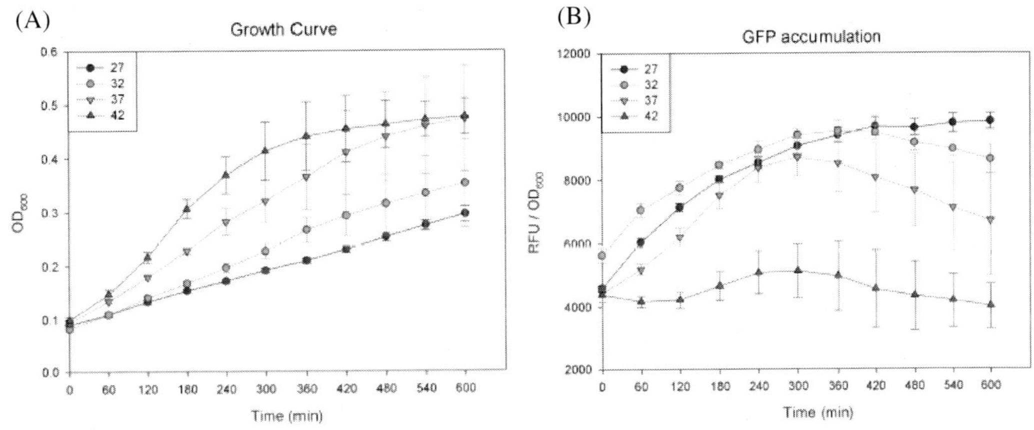

Figure 8.16. Growth curve and GFP expression of the PR4 circuit at different temperatures. *E. coli* DH5α containing the PR4 circuit cultured at different temperatures (27, 32, 37, and 42 °C). (A) The growth curves of the genetic circuit at different temperatures. (B) The GFP expressions of the genetic circuit at different temperatures.

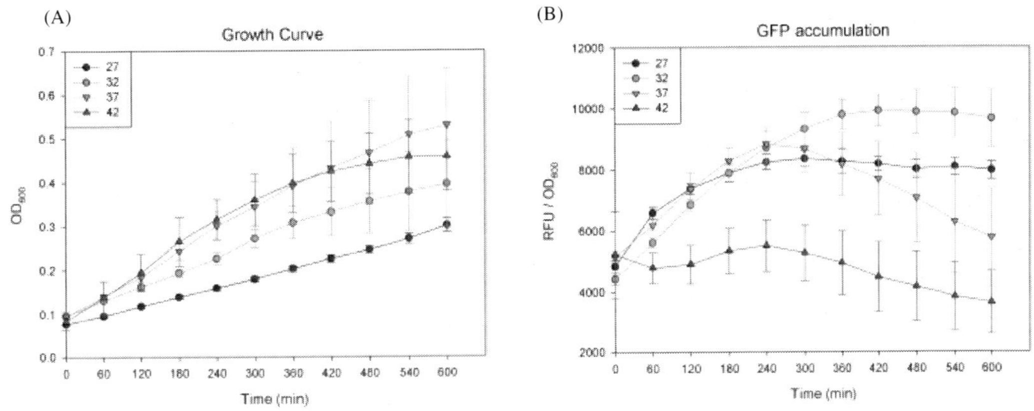

Figure 8.17. Growth curve and GFP expression of the J14 circuit at different temperatures. *E. coli* DH5α containing the PR4 circuit cultured at different temperatures (27, 32, 37, and 42 °C). (A) The growth curves of the genetic circuit at different temperatures. (B) The GFP expressions of the genetic circuit at different temperatures.

The optimized expression of the exogenous proteins can reduce the toxicity, stress, and negative feedback intermediates to the bacterial cells. We further introduced the temperature control scheme into the isobutanol metabolic pathway. With the embedding of our 2010 iGEM circuit to control the enzymes in the isobutanol biosynthetic pathway, we must only change the temperature in order to regulate the expression of the key enzymes in the isobutanol metabolic pathway to effectively control the final production. The roles of the required enzymes, *i.e.*, Alss, Ilvc, and IlvD, are used to accumulate the precursor 2-ketoisovalerate in the isobutanol pathway, where KivD is the key enzyme, which needs to be regulated carefully by our temperature-controllable circuit to produce isobutanol, since the excess accumulation of isobutanol is toxic and harmful to *E. coli* growth and survival [43].

For the temperature tests at 27, 32, 37, and 42 °C, the maximal yield of isobutanol was achieved at 42 °C (0.8%; 100 mM). However, our previous experiments found that the fluorescence expression of the 2010 iGEM circuit at 42 °C is less than that at 27 °C (approximately 32 times lower). From this, it was observed that a low concentration of KivD may conversely create more isobutanol. Bond-Watts et al. have found that if more active promoters are used to synthesize enzymes in the n-butanol production process, then the majority of these enzymes will become insoluble proteins such that the activity of the enzymes is lost, which leads to a slowdown of production.[56] Furthermore, the maximum activity of KivD is at 45 °C [44]. This means that KivD has a better enzyme activity as the temperature increases in this experiment, and the concentration of the 2-ketoisovalerate, the KivD substrate, should not be accumulated to the excess level required for the KivD catalysis capacity. Although the enzyme expression became higher at 27 °C, the nutrient source was significantly depleted, and the enzymatic activity was also decreased. Consequently, the yield of isobutanol reached the highest performance at 42 °C in this synthetic genetic system. However, when the temperature was higher than 42 °C, the isobutanol yield was decreased. The reason for this is the cell being unable to tolerate the higher temperature and declining propagation with increasing temperature over 42 °C (Figure 8.18A).

In this chapter, several factors have also been found to have significant influence on the production of isobutanol. (1) The bacterial strains affected the isobutanol yield. For example, our data has shown that JM109 and MG41655 could not yield detectable isobutanol in the medium, and that EP1300 and DH106 could barely produce 0.1% isobutanol. However, DH5α produced the highest level of isobutanol at approximately 0.4% (Figure 8.18B). (2) The content of the culture medium was also related. In our experiment, when a specific trace metal mix was added into the M9 culture medium, the production of isobutanol could increase by up to approximately 10 times. When we further added a 0.5% yeast extract into the culture medium, not only did the growth rate of the bacteria increase significantly, but the quantity of isobutanol produced was also doubled. The effect of yeast extract on the production of isobutanol has also been mentioned previously [43]. (3) The influence of the amount of each key enzyme of the isobutanol biosynthetic pathway was involved. We placed LacI-regulated promoters to express the three key enzymes, Alss, Ilvc, and IlvD, in our isobutanol biosynthetic pathway. This promoter has a leaky characteristic expression in *E. coli*, and would be induced to have a high expression by the inducer IPTG. However, when IPTG was added to induce a higher expression of these enzymes in our experiment, the amount of isobutanol produced was conversely decreased (data not shown). This may be due to the previously mentioned reason that too much exogenous protein expression causes the perturbation of cell growth and depletes the nutrient source. Therefore, how to produce adequate quantities of each enzyme to avoid over-expression is an important subject.

In a previous study, the researchers also removed the unnecessary genes in *E. coli* to achieve the highest yield of isobutanol after 24 h of continuous culturing.[43] In our study, we could further apply the optimized enzyme expressions without chemical inducers but with simulations to predict the maximal product yield, and this could be a basis of utilization for the other biosynthetic pathway.

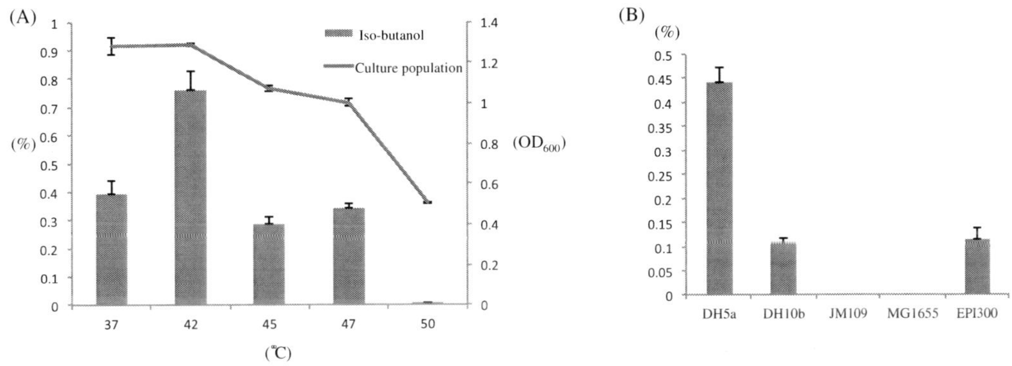

Figure 8.18. The exhibition of the isobutanol circuit at higher temperature and in different *E. coli* strains. (A) The isobutanol yield and cell culture population of the isobutanol circuit in DH5α at and over 37 °C after 24-h incubation. The bar chart shows the isobutanol percentage (v/v, left ordinate) in the culture medium, while the curve represents the cell population (OD_{600}, right ordinate). (B) In comparison with isobutanol produced by DH5α, DH10b, JM109, MG1655, and EPI300. The cell was cultured as described, and the cell media was collected for GC-FID analysis to determine the quantity of isobutanol.

8.5. Conclusion

In this chapter, systematic metabolic engineering design methods based on the steady state of the S-system model and the discrete-time stochastic system are introduced to improve metabolic production. Further, a temperature-controllable enzyme expression circuit is integrated into a biosynthetic isobutanol pathway to improve isobutanol yields in *E. coli*, which otherwise would lack several enzymes and therefore could not produce isobutanol. The robust metabolic engineering circuit design methods proposed hold potential for systematic designs in metabolic engineering, biofuel production, and drug design.

8.6. Appendix

8.6.1. Proof of Proposition 8.2

We choose the Lyapunov function $V(X(t))=E\{X^T(t)PX(t)\}$.

$$\begin{aligned}\Delta V &\triangleq V(t+1)-V(t)\\ &= E\left\{X^T(t+1)PX(t+1)-X^T(t)PX(t)\right\}\\ &= E\left\{X^T(t)\left(A^TPA-P+\sum_{i=1}^{L}\sigma_i^2A_i^TPA_i\right)X(t)+X^T(t)A^TPB_vv(t)\right.\\ &\left.\quad +v^T(t)B_v^TPAX(t)+v^T(t)B_v^TPB_vv(t)\right\}\end{aligned}$$

If the following inequality holds:

$$E\left\{X^T(t)\left(A^TPA-P+\sum_{i=1}^{L}\sigma_i^2A_i^TPA_i\right)X(t)+X^T(t)A^TPB_vv(t)\right.$$
$$\left.+v^T(t)B_v^TPAX(t)+v^T(t)B_v^TPB_vv(t)+y^T(t)y(t)-\rho^2v^T(t)v(t)\right\}\leq 0 \tag{8.60}$$

then

$$\Delta V\leq -E\left\{y^T[k]y[k]-\rho^2v^T[k]v[k]\right\}$$

So we sum the above inequality from 0 to ∞, with the initial condition $X(0)$, to get

$$\sum_{t=0}^{\infty}\left(V(X(t+1))-V(X(t))\right)+y^T(t)y(t)-\rho^2v^T(t)v(t)\leq 0$$
$$\Rightarrow V(X(\infty))-V(X(0))+\sum_{t=0}^{\infty}\left(y^T(t)y(t)-\rho^2v^T(t)v(t)\right)\leq 0 \tag{8.61}$$

Because $V(X(\infty))\geq 0$, if the initial condition of $V(X(0))=0$, then we can get $\Sigma_{t=0}^{\infty}$ $(y^T(t)y(t)-\rho^2v^T(t)v(t))\leq 0$, which can be rewritten as (8.38), i.e.,

$$\|y(t)\|_2^2\leq\rho^2\|v(t)\|_2^2 \text{ or } \frac{\|y(t)\|_2^2}{\|v(t)\|_2^2}\leq\rho^2$$

If the $V(X(0))\neq 0$, then the inequality in (8.61) should be modified as

$$\|y(t)\|_2^2\leq X^T(0)PX(0)+\rho^2\|v(t)\|_2^2$$

The above results hold only when the inequality (8.60) holds, i.e., the following inequality holds:

$$E\left\{X^T(t)\left(A^TPA-P+\sum_{i=1}^{L}\sigma_i^2A_i^TPA_i\right)X(t)+X^T(t)A^TPB_vv(t)+v^T(t)B_v^TPAX(t)\right.$$
$$\left.+v^T(t)B_v^TPB_vv(t)+y^T(t)y(t)-\rho^2v^T(t)v(t)\right\}\leq 0$$
$$\Rightarrow E\left\{X^T(t)\left(A^TPA-P+\sum_{i=1}^{L}\sigma_i^2A_i^TPA_i+C^TC\right)X(t)+X^T(t)A^TPB_vv(t)\right.$$

$$+v^T(t)B_v^T PAX(t)+v^T(t)\left(B_v^T PB_v-\rho^2 I\right)v(t)\Big\}\le 0$$

$$\Rightarrow E\left\{\begin{bmatrix} X(t) \\ v(t)\end{bmatrix}^T\begin{bmatrix} A^T PA-P+\sum_{i=1}^{L}\sigma_i^2 A_i^T PA_i+C^T C & A^T PB_v \\ B_v^T PA & B_v^T PB_v-\rho^2 I\end{bmatrix}\begin{bmatrix} X(t) \\ v(t)\end{bmatrix}\right\}\le 0$$

Thus, if the matrix inequality in (8.39) holds, then the robust filtering in (8.38) holds.

8.6.2. Proof of Proposition 8.3

With a similar procedure, we have the equation:

$$\Delta V=E\left\{\left(\frac{\partial V(X(t))}{\partial X(t)}\right)^T\left(f(X(t))-X(t)\right)+\frac{1}{2}\left(\frac{\partial V(X(t))}{\partial X(t)}\right)^T\left(\sum_{i=1}^{L}f_i(X(t))n_i(t)\right)\right.$$

$$+\frac{1}{2}\left(\sum_{i=1}^{L}f_i(X(t))n_i(t)\right)^T\left(\frac{\partial V(X(t))}{\partial X(t)}\right)+\frac{1}{2}\left(\frac{\partial V(X(t))}{\partial X(t)}\right)^T B_v v(t)$$

$$\left.+\frac{1}{2}\left(B_v v(t)\right)^T\left(\frac{\partial V(X(t))}{\partial X(t)}\right)\right\}$$

$$\le E\left\{\left(\frac{\partial V(X(t))}{\partial X(t)}\right)^T\left(f(X(t))-X(t)\right)+\frac{1}{2}\left(\frac{\partial V(X(t))}{\partial X(t)}\right)^T\left(\frac{\partial V(X(t))}{\partial X(t)}\right)\right.$$

$$+\frac{1}{2}\sum_{i=1}^{L}\sigma_i^2 f_i^T(X(t))f_i(X(t))+\frac{1}{2\rho^2}\left(\frac{\partial V(X(t))}{\partial X(t)}\right)^T B_v B_v^T\left(\frac{\partial V(X(t))}{\partial X(t)}\right)$$

$$\left.+\frac{1}{2}\rho^2 v^T(t)v(t)+\frac{1}{2}\left(y^T(t)y(t)-y^T(t)y(t)\right)\right\}$$

If the following inequality holds:

$$\left(\frac{\partial V(X(t))}{\partial X(t)}\right)^T\left(f(X(t))-X(t)\right)+\frac{1}{2}\left(\frac{\partial V(X(t))}{\partial X(t)}\right)^T\left(\frac{\partial V(X(t))}{\partial X(t)}\right)$$

$$+\frac{1}{2}\sum_{i=1}^{L}\sigma_i^2 f_i^T(X(t))f_i(X(t))+\frac{1}{2\rho^2}\left(\frac{\partial V(X(t))}{\partial X(t)}\right)^T B_v B_v^T\left(\frac{\partial V(X(t))}{\partial X(t)}\right) \tag{8.62}$$

$$+\frac{1}{2}y^T(t)y(t)\le 0$$

then we have

$$\begin{aligned}\Delta V &= V\left(X\left(t+1\right)-V\left(X\left(t\right)\right)\right)\\ &\leq \frac{1}{2}\left(-y^T\left(t\right)y\left(t\right)+\rho^2 v^T\left(t\right)v\left(t\right)\right)\end{aligned}$$

With a similar procedure in the Appendix, to sum the above inequality from 0 to ∞ with the initial condition $X(0) = 0$ we could get the inequality in (8.38).

From the above analysis, we conclude that if (8.62) holds, then the robust desired noise attenuation level ρ in (8.38) holds for the nonlinear perturbative regulatory network in (8.45).

8.6.3. Methods of Synthetic Gene Circuit for Isobutanol in 8.4

Reagents

All restriction enzymes and DNA ligation kits were purchased from New England Biolabs. The chemicals used in this chapter were ordered from Sigma-Aldrich. The oligonucleotides were ordered from Integrated DNA Technologies.

Bacterial Strains, Plasmids, and Genetic Manipulation

The *E. coli* strain, MG1655 (ΔrecA), from Epicentre Technologies (Madison, WI) was used for the fluorescence-measurable experiments, and strain DH5α from Yeastern Biotech (ECOS) was used in all construction procedures.

The DNA parts used in this chapter are listed in Table 8.1. For *alss*, the genomic DNA of *Bacillus subtilis* was used as a PCR template with a pair of primers, alss-PF and alss-PR. The PCR products were ligated with pGEMT, creating alss/TA. To clone *ilvC*, the genomic DNA of *E. coli* BW25113 was used as a PCR template with a pair of primers, ilvC-PF and ilvC-PR. The PCR products were also ligated with pGEMT, creating ilvC/TA. To clone *ilvD*, the genomic DNA of *E. coli* BW25113, was used as a PCR template with a pair of primers, ilvD-PF and ilvD-PR; the PCR products were then ligated with pGEMT, creating ilvD/TA. Finally, to clone *kivd*, the genomic DNA of *Lactococcus lactis* (ATCC) was used as a PCR template with a pair of primers, kivd-PF and kivd-PR, and the PCR products were then ligated with pGEMT, creating kivd/TA.

Medium and Cultivation

The M9 medium contained the following components: Na_2HPO_4 (34 g/L), K_2HPO_4 (15 g/L), NaCl (2.5 g/L), NH_4Cl (5 g/L), casamino acids (10 g/L), and Trace Metal Mix A5. The Trace Metal Mix A5 contained the following components: H_3BO_3 (28.6 mg/L), $MnCl_2 \cdot 4H_2O$ (18.1 mg/L), $ZnSO_4 \cdot 7H_2O$ (2.22 mg/L), $Na_2MoO_4 \cdot 2H_2O$ (3.9 mg/L), $CuSO_4 \cdot 5H_2O$ (0.79 mg/ L), and $Co(NO_3)_2 \cdot 6H_2O$ (0.49 mg/ L).

For Fluorescence Measurement

Bacteria were grown in a 15-ml test tube at 37 °C and 150 rpm, with a culture medium supplemented with the M9 medium, 0.2% casamino acids, 1 mM $MgSO_4$, 2 M thiamine, 10 mM glucose, and 50 ng/ml Kanamycin. The culture was grown overnight and measured by a photospectrometer (Thermo, HELIOS EPSILON) at 600 nm and diluted 500 folds with fresh M9 medium. The GFP fluorescence level of the diluted culture was measured by the microplate reader (BioTek, Synergy™ H1, where the GFP settings were 490/510 nm for excitation and emission) until the cultural OD_{600} reached 0.2.

For Gas Chromatography Analysis

For isobutanol production, 0.5% (v/v) of the overnight culture was inoculated into 5 ml of fresh culture medium in a 15-ml test tube at 37 °C for 1–2 h. After the OD_{600} reached 0.2, the cells were transferred to the rotary shaker (150 rpm.) at the experimental temperature. A number of samples were taken from 0 to 24 hours and the media containing isobutanol was quantified with a gas chromatography–flame ionization detector (GC–FID) (CLARUS 500, Perkin Elmer, equipped with the flame ionization detector).

Intermediate Analysis

The separation of isobutanol and isobutyraldehyde was carried out using a GC equipped with a Capillary Column (60 m, 0.32-mm internal diameter, 1-µm film thickness; SUPELCO). The GC oven temperature was initially held at 40 °C for 1 min and raised with a gradient of 20 °C min^{-1} until reaching a maximum at 100 °C and held for 4 min. It was then raised with a gradient of 15 °C min^{-1} until 130 °C and held for 4 min. Helium was used as the carrier gas with an inlet pressure of 14 p.s.i. The injector and detector were maintained at 220 °C. A 1-µl sample was injected in the splitless injection mode [57]. Isobutanol or isobutyraldehyde was used as the internal standard (Figure 8.19).

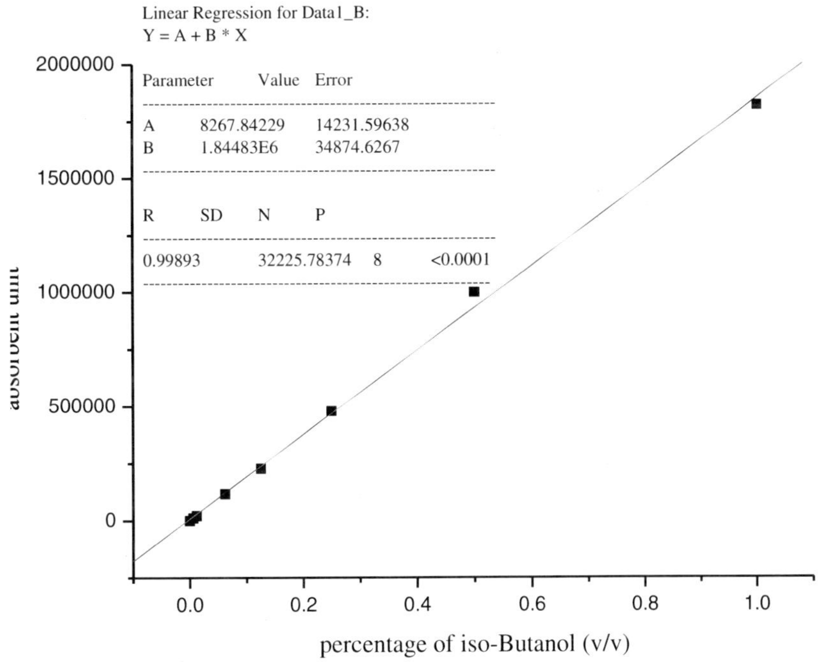

Figure 8.19. Standard curve of isobutanol. The different diluted concentrations of isobutanol (1–0.00625%, v/v) to quantitate the purity.

The intermediate, 2-ketoisovalerate, was analyzed using high-performance liquid chromatography (HPLC) (Waters PDA 996) equipped with a GRACE 201TP54 column. The mobile phase contained 15 mM of ammonium acetate, 0.5% (v/v) methanol, and 0.1% (v/v) H_3PO_4, and the flow rate was 0.8 ml/min [58]. The 2-ketoisovalerate was used as the standard.

Supporting Information Available

The protein expression profiles of the different circuits, and the proofs for the cloning genes and product analysis. Comparisons of isobutanol yield in different *E. coli* strains and at higher temperature. This material is available free of charge via the Internet at http://pubs.acs.org.

REFERENCES

[1] Kim, PJ; Lee, DY; Kim, TY; Lee, KH; Jeong, H; Lee, SY; et al. "Metabolite essentiality elucidates robustness of Escherichia coli metabolism," *Proc Natl Acad Sci U S A*, vol. 104, 13638-42, Aug 21 2007.

[2] Segre, D; Vitkup, D; Church, GM. "Analysis of optimality in natural and perturbed metabolic networks," *Proc Natl Acad Sci U S A*, vol. 99, 15112-7, Nov 12 2002.

[3] Park, JH; Lee, SY; Kim, TY; Kim, HU. "Application of systems biology for bioprocess development," *Trends Biotechnol*, vol. 26, 404-12, Aug 2008.

[4] Chandran, D; Copeland, WB; Sleight, SC; Sauro, HM. "Mathematical modeling and synthetic biology," *Drug Discovery Today: Disease Models*, vol. 5, 299-309, 2008.

[5] Shiue, E; Prather, KLJ. "Synthetic biology devices as tools for metabolic engineering," *Biochemical Engineering Journal*, vol. 65, 82-89, Jun 15 2012.

[6] Ghosh, S; Matsuoka, Y; Asai, Y; Hsin, KY; Kitano, H. "Software for systems biology: from tools to integrated platforms," *Nature Reviews Genetics*, vol. 12, 821-832, Dec 2011.

[7] Chen, BS; Wu, WS; Wang, YC; Li, WH. "On the Robust Circuit Design Schemes of Biochemical Networks: Steady-State Approach," *IEEE Transactions on Biomedical Circuits and Systems*, vol. 1, 91-104, Jun 2007.

[8] Voit, EO. *Computational analysis of biochemical systems : a practical guide for biochemists and molecular biologists*. New York: Cambridge University Press, 2000.

[9] Tsai, KY; Wang, FS. "Evolutionary optimization with data collocation for reverse engineering of biological networks," *Bioinformatics*, vol. 21, 1180-1188, Apr 1 2005.

[10] Chen, BS; Chen, PW. "Robust Engineered Circuit Design Principles for Stochastic Biochemical Networks With Parameter Uncertainties and Disturbances," *IEEE Transactions on Biomedical Circuits and Systems*, vol. 2, 114-132, Jun 2008.

[11] Boyd, SP. *Linear matrix inequalities in system and control theory*. Philadelphia: Society for Industrial and Applied Mathematics, 1994.

[12] Chen, BS; Chen, WH; Wu, HL. "Robust H-2/H-∞ Global Linearization Filter Design for Nonlinear Stochastic Systems," *IEEE Transactions on Circuits and Systems I-Regular Papers*, vol. 56, 1441-1454, Jul 2009.

[13] Chen, BS; Tseng, CS; Uang, HJ. "Robustness design of nonlinear dynamic systems via fuzzy linear control," *IEEE Transactions on Fuzzy Systems*, vol. 7, 571-585, Oct 1999.
[14] Chen, BS; Tseng, CS; Uang, HJ. "Mixed H-2/H-infinity fuzzy output feedback control design for nonlinear dynamic systems: An LMI approach," *IEEE Transactions on Fuzzy Systems*, vol. 8, 249-265, Jun 2000.
[15] Hasty, J; McMillen, D; Collins, JJ. "Engineered gene circuits," *Nature*, vol. 420, 224-230, Nov 14 2002.
[16] Kitano, H. "Biological robustness," *Nat Rev Genet*, vol. 5, 826-37, Nov 2004.
[17] Goulian, M. "Robust control in bacterial regulatory circuits," *Current Opinion in Microbiology*, vol. 7, 198-202, Apr 2004.
[18] Atsumi, S; Liao, JC. "Metabolic engineering for advanced biofuels production from Escherichia coli," *Current opinion in biotechnology*, vol. 19, 414-9, Oct 2008.
[19] Jin, C; Yao, MF; Liu, HF; Lee, CFF; Ji, J. "Progress in the production and application of n-butanol as a biofuel," *Renewable & Sustainable Energy Reviews*, vol. 15, 4080-4106, Oct 2011.
[20] Yan, Y; Liao, JC. "Engineering metabolic systems for production of advanced fuels," *Journal of industrial microbiology & biotechnology*, vol. 36, 471-9, Apr 2009.
[21] Callura, JM; Dwyer, DJ; Isaacs, FJ; Cantor, CR; Collins, JJ. "Tracking, tuning, and terminating microbial physiology using synthetic riboregulators," *Proceedings of the National Academy of Sciences of the United States of America*, vol. 107, 15898-903, Sep 7 2010.
[22] Friedland, AE; Lu, TK; Wang, X; Shi, D; Church, G; Collins, JJ. "Synthetic gene networks that count," *Science*, vol. 324, 1199-202, May 29 2009.
[23] Kobayashi, H; Kaern, M; Araki, M; Chung, K; Gardner, TS; Cantor, CR; et al. "Programmable cells: Interfacing natural and engineered gene networks," *Proceedings of the National Academy of Sciences of the United States of America*, vol. 101, 8414-8419, Jun 1 2004.
[24] Salis, HM; Mirsky, EA; Voigt, CA. "Automated design of synthetic ribosome binding sites to control protein expression," *Nature biotechnology*, vol. 27, 946-50, Oct 2009.
[25] Tabor, JJ; Levskaya, A; Voigt, CA. "Multichromatic control of gene expression in Escherichia coli," *Journal of molecular biology*, vol. 405, 315-24, Jan 14 2011.
[26] Zhang, K; Li, H; Cho, KM; Liao, JC. "Expanding metabolism for total biosynthesis of the nonnatural amino acid L-homoalanine," *Proceedings of the National Academy of Sciences of the United States of America*, vol. 107, 6234-9, Apr 6 2010.
[27] Jiang, X; Zhang, H; Yang, J; Liu, M; Feng, H; Liu, X; et al. "Induction of gene expression in bacteria at optimal growth temperatures," *Appl Microbiol Biotechnol*, vol. 97, 5423-31, Jun 2013.
[28] Ishii, N; Nakahigashi, K; Baba, T; Robert, M; Soga, T; Kanai, A; et al. "Multiple high-throughput analyses monitor the response of E. coli to perturbations," *Science*, vol. 316, 593-7, Apr 27 2007.
[29] Digel, I; Kayser, P; Artmann, GM. "Molecular processes in biological thermosensation," *Journal of biophysics*, vol. 2008, 602870, 2008.
[30] Kaempfer, R. "RNA sensors: novel regulators of gene expression," *EMBO reports*, vol. 4, 1043-7, Nov 2003.
[31] Klinkert, B; Narberhaus, F. "Microbial thermosensors," *Cellular and molecular life sciences : CMLS*, vol. 66, 2661-76, Aug 2009.

[32] Narberhaus, F; Waldminghaus, T; Chowdhury, S. "RNA thermometers," *FEMS microbiology reviews*, vol. 30, 3-16, Jan 2006.

[33] Chowdhury, S; Maris, C; Allain, FH; Narberhaus, F. "Molecular basis for temperature sensing by an RNA thermometer," *The EMBO journal*, vol. 25, 2487-97, Jun 7 2006.

[34] Chowdhury, S; Ragaz, C; Kreuger, E; Narberhaus, F. "Temperature-controlled structural alterations of an RNA thermometer," *The Journal of biological chemistry*, vol. 278, 47915-21, Nov 28 2003.

[35] Neupert, J; Karcher, D; Bock, R. "Design of simple synthetic RNA thermometers for temperature-controlled gene expression in Escherichia coli," *Nucleic acids research*, vol. 36, e124, Nov 2008.

[36] Waldminghaus, T; Heidrich, N; Brantl, S; Narberhaus, F. "FourU: a novel type of RNA thermometer in Salmonella," *Molecular microbiology*, vol. 65, 413-24, Jul 2007.

[37] Hurme, R; Berndt, KD; Normark, SJ; Rhen, M. "A proteinaceous gene regulatory thermometer in Salmonella," *Cell*, vol. 90, 55-64, Jul 11 1997.

[38] Rinnenthal, J; Klinkert, B; Narberhaus, F; Schwalbe, H. "Direct observation of the temperature-induced melting process of the Salmonella fourU RNA thermometer at base-pair resolution," *Nucleic acids research*, vol. 38, 3834-47, Jun 2010.

[39] Johansson, J; Mandin, P; Renzoni, A; Chiaruttini, C; Springer, M; Cossart, P. "An RNA thermosensor controls expression of virulence genes in Listeria monocytogenes," *Cell*, vol. 110, 551-61, Sep 6 2002.

[40] Davis, SC; Diegel, SW. United States. Dept. of Energy. Office of Energy Efficiency and Renewable Energy., United States. Dept. of Energy. Office of Energy Efficiency and Renewable Energy. Office of Planning Budget Formulation and Analysis., and Oak Ridge National Laboratory., *Transportation energy data book : edition 31*. Oak Ridge, Tenn.: Oak Ridge National Laboratory, 2012.

[41] Trinh, CT; Li, J; Blanch, HW; Clark, DS. "Redesigning Escherichia coli metabolism for anaerobic production of isobutanol," *Appl Environ Microbiol*, vol. 77, 4894-904, Jul 2011.

[42] Lee, YY; Hsu, CY; Lin, LJ; Chang, CC; Cheng, HC; Yeh, TH; et al. "Systematic design methodology for robust genetic transistors based on I/O specifications via promoter-RBS libraries," *BMC Syst Biol*, vol. 7, p. 109, 2013.

[43] Atsumi, S; Hanai, T; Liao, JC. "Non-fermentative pathways for synthesis of branched-chain higher alcohols as biofuels," *Nature*, vol. 451, 86-9, Jan 3 2008.

[44] de la Plaza, M; Fernandez de Palencia, P; Pelaez, C; Requena, T. "Biochemical and molecular characterization of alpha-ketoisovalerate decarboxylase, an enzyme involved in the formation of aldehydes from amino acids by Lactococcus lactis," *FEMS microbiology letters*, vol. 238, 367-74, Sep 15 2004.

[45] Atsumi, S; Wu, TY; Eckl, EM; Hawkins, SD; Buelter, T; Liao, JC. "Engineering the isobutanol biosynthetic pathway in Escherichia coli by comparison of three aldehyde reductase/alcohol dehydrogenase genes," *Applied microbiology and biotechnology*, vol. 85, 651-7, Jan 2010.

[46] Zhang, K; Sawaya, MR; Eisenberg, DS; Liao, JC. "Expanding metabolism for biosynthesis of nonnatural alcohols," *Proc Natl Acad Sci U S A*, vol. 105, 20653-8, Dec 30 2008.

[47] Krause, FS; Blombach, B; Eikmanns, BJ. "Metabolic engineering of Corynebacterium glutamicum for 2-ketoisovalerate production," *Appl Environ Microbiol*, vol. 76, 8053-61, Dec 2010.

[48] Clomburg, JM; Gonzalez, R. "Biofuel production in Escherichia coli: the role of metabolic engineering and synthetic biology," *Applied microbiology and biotechnology*, vol. 86, 419-34, Mar 2010.

[49] Goeddel, DV; Kleid, DG; Bolivar, F; Heyneker, HL; Yansura, DG; Crea, R; et al. "Expression in Escherichia coli of chemically synthesized genes for human insulin," *Proceedings of the National Academy of Sciences of the United States of America*, vol. 76, 106-10, Jan 1979.

[50] Huffer, S; Roche, CM; Blanch, HW; Clark, DS. "Escherichia coli for biofuel production: bridging the gap from promise to practice," *Trends in biotechnology*, vol. 30, 538-45, Oct 2012.

[51] Pfeifer, BA; Admiraal, SJ; Gramajo, H; Cane, DE; Khosla, C. "Biosynthesis of complex polyketides in a metabolically engineered strain of E. coli," *Science*, vol. 291, 1790-2, Mar 2 2001.

[52] Berrow, NS; Bussow, K; Coutard, B; Diprose, J; Ekberg, M; Folkers, GE; et al. "Recombinant protein expression and solubility screening in Escherichia coli: a comparative study," *Acta crystallographica. Section D, Biological crystallography*, vol. 62, 1218-26, Oct 2006.

[53] Milanowska J; Gruszecki, WI. "Heat-induced and light-induced isomerization of the xanthophyll pigment zeaxanthin," *Journal of photochemistry and photobiology. B, Biology*, vol. 80, 178-86, Sep 1 2005.

[54] van den Berg, H; Faulks, R, Granado, HF; Hirschberg, J; Olmedilla, B; Sandmann, G; et al. "The potential for the improvement of carotenoid levels in foods and the likely systemic effects," *Journal of the Science of Food and Agriculture*, vol. 80, 880-912, May 15 2000.

[55] Spiess, C; Beil, A; Ehrmann, M. "A temperature-dependent switch from chaperone to protease in a widely conserved heat shock protein," *Cell*, vol. 97, 339-47, Apr 30 1999.

[56] Bond-Watts, BB; Bellerose, RJ; Chang, MC. "Enzyme mechanism as a kinetic control element for designing synthetic biofuel pathways," *Nature chemical biology*, vol. 7, 222-7, Apr 2011.

[57] Machado, HB; Dekishima, Y; Luo, H; Lan, EI; Liao, JC. "A selection platform for carbon chain elongation using the CoA-dependent pathway to produce linear higher alcohols," *Metabolic engineering*, vol. 14, 504-11, Sep 2012.

[58] Buescher, JM; Moco, S; Sauer, U; Zamboni, N. "Ultrahigh performance liquid chromatography-tandem mass spectrometry method for fast and robust quantification of anionic and aromatic metabolites," *Anal Chem*, vol. 82, 4403-12, Jun 1 2010.

Chapter 9

SYSTEMS BIOLOGY APPROACH BY STOCHASTIC DYNAMIC NASH GAMES

ABSTRACT

The dynamic Nash game has applications in many scientific disciplines including applied mathematics, economics, engineering, aeronautics, sociology, and politics. In this chapter, nonlinear stochastic dynamic Nash games are introduced for robust control of immune systems, robust synthetic network design, and evolution of biological networks from the minimax game perspective. In these problems, the biological networks need to maintain their phenotypes in the face of uncertain environmental disturbances. In this situation, the biological network and an uncertain disturbance are considered players. The network wants to minimize the influence of the disturbance while the disturbance wants to maximize its effect on the network (because of the uncertainty of disturbance and considering its worst-case effect on the safe side), thus there exists a conflict of interest. This is the so-called worst-case design against uncertain external disturbance. Based on nonlinear stochastic game theory, a complex HJI needs to be solved for such minimax robust design problems. In this chapter, the T-S fuzzy model is introduced to interpolate several local linear systems to approximate a nonlinear biological network. This allows the HJI to be replaced by a set of LMIs, which can be readily solved. In this situation, nonlinear stochastic Nash game problems in systems biology can be addressed efficiently.

INTRODUCTION

Game theory studies rational decision making in strategic games. In game theory, each of the players can choose their strategies to maximize their own profit. Each player receives a certain payoff depending on their choices and tries to maximize their own return. A situation in which neither of the players can improve his payoff by a unilateral change of strategy is called a Nash equilibrium. When game theory is applied to deal with Nash equilibrium strategies in dynamic systems, it is called a dynamic Nash game.

In this chapter, a stochastic Nash game with a systems biology approach incorporating intrinsic and external noise will be considered. As a first example, a robust control of immune response is proposed for therapeutic enhancement that matches a prescribed immune response with uncertain initial states and environmental disturbances, including continuous intrusion of exogenous pathogens. The worst-case effect of all possible environmental disturbances and an uncertain initial state is minimized for the enhanced immune system. A robust control is designed to track a prescribed immune model response from a minimax matching perspective. This minimax matching problem can be transformed into an equivalent dynamic game problem. Another example in robust synthetic biology is the engineering of artificial genetic circuits in a host cell with uncertain initial conditions and extracellular environmental disturbances. This robust genetic circuit design could be constructed from the minimax perspective, which can be transformed into an equivalent stochastic game problem. The third example involves the evolutionary game problem of biological networks. Biological evolution is a physical, genetic, or behavioral change in a biological network population over time because of natural selection. Network robustness is the persistence of phenotypic traits under genetic variation and environmental disturbance caused by environmental changes. Since the worst-case effect of environmental changes on network phenotype fitness is inverse to evolution level, the maximum fitness of genetic networks with the desired phenotypic trait can be formulated as a stochastic minimax evolutionary game problem. These three stochastic dynamic game problems for biological networks will be discussed in the following sections.

9.1. Robust Model Matching Control of Immune Systems under Environmental Disturbances: Dynamic Game Approach

9.1.1. Introduction of Robust Control of Immune systems

A dynamic response of the immune system, which includes innate immune system and adaptive immune system, is induced by infectious microbes or environmental disturbances. The innate immune system provides a tactical response, signaling the presence of 'non-self' organisms and activating B cells to produce antibodies to bind to the intruders' epitopic sites. The antibodies identify targets for scavenging cells that engulf and consume the microbes, reducing them to non-functioning units [1]. The antibodies can also stimulate the production of cytokines, complement factors and acute-phase response proteins that either damage an intruder's plasma membrane directly or trigger the second phase of immune response. The innate immune system protects against many extracellular bacteria or free viruses found in blood plasma, lymph, tissue fluid, or interstitial space between cells, but it cannot clean out microbes that burrow into cells, such as viruses, intracellular bacteria, and protozoa [1-3].

Activated by the innate immune response, the adaptive immune system could provide strategic response to invading microbe and yield protective cells. These protective cells could remember specific antigens and produce antibodies to counter the antigens, and seek for epitopes of antigens on the surfaces of infected cells. It is found that adaptive immune mechanisms depend on the actions of B- and T-lymphocytes that become dedicated to a single antibody type through clonal selection. Meanwhile, killer T-cells (or cytotoxic T-lymphocytes) bind to infected cells and kill them by initiating programmed cell death (apoptosis). In addition, helper T-cells assist naive B-cells in maturing into plasma cells that produce the needed antibody type. Finally, immune cells with narrowly focused memory are generated, ready to respond rapidly if invading microbes with the same antigen epitopes are encountered again. Elements of the innate and adaptive immune systems are shared, and response mechanisms are coupled, even though distinctive modes of operation can be identified [1-3].

For clinical treatment of infection, the current therapeutic methods focus on killing the invading microbes, neutralizing their response, and providing palliative or healing care to affected organs of the body. Few biological or chemical agents have adverse side effects; for example, an agent that kills a virus may also damage healthy 'self' cells. A critical function of drug discovery and development is hence to identify new compounds that have maximum intended efficacy with minimal side effects on the general population. These examples include antibiotics as microbe killers; interferons as microbe neutralizers; interleukins, antigens from killed (i.e., non-toxic) pathogens, and pre-formed and monoclonal antibodies as immunity enhancers; and anti-inflammatory and anti-histamine compounds as palliative drugs [1].

Recently, there are many models of immune response to infection [4-7] with special emphasis on the human-immunodeficiency virus (HIV) [8-11]. Some papers have discussed immune defense models with moving target strategy [12]. Norbert Wiener [13] and Richard Bellman [14] appreciated and anticipated the application of mathematical analysis to treatment in a broad sense, and Swan surveys early optimal control applications to biomedical problems [15]. Notably, Kirschner [16] offers an optimal control approach to HIV treatment, and intuitive control approaches are presented [17-21].

The dynamics of drug response (pharmacokinetics) have been modeled in several works [22, 23] and control theory is applied to drug delivery in other studies [22, 24-29]. Recently, Stengel [30] presented a simple model for the response of the innate immune system to infection and therapeutic intervention, reviewed the prior method and results of optimization, and introduced a significant extension to the optimal control of enhancing immune response by solving a two-point boundary-value problem via an iterative method. Their results show not only the progression from an initially life-threatening state to a controlled or cured condition but also the optimal history of therapeutic agents that produces that condition. In their study, the therapeutic method is extended by adding linear-optimal feedback control to the nominal optimal solution. However, the performance of quadratic optimal control for immune systems may be decayed by the continuous exogenous pathogens input, which is considered as an environmental disturbance of the immune system. Furthermore, some overshoots may occur in the optimal control process and may lead to organ failure because the quadratic optimal control only minimizes a quadratic cost function that is only the integration of squares of states and allows the existence of overshoot [31]. A series dynamic optimization method is therefore proposed to design the optimal schedule for host defense, immune memory and post-infection pathogen levels in mammals [32-35].

In this section, a robust model matching control of immune response is proposed for therapeutic enhancement to match a desired immune response under uncertain exogenous pathogens input, environmental disturbances and uncertain initial states. Because of the uncertainties of these factors mentioned above, in order to attenuate their detrimental effects, their worst-case effects should be considered in the matching control procedure from the robust design perspective. The worst-case effect of all possible uncertain factors on the matching error to a desired immune response is minimized for the enhanced immune systems, i.e., the proposed robust model matching control is designed from the minimax matching perspective. This minimax matching could be transformed to an equivalent dynamic game problem [36]. The exogenous pathogens input is considered as a player to maximize (worsen) the matching error, while the therapeutic control agent is considered as another player to minimize the matching error. Since the innate immune system is highly nonlinear, it is not easy to solve the robust model matching control problem by the nonlinear dynamic game method directly. Recently, fuzzy systems have been employed to efficiently approximate nonlinear dynamic systems to solve the nonlinear control problem [37-40]. A fuzzy model is proposed to interpolate several linearized immune systems at different operating points to approximate the innate immune system via smooth fuzzy membership functions. Then, with the help of fuzzy approximation method, a fuzzy dynamic game scheme is developed so that the minimax matching control of immune systems could be easily solved by the linear dynamic game method, which can be subsequently solved by a constrained optimization scheme via the linear matrix inequality (LMI) technique [41] with the help of Robust Control Toolbox in Matlab. Because the fuzzy dynamic model can approximate any nonlinear dynamic system, the proposed model matching method via fuzzy game theory can be applied to the robust control design of any model of immune system that can be Takagi-Sugeno (T-S) fuzzy interpolated. Finally, the computational simulation examples considering the side effect of agents are given to illustrate the design procedure and to confirm the efficiency and efficacy of the proposed minimax match control method for immune systems.

9.1.2. Model of Immune Response

For the principal goals to study the general course of a disease and to clarify some observational results, a simple four-nonlinear, ordinary differential equation for the dynamic model of infectious disease is introduced as the following equations to describe rates of change of pathogen, immune cell and antibody concentrations and of an indicator of organic health [4, 30] A more general dynamic model will be given next in sequel [42].

$$\begin{aligned}
\dot{x}_1 &= (a_{11} - a_{12}x_3)x_1 + b_1u_1 + w_1 \\
\dot{x}_2 &= a_{21}(x_4)a_{22}x_1x_3 - a_{23}(x_2 - x_2^*) + b_2u_2 + w_2 \\
\dot{x}_3 &= a_{31}x_2 - (a_{32} + a_{33}x_1)x_3 + b_3u_3 + w_3 \\
\dot{x}_4 &= a_{41}x_1 - a_{42}x_4 + b_4u_4 + w_4 \\
a_{21}(x_4) &= \begin{cases} \cos(\pi\, x_4), & 0 \le x_4 \le 1/2 \\ 0, & 1/2 \le x_4 \end{cases}
\end{aligned} \tag{9.1}$$

where x_1 denotes the concentration of a pathogen that expresses a specific foreign antigen; x_2 denotes the concentration of immune cells that are specific to the foreign antigen; x_3 denotes the concentration of antibodies that bind to the foreign antigen; x_4 denotes the characteristic of a damaged organ [x_4=0:healthy, $x_4 \geq 1$:dead]. The combined therapeutic control agents and the exogenous inputs are described as follows: $u_1$denotes the pathogen killer's agent; u_2 denotes the immune cell enhancer; u_3 denotes the antibody enhancer; u_4 denotes the organ healing factor (or health enhancer); and w_1 denotes the rate of continuing introduction of exogenous pathogens. w_2~w_4 denote the environmental disturbances or unmodeled errors and residues.

$a_{21}(x_4)$ is a nonlinear function that describes the mediation of immune cell generation by the damaged cell organ. And if there is no antigen, then the immune cell maintains the steady equilibrium value of x_2^*. The parameters have been chosen to produce a system that recovers naturally from the pathogen infections (without treatment) as a function of initial conditions during a period of times. For the benchmark example in (9.1), both parameters and time units are abstractions, as no specific disease is addressed. The state and control are always positive because concentrations cannot go below zero, and organ death is indicated when $x_4 \geq 1$. The structural relationship of system variables in (9.1) is illustrated in Figure 9.1. Organ health mediates immune cell production, inferring a relationship between immune response and fitness of the individual. Antibodies bind to the attacking antigens, thereby killing pathogenic microbes directly, activating complement proteins, or triggering an attack by phagocytic cells, e.g. macrophages and neutrophils. Each element of the state is subject to independent control, and new microbes may continue to enter the system.

Several typical uncontrolled responses to increasing levels of initial pathogen concentration under sub-clinical, clinical, chronic, and lethal conditions are shown in Figure 9.2 [30]. In general, the sub-clinical response would not require medical examination, while the clinical case warrants medical consultation but is self-healing without intervention. Pathogen concentration stabilizes at non-zero values in the chronic case, which is characterized by permanently degraded organ health, and it diverges in the lethal case and expire the organ. The 'lethal' simulation of Figure 9.2 is allowed to continue to past the point at which x_4 exceeds one [1]. Finally, a more general disease dynamic model could be represented as

$$\dot{x}(t) = f(x(t)) + Bu(t) + Dw(t), \quad x(0) = x_0 \tag{9.2}$$

where $x(t) \in \mathbb{R}^{n\times 1}$ is the state vector, $u(t) \in \mathbb{R}^{m\times 1}$ is the control agents input; $w(t) \in \mathbb{R}^{n\times 1}$ includes exogenous pathogens and environmental disturbances or uncertainty vector. $f(x(t))$ denotes all possible nonlinear interactions in the immune system.

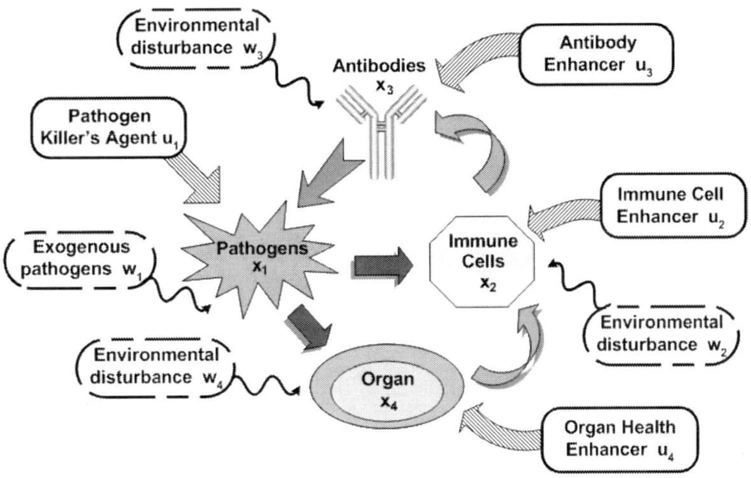

Figure 9.1. Innate and enhanced immune responses to a pathogenic attack under exogenous pathogens and environmental disturbances in (9.1).

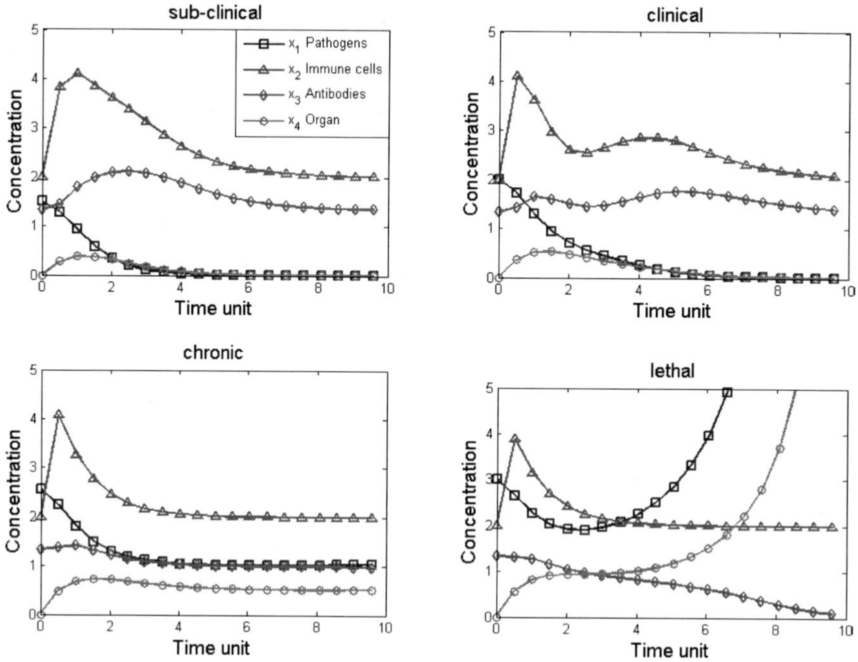

Figure 9.2. Native immune responses to pathogens which are under sub-clinical, clinical, chronic, and lethal conditions [30].

9.1.3. Robust Therapeutic Control of Immune Response

The optimal control is to specify $u(t)$ such that the following cost function is minimized [30].

$$J = \frac{1}{2}\left\{ x^T(t_f)Px(t_f) + \int_0^{t_f} [x^T(t)Qx(t) + u^T(t)Ru(t)]dt \right\} \tag{9.3}$$

where P, Q, and R are weighting matrices to be specified by designer. Because the quadratic control is only to minimize J in (9.3), i.e., the integration of $x^T(t)Qx(t)+u^T(t)Ru(t)$ to be minimized, a control leading to large overshoot of $x(t)$ but with small integration of $x^T(t)Qx(t)$ may be specified in the quadratic optimal control design [31]. This therapeutic control will lead to organ failure because $x_4(t)\geq 1$. Furthermore, the cost function does not include exogenous pathogens and environmental disturbances $w(t)$, which may degrade the performance of the quadratic optimal control. Therefore, it is more appealing to prescribe a desired time response of the disease dynamic in (9.2) beforehand. Next, we design therapeutic control agents $u(t)$ to optimally track the desired time response and at the same time the influence of exogenous pathogens and environmental disturbances $w(t)$ on the tracking should be eliminated as much as possible.

Consider a reference model of immune system with a desired time response prescribed as follows

$$\dot{x}_r(t) = A_r x_r(t) + r(t) \tag{9.4}$$

where $x_r(t) \in \mathbb{R}^{n\times 1}$ is the reference state vector; $A_r \in \mathbb{R}^{n\times n}$ is a specific asymptotically stable matrix and $r(t)$ is a desired reference signal. It is assumed that $x_r(t)$, $\forall t>0$ represents a desired immune response for equation (9.2) to follow, i.e., the therapeutic control is to specify $u(t)$ such that the tracking error $\tilde{x}(t)=x(t)-x_r(t)$ must be as small as possible under the influence of uncertain exogenous pathogens and environmental disturbances $w(t)$. Since the exogenous pathogens and environmental disturbances $w(t)$ and the initial state $x(0)$ are uncertain and reference signal $r(t)$ could be arbitrarily assigned, the robust model matching control design should be specified so that the worst-case effect of three uncertainties $w(t)$, $x(0)$ and $r(t)$ on the tracking error could be minimized, i.e., both the minimax matching and robustness against uncertainties $w(t)$, $x(0)$ and $r(t)$ should be achieved simultaneously [36, 41].

$$\min_{u(t)} \max_{w(t),r(t)} \frac{\int_0^{t_f} (\tilde{x}^T(t)Q\tilde{x}(t) + u^T(t)Ru(t))dt}{\int_0^{t_f} (w^T(t)w(t) + r^T(t)r(t))dt + \tilde{x}^T(0)\tilde{x}(0)} \le \rho^2, \quad \forall \tilde{x}(0) \tag{9.5}$$

where the weighting matrices Q and R are assumed diagonal as follows

$$Q = \begin{bmatrix} q_{11} & 0 & 0 & 0 \\ 0 & q_{22} & 0 & 0 \\ 0 & 0 & q_{33} & 0 \\ 0 & 0 & 0 & q_{44} \end{bmatrix}, R = \begin{bmatrix} r_{11} & 0 & 0 & 0 \\ 0 & r_{22} & 0 & 0 \\ 0 & 0 & r_{33} & 0 \\ 0 & 0 & 0 & r_{44} \end{bmatrix}$$

The diagonal element q_{ii} of Q denotes the punishment on the corresponding tracking error and the diagonal element r_{ii} of R denotes the relative therapeutic cost. Since the worst-case effect of $w(t)$, $r(t)$ and uncertain initial state $x(0)$ on tracking error $\tilde{x}(t)$ and control $u(t)$ is minimized from the energy point of view, the minimax problem of equation (9.5) is suitable for the minimax matching problem under unknown initial $x(0)$, uncertain environmental disturbances $w(t)$ and changeable reference $r(t)$, which are always met in practical design cases. Because it is not easy to solve the Nash dynamic game problem in (9.5) subject to (9.2) and (9.4) directly, we provide an upper bound ρ^2 of the minimax problem.

Remark 9.1: Actually, the design idea is the same as the model adaptive control (MRAC) [43]. The desired time response in (9.4) is the model reference in Åström and Wittenmark. The difficult of the model reference control design of immune system is that all the immune systems are nonlinear and external disturbances are uncertain. Therefore, the minimax game theory in (9.5) and fuzzy interpolation method are employed to simplify the design procedure of the nonlinear MRAC design problem of immune systems in the next approach.

We will first solve the above sub-minimax problem and then decrease the upper bound ρ^2 as small as possible to get the real minimax problem. Since the denominator in (9.5) is independent of $u(t)$ and is not zero, equation (9.5) is equivalent to [36, 41, 44].

$$\min_{u(t)} \max_{w(t),r(t)} \int_0^{t_f} (\tilde{x}^T(t)Q\tilde{x}(t) + u^T(t)Ru(t) - \rho^2 w^T(t)w(t) - \rho^2 r^T(t)r(t))dt \le \rho^2 \tilde{x}^T(0)\tilde{x}(0), \quad \forall \tilde{x}(0) \tag{9.6}$$

Let us denote

$$\min_{u(t)} \max_{w(t),r(t)} J(u(t), w(t), r(t))$$
$$= \min_{u(t)} \max_{w(t),r(t)} \int_0^{t_f} (\tilde{x}^T(t)Q\tilde{x}(t) + u^T(t)Ru(t) - \rho^2 w^T(t)w(t) - \rho^2 r^T(t)r(t))dt$$

From the above analysis, the dynamic game problem in (9.5) or (9.6) is equivalent to finding the worst-case disturbance $w^*(t)$ and reference signal $r^*(t)$ which maximize $J(u(t),w(t),r(t))$ and then a minimax control $u^*(t)$ which minimizes $J(u(t),w^*(t),r^*(t))$ such that the minimax value $J(u^*(t),w^*(t),r^*(t))$ is less than $\rho^2\tilde{x}^T(0)\tilde{x}(0)$, i.e.

$$J(u^*(t), w^*(t), r^*(t)) = \min_{u(t)} J(u(t), w^*(t), r^*(t))$$
$$= \min_{u(t)} \max_{w(t),r(t)} J(u(t), w(t), r(t)) \le \rho^2 \tilde{x}^T(0)\tilde{x}(0), \forall \tilde{x}(0) \tag{9.7}$$

Hence, if there exist $u^*(t)$, $w^*(t)$, and $r^*(t)$ such that minimax matching problem in (9.7) is solved, then they can satisfy the robust model matching performance in (9.5) as well. Therefore, the first step of robust matching control design of therapeutic agents for immune systems is to solve the following dynamic game problem.

$$\min_{u(t)} \max_{w(t),r(t)} J(u(t), w(t), r(t)) \tag{9.8}$$

subject to the disease dynamic model in (9.2) and the desired reference model in (9.4). After that, the next step is to check whether the condition $J(u^*(t),w^*(t),r^*(t)) \le \rho^2\tilde{x}^T(0)\tilde{x}(0)$ is satisfied or not for any $\tilde{x}(0)$.

In general, it is not easy to solve the minimax matching problem directly; it should be transformed to an equivalent minimax regulation problem. Let us denote

$$F(\bar{x}(t)) = \begin{bmatrix} f(x(t)) \\ A_r x_r(t) \end{bmatrix}, \bar{x}(t) = \begin{bmatrix} x(t) \\ x_r(t) \end{bmatrix} \in \mathbb{R}^{2n\times 1}, u(t) \in \mathbb{R}^{m\times 1}, \text{ and } v(t) = \begin{bmatrix} w(t) \\ r(t) \end{bmatrix} \in \mathbb{R}^{2n\times 1}$$

Then we can rewrite the minimax matching problem as

$$\min_{u(t)} \max_{v(t)} J(u(t), v(t))$$
$$= \min_{u(t)} \max_{v(t)} \int_0^{t_f} (\bar{x}^T(t)\mathbf{Q}\bar{x}(t) + u^T(t)Ru(t) - \rho^2 v^T(t)v(t))dt, \quad \forall \bar{x}(0) \tag{9.9}$$

subject to the following augmented system of (9.2) and (9.4)

$$\dot{\bar{x}}(t) = F(\bar{x}(t)) + \mathbf{B}u(t) + Cv(t) \tag{9.10}$$

where

$$\mathbf{Q} = \begin{bmatrix} Q & -Q \\ -Q & Q \end{bmatrix}, \mathbf{B} = \begin{bmatrix} B \\ 0 \end{bmatrix}, C = \begin{bmatrix} D & 0 \\ 0 & I \end{bmatrix}, \mathbf{I} = \begin{bmatrix} I & -I \\ -I & I \end{bmatrix}$$

in which I denotes the 4-by-4 identity matrix. Then the minimax matching problem in (9.8) is equivalent to the following minimax regulation problem of the augmented system in (9.10).

$$\begin{aligned} &\min_{u(t)} \max_{v(t)} J(u(t), v(t)) \\ &= \min_{u(t)} \max_{v(t)} \int_0^{t_f} (\bar{x}^T(t)\mathbf{Q}\bar{x}(t) + u^T(t)Ru(t) - \rho^2 v^T(t)v(t))dt \le \rho^2 \bar{x}^T(0)\mathbf{I}\bar{x}(0), \forall \tilde{x}(0) \end{aligned} \tag{9.11}$$

subject to (9.10).

Proposition 9.1 [42]: The dynamic game problem for robust matching control of immune response in (9.11) could be solved by the following minimax matching control $u^*(t)$ and the worst-case disturbance $v^*(t)$

$$u^*(t) = -\frac{1}{2}R^{-1}\mathbf{B}^T \frac{\partial V(\bar{x}(t))}{\partial \bar{x}(t)} \tag{9.12}$$

$$v^*(t) = \frac{1}{2\rho^2}C^T \frac{\partial V(\bar{x}(t))}{\partial \bar{x}(t)} \tag{9.13}$$

where $V(\bar{x}(t))>0$ is positive solution of the following Hamilton-Jacobi inequality(HJI)

$$\begin{aligned} &\left(\frac{\partial V(\bar{x}(t))}{\partial \bar{x}(t)}\right)^T F(\bar{x}(t)) + \bar{x}^T(t)\mathbf{Q}\bar{x}(t) - \frac{1}{4}\left(\frac{\partial V(\bar{x}(t))}{\partial \bar{x}(t)}\right)^T \mathbf{B}R^{-1}\mathbf{B}^T \frac{\partial V(\bar{x}(t))}{\partial \bar{x}(t)} \\ &+ \frac{1}{4\rho^2}\left(\frac{\partial V(\bar{x}(t))}{\partial \bar{x}(t)}\right)^T CC^T \frac{\partial V(\bar{x}(t))}{\partial \bar{x}(t)} < 0 \end{aligned} \tag{9.14}$$

with

$$V(\bar{x}(0)) \le \rho^2 \bar{x}^T(0)\mathbf{I}\bar{x}(0), \quad \forall \bar{x}(0) \tag{9.15}$$

Proof: see 9.5.1.

Since ρ^2 is the upper bound of Nash game problem in (9.5), based on the analysis above, the minimax matching control $u^*(t)$ and the worst-case disturbance $v^*(t)$ still need to minimize the upper bound ρ^2 as follows [42]

$$\rho_0^2 = \min_{V(\bar{x}(t))>0} \rho^2 \tag{9.16}$$

subject to (9.14) and (9.15).

After solving a $V(\bar{x}(t))$ and ρ_0^2 from the constrained optimization in (9.16), we substitute this solution $V(\bar{x}(t))$ to obtain the minimax matching control $u^*(t)$ in (9.12).

9.1.4. Robust Model Matching Control of Innate Immune System via Fuzzy Interpolation Method

Because it is very difficult to solve the nonlinear HJI in (9.14), no simple approach is available to solve the constrained optimization problem in (9.16) for robust model matching control of innate immune system. Recently [37, 38, 45], the fuzzy T-S model has been widely applied to approximate the nonlinear system via interpolating several linearized systems at different operating points so that the nonlinear dynamic game problem could be transformed to a fuzzy dynamic game problem. Using such approach, the HJI in (9.14) can be replaced by a set of linear matrix inequalities (LMI). In this situation, the nonlinear dynamic game problem in (9.5) could be easily solved by fuzzy dynamic game method for the design of robust model matching control for innate immune response systems.

Suppose the augmented system in (9.10) can be represented by the Takagi-Sugeno (T-S) fuzzy model [45]. The T-S fuzzy model is a piecewise interpolation of several linearized models through membership functions. The fuzzy model is described by fuzzy If-Then rules and will be employed to deal with the nonlinear dynamic game problem for robust model matching control to achieve a desired immune response under exogenous pathogens input and environmental disturbances. The i-th rule of fuzzy model for nonlinear system in (9.10) is of the following form [37, 38].

Rule *i*:

$$\begin{aligned} &\text{If } x_1(t) \text{ is } F_{i1}, \ \ldots, \text{ and } x_g(t) \text{ is } F_{ig} \\ &\text{then } \dot{\bar{x}} = \mathbf{A}_i \bar{x}(t) + \mathbf{B}u(t) + Cv(t), \quad i = 1, \ldots L \end{aligned} \tag{9.17}$$

in which $\mathbf{A}_i = \begin{bmatrix} A_i & 0 \\ 0 & A_r \end{bmatrix}, \mathbf{B} = \begin{bmatrix} B \\ 0 \end{bmatrix}, C = \begin{bmatrix} D & 0 \\ 0 & I \end{bmatrix}$, and F_{ij} is the fuzzy set; $\mathbf{A}_i$, $\mathbf{B}$, and C are known constant matrices; L is the number of If-Then rules, g is the number of premise variables and $x_1(t)$, …, $x_g(t)$ are the premise variables. The fuzzy system is inferred as follows [37, 38, 45].

$$\dot{\bar{x}}(t) = \frac{\sum_{i=1}^{L} \mu_i(x(t))[\mathbf{A}_i \bar{x}(t) + \mathbf{B}u(t) + Cv(t)]}{\sum_{i=1}^{L} \mu_i(x(t))}$$
$$= \sum_{i=1}^{L} h_i(x(t))[\mathbf{A}_i \bar{x}(t) + \mathbf{B}u(t) + Cv(t)] \tag{9.18}$$

where $\mu_i(x(t)) = \prod_{j=1}^{g} F_{ij}(x_j(t))$, $h_i(x(t)) = \mu_i(x(t)) / \sum_{i=1}^{L} \mu_i(x(t))$, $x(t)=[x_1(t), \ldots, x_g(t)]$, and $F_{ij}(x_j(t))$ is the grade of membership of $x_j(t)$ in F_{ij}.

We assume

$$\mu_i(x(t)) \geq 0 \text{ and } \sum_{i=1}^{L} \mu_i(x(t)) > 0 \tag{9.19}$$

Therefore, we get

$$h_i(x(t)) \geq 0 \text{ and } \sum_{i=1}^{L} h_i(x(t)) = 1 \tag{9.20}$$

The T-S fuzzy model in (9.18) is to interpolate L linear systems to approximate the nonlinear system in (9.10) via the fuzzy basis function $h_i(x(t))$. We specify the parameter $\mathbf{A}_i$ easily so that $\sum_{i=1}^{L} h_i(x(t))\mathbf{A}_i \bar{x}(t)$ in (18) can approximate $F(\bar{x}(t))$ in (9.10) by the fuzzy identification method [45].

After the nonlinear system in (9.10) is approximated as the T-S fuzzy system in (9.18), the nonlinear dynamic game problem in (9.10) and (9.11) is replaced by solving the fuzzy dynamic game problem in (9.18) and (9.11).

Proposition 9.2 [42]: The minimax control and the worst-case disturbance for the fuzzy dynamic game problem in (9.11) subject to (9.18) are solved respectively as follows.

$$u^*(t) = -R^{-1}\mathbf{B}^T P\bar{x}(t) \text{ and } v^*(t) = \frac{1}{\rho^2} C^T P\bar{x}(t) \tag{9.21}$$

where P is the positive definite symmetric matrix solution of the following Riccati-like inequality

$$P\mathbf{A}_i + \mathbf{A}_i^T P + \mathbf{Q} - P^T \mathbf{B} R^{-1} \mathbf{B}^T P + \frac{1}{\rho^2} P^T C C^T P \leq 0, \quad i = 1, \ldots, L$$
$$P \leq \rho^2 \mathbf{I} \tag{9.22}$$

Proof: see 9.5.2.

By fuzzy approximation, obviously, the HJI in (9.14) can be approximated by a set of algebraic inequalities in (9.22).

Since ρ^2 is the upper bound of minimax Nash game problem in (9.5), the minimax game problem still needs to minimize ρ^2 as follows

$$\rho_0^2 = \min_{P>0} \rho^2 \tag{9.23}$$

subject to (9.22).

In order to solve the above constrained optimization in (9.23) by the conventional LMI method, we let $W=P^{-1}>0$. Then the equation (9.22) can be equivalent to

$$\mathbf{A}_i W + W\mathbf{A}_i^T + W\mathbf{Q}W - \mathbf{B}R^{-1}\mathbf{B}^T + \frac{1}{\rho^2}CC^T \leq 0, \quad i=1,\cdots,L$$

or

$$\mathbf{A}_i W + W\mathbf{A}_i^T + W\begin{bmatrix} Q & -Q \\ -Q & Q \end{bmatrix} W - \mathbf{B}R^{-1}\mathbf{B}^T + \frac{1}{\rho^2}CC^T \leq 0, i=1,\cdots,L$$

or

$$\mathbf{A}_i W + W\mathbf{A}_i^T + W\begin{bmatrix} Q^{1/2} \\ -Q^{1/2} \end{bmatrix} I \begin{bmatrix} Q^{1/2} & -Q^{1/2} \end{bmatrix} W - \mathbf{B}R^{-1}\mathbf{B}^T + \frac{1}{\rho^2}CC^T \leq 0, i=1,\cdots,L$$

By the schur complements [41], the constrained optimization in (9.22) and (9.23) is equivalent to the following LMI-constrained optimization [42]:

$$\rho_0^2 = \min_{W>0} \rho^2 \tag{9.24}$$

subject to

$$\begin{bmatrix} \mathbf{A}_i W + W\mathbf{A}_i^T - \mathbf{B}R^{-1}\mathbf{B}^T + \frac{1}{\rho^2}CC^T & W\begin{bmatrix} Q^{1/2} \\ -Q^{1/2} \end{bmatrix} \\ \begin{bmatrix} Q^{1/2} & -Q^{1/2} \end{bmatrix} W & -I \end{bmatrix} \leq 0, i=1,\ldots,L \tag{9.25}$$

$$\rho^2 W \geq I$$

Remark 9.2

(1) The fuzzy basis function $h_i(x(t))$ in (9.18) and (9.20) can be replaced by other interpolation functions, for example, cubic spline functions.
(2) By fuzzy approximation, the HJI in (9.14) of nonlinear dynamic game problem is replaced by a set of inequalities in (9.22), which can be easily solved by LMI-constrained optimization in (9.25).
(3) The constrained optimization to solve ρ_0 and $W=P^{-1}$ in (9.24) and (9.25) can be easily solved by decreasing ρ^2 until there exists no $W>0$ solution in (9.25). After solving W and then $P=W^{-1}$ from the constrained optimization problem in (9.24) and (9.25), the minimax control can be obtained from (9.21).
(4) The solution $W>0$ in LMI-constrained optimization (9.25) can be solved by Robust Control Toolbox in Matlab efficiently.
(5) If the conventional quadratic optimal control in (9.3) is considered [30], i.e., the effect of disturbance is not considered in the design procedure, the optimal tracking control problem is equivalent to letting $\rho^2=\infty$ in (9.5) [31]. Then the optimal control design $u^*(t)=-R^{-1}\mathbf{B}P\bar{x}(t)$ can be solved by a common positive definite symmetric matrix P from the equation (9.22) with $\rho^2=\infty$, i.e., solving a common positive definite symmetric matrix $P>0$ from the following constrained inequalities $P\mathbf{A}_i+\mathbf{A}_i^TP+\mathbf{Q}-P^T\mathbf{B}R^{-1}\mathbf{B}^TP\leq 0$, $i=1, \ldots, L$, [41]. In order to solve the optimal tracking control by LMI technique, the optimal tracking control is equivalent to solving a common $W=P^{-1}$ from the following constrained inequalities

$$\mathbf{A}_iW+W\mathbf{A}_i^T+W\begin{bmatrix} Q^{1/2} \\ -Q^{1/2} \end{bmatrix} I\begin{bmatrix} Q^{1/2} & -Q^{1/2} \end{bmatrix}W-\mathbf{B}R^{-1}\mathbf{B}^T\leq 0, i=1,\ldots,L \tag{9.26}$$

or equivalently,

$$\begin{bmatrix} \mathbf{A}_iW+W\mathbf{A}_i^T-\mathbf{B}R^{-1}\mathbf{B}^T & W\begin{bmatrix} Q^{1/2} \\ -Q^{1/2} \end{bmatrix} \\ \begin{bmatrix} Q^{1/2} & -Q^{1/2} \end{bmatrix}W & -I \end{bmatrix}\leq 0, i=1,\ldots,L \tag{9.27}$$

which is equivalent to (9.25) with $\rho^2=\infty$.

According to the analysis above, the robust model matching control of innate immune system via fuzzy interpolation method is summarized as follows.

Design Procedure:

1) Give a desired reference model in (9.4) of immune system.
2) Select membership functions and construct fuzzy plant rules in (9.17).

3) Give weighting matrices Q and R in (9.5).
4) Solve the LMI-constrained optimization in (9.25) to obtain W (thus $P=W^{-1}$ can also be obtained) and ρ_0^2.
5) Construct the controller under the worst-case disturbance in (9.21).

Remark 9.3: The software packages such as Robust Control Toolbox in Matlab can be employed to solve the LMI-constrained optimization problem in (25) easily.

9.1.5. Computational Simulation Examples

Example 9.1: We consider the innate immune system in (9.1) and in Figure 9.1. The values of the parameters are in the Table 9.1. The environmental disturbances w_1~w_4 are unknown but bounded signals. Under infectious situation, the microbes infect the organ not only by an initial concentration at the beginning but also by the continuous pathogens input. For the convenience of computer simulation, suppose the continuous pathogens input to the immune system is viewed as an environmental disturbance w_1 shown in Figure 9.3. For the convenience of simulation, w_2~w_4 are assumed zero mean white noises with standard deviations all equal to 2. The stochastic noises of immune systems are mainly due to measurement errors, modeling errors and process noises [46]. The dynamic model of innate immune system under exogenous pathogens input and environmental disturbances is controlled by a combined therapeutic control shown as follows [30].

$$\begin{aligned}
\dot{x}_1 &= (1-x_3)x_1 - u_1 + 3w_1 \\
\dot{x}_2 &= a_{21}(x_4)3x_1x_3 - (x_2-2) + u_2 + 3w_2 \\
\dot{x}_3 &= x_2 - (1.5+x_1)x_3 + u_3 + 3w_3 \\
\dot{x}_4 &= 0.5x_1 - x_4 - u_4 + 3w_4 \\
a_{21}(x_4) &= \begin{cases} \cos(\pi\, x_4), & 0 \le x_4 \le 1/2 \\ 0, & 1/2 \le x_4 \end{cases}
\end{aligned} \tag{9.28}$$

with the set of the initial condition $x(0)=[x_1(0)\ x_2(0)\ x_3(0)\ x_4(0)]^T=[3\ 3.1\ 1\ 0.98]^T$.

In this example, therapeutic controls u_1~u_4 are combined to enhance the immune system. Our reference model design objective is that system matrix A_r and $r(t)$ should be specified beforehand so that its transient responses and steady state of reference system for innate immune response system are desired. If the real parts of eigenvalues of A_r are more negative (i.e., more robust stable), the tracking system will be more robust to environmental disturbances. After some numerical simulations for clinical treatment, the desired reference signals are obtained by the following reference model. (see Figure 9.4)

$$\dot{x}_r(t) = \begin{bmatrix} -12 & 0 & 0 & 0 \\ 0 & -2.3 & 0 & 0 \\ 0 & 0 & -10 & 0 \\ 0 & 0 & 0 & -3 \end{bmatrix} \cdot x_r(t) + B_r \cdot u_{step}(t) \tag{9.29}$$

where $B_r=[0\ 4.6\ 13.333\ 0]^T$ and $u_{step}(t)$ is the unit step function.

From the investigation of the uncontrolled innate immune response (lethal case) in Figure 9.5, the pathogen concentration is increasing rapidly and causes organ failure. We try to administrate a treatment after a period of pathogens infection to enhance the immune system. The cutting line (black solid line) in Figure 9.5 is a proper time to take drugs. Suppose the set of the initial condition of the desired reference model is about $x_r(0)=[2.9\ 3.2\ 1.1\ 0.9]^T$. The time response of the desired reference model in (9.29) is shown in Figure 9.4.

Table 9.1. Model parameters of dynamic innate immune system in (9.28) [1, 47]

Parameter	z	Value	Description
a_{11}		1	Pathogens reproduction rate coefficient
a_{12}		1	The suppression by pathogens coefficient
a_{22}		3	Immune reactivity coefficient
a_{23}		1	The mean immune cell production rate coefficient
x_2		2	The steady-state concentration of immune cells
a_{31}		1	Antibodies production rate coefficient
a_{32}		1.5	The antibody mortality coefficient
a_{33}		0.5	The rate of antibodies suppress pathogens
a_{41}		0.5	The organ damage depends on the pathogens damage possibilities coefficient
a_{42}		1	Organ recovery rate
b_1		−1	Pathogen killer's agent coefficient
b_2		1	Immune cell enhancer coefficient
b_3		1	Antibody enhancer coefficient
b_4		−1	Organ health enhancer coefficient

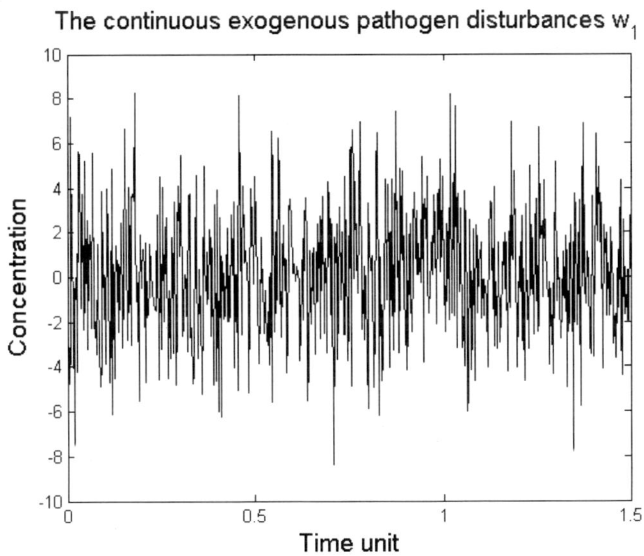

Figure 9.3. The continuous exogenous pathogens intrusion in (9.28).

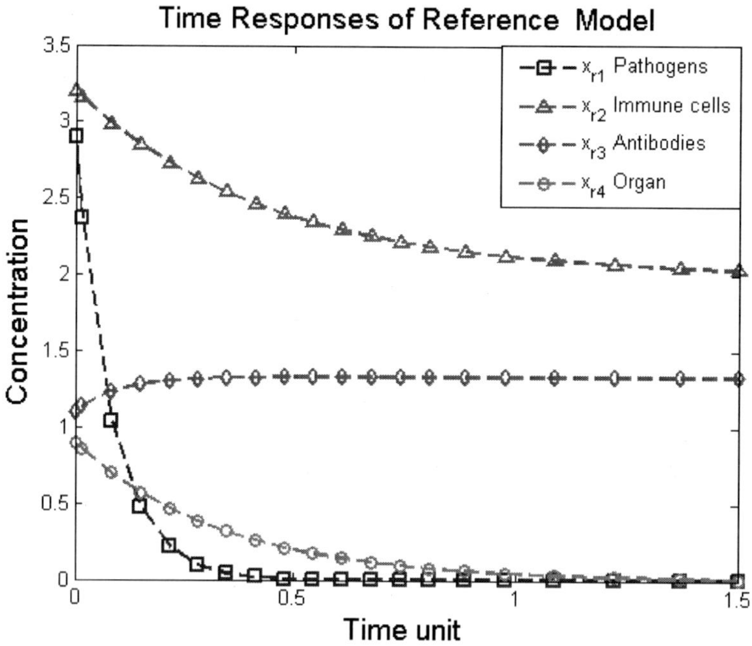

Figure 9.4. The desired reference model with four desired states in (9.29): pathogens (x_{r1}, blue, dashed square line), immune cells (x_{r2}, green, dashed triangle line), antibodies (x_{r3}, red, dashed diamond line) and organ (x_{r4}, magenta, dashed circle line). The initial conditions of the reference model are $x_r(0)$=[2.9 3.2 1.1 0.9]^T.

To minimize the design effort and complexity for this nonlinear innate immune system in (9.28), we employ the T-S fuzzy model to construct fuzzy rules to approximate the nonlinear innate immune system with the innate immune system's state variables as premise variables in the following [42].

Rule *i*:

$$\begin{aligned}&\text{If } x_1(t) \text{ is } F_{i1}, x_2(t) \text{ is } F_{i2}, x_3(t) \text{ is } F_{i3}, \text{ and } x_4(t) \text{ is } F_{i4}\\ &\text{then } \dot{\bar{x}} = \mathbf{A}_i\bar{x}(t)+\mathbf{B}u(t)+Cv(t), \quad i=1,\dots L\end{aligned} \tag{9.30}$$

where $\bar{x}=[x_1\ x_2\ x_3\ x_4\ x_{r1}\ x_{r2}\ x_{r3}\ x_{r4}]^T$, $u=[u_1\ u_2\ u_3\ u_4]^T$, $v=[w_1\ w_2\ w_3\ w_4\ r_1\ r_2\ r_3\ r_4]^T$, the parameters $\mathbf{A}_i$, $\mathbf{B}$, C are described in the 9.5.3 and the number of the fuzzy rules is L=16. To construct the fuzzy model in (9.30), we need to find the operating points of the innate immune response. Suppose the operating points for x_1 are at $\bar{x}_{11}$=0, and $\bar{x}_{12}$=4. Similarly, the operating points of x_2, x_3, and x_4 are at $\bar{x}_{21}$=0, $\bar{x}_{22}$=10, $\bar{x}_{31}$=0, $\bar{x}_{32}$=5, $\bar{x}_{41}$=0, and $\bar{x}_{42}$=1, respectively. For the convenience of design, triangle type membership functions are taken for Rule 1 through Rule 16. We create two triangle type membership functions for each state at these operating points (see Figure 9.6). In order to accomplish the robust matching performance, we should tune up a set of the weighting matrices Q and R of the cost function in (9.11) as follows

$$Q=\begin{bmatrix}1&0&0&0\\0&1&0&0\\0&0&1&0\\0&0&0&1\end{bmatrix}, R=\begin{bmatrix}0.003&0&0&0\\0&0.003&0&0\\0&0&0.003&0\\0&0&0&0.003\end{bmatrix}$$

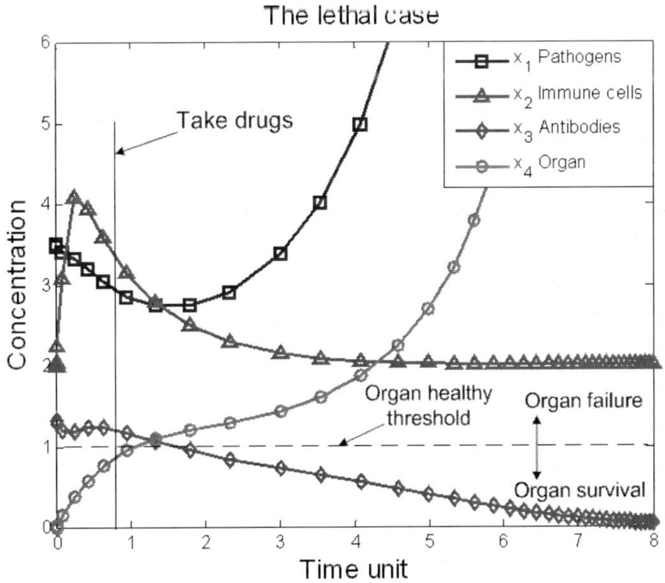

Figure 9.5. The uncontrolled immune responses (lethal case) in (9.28) are shown to increase the level of pathogen concentration at the beginning of the time period. In this case, we try to administrate a treatment after a short period of pathogens infection. The cutting line (black solid line) is an optimal time point to give drugs. The organ will survive or fail based on the organ health threshold (horizontal dashed line) [x_4<1: survival, x_4≥1: **failure**].

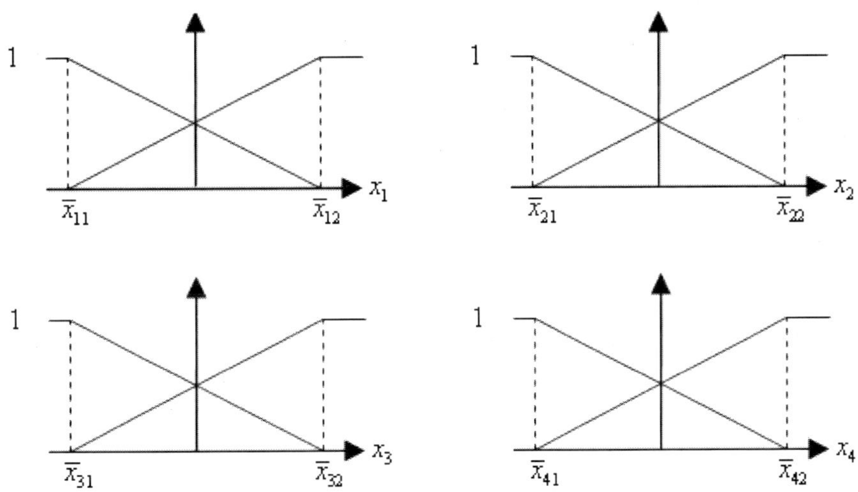

Figure 9.6. Membership functions for four states x_1, x_2, x_3, and x_4.

After specifying the desired reference model, we need to solve the constrained optimization in (9.24) for the robust minimax control in (9.21) by employing Matlab Robust Control Toolbox. Finally, we obtain a minimum attenuation level ρ_0^2=0.98 and a common positive definite symmetric matrix P for equation (9.22) as follows

$$P = \begin{bmatrix} 0.43313 & 0 & 0 & 0 & -0.43313 & 0 & 0 & 0 \\ 0 & 0.56172 & 0 & 0 & 0 & -0.56172 & 0 & 0 \\ 0 & 0 & 0.42678 & 0 & 0 & 0 & -0.42678 & 0 \\ 0 & 0 & 0 & 0.28482 & 0 & 0 & 0 & -0.28482 \\ -0.43313 & 0 & 0 & 0 & 0.50151 & 0 & 0 & 0 \\ 0 & -0.56172 & 0 & 0 & 0 & 0.62738 & 0 & 0 \\ 0 & 0 & -0.42678 & 0 & 0 & 0 & 0.49526 & 0 \\ 0 & 0 & 0 & -0.28482 & 0 & 0 & 0 & 0.34567 \end{bmatrix}$$

Figure 9.7 and Figure 9.8 present the simulation results for the robust model matching control. Figure 9.7 shows the responses of the controlled immune system by minimax model matching control with the concentrations of the pathogens x_1, immune cells x_2, antibodies x_3 and organ index x_4 to track the desired reference states x_{r1}, x_{r2}, x_{r3}, and x_{r4}, respectively. From the simulation results, the tracking performance of the robust model matching control via T-S fuzzy interpolation is quite satisfactory. The Figure 9.8 shows the four combined therapeutic control signals. Obviously, from Figure 9.8 and Figure 9.9, it is seen that the effect of stochastic external disturbances on the reference model tracking of immune system is attenuated significantly by the proposed robust therapeutic control design.

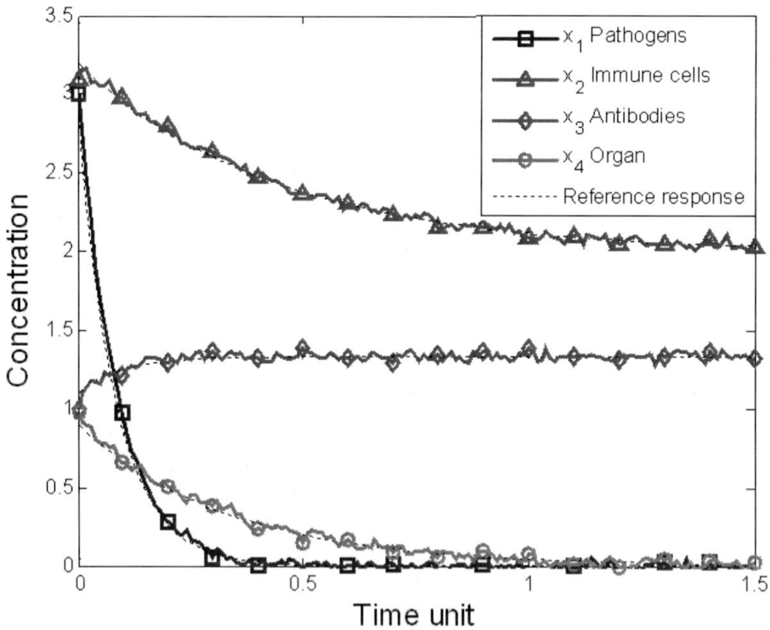

Figure 9.7. The tracking of innate immune system to the desired reference model by the robust minimax matching control under the continuous exogenous pathogens and environmental disturbances.

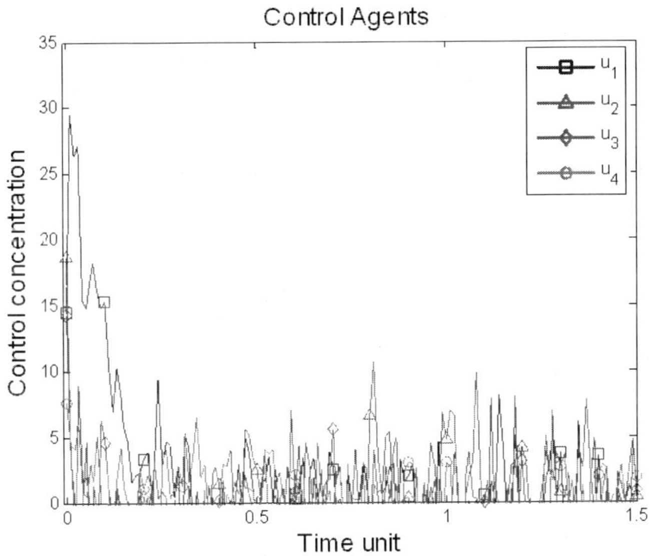

Figure 9.8. The minimax controls in the simulation example. The drug controls u_1 (blue, solid square line) for pathogens, u_2 for immune cells (green, solid triangle line), u_3 for antibodies (red, solid diamond line) and u_4 for organ (magenta, solid circle line).

Example 9.2: It is sometime unavoidable for drugs to have adverse side effects. Limiting the impact of side effects should be considered in this therapeutic control design. Suppose the pathogen killer's agent u_1 (i.e., drugs or antibiotics) also degrades organ health. The modified model of equation (9.28) of the organ state x_4 becomes

$$\dot{x}_4 = 0.5x_1 - x_4 - u_4 + 0.2u_1 + 3w_4 \tag{9.31}$$

i.e., an increase of 0.2 side effect of the pathogen killer's agent u_1 on the organ state x_4. The detailed design of fuzzy approximation is similar to the previous procedure and the parameters are the same except **B**, i.e.,

$$\mathbf{B}^T = \begin{bmatrix} -1 & 0 & 0 & 0.2 & 0 & 0 & 0 & 0 \\ 0 & 1 & 0 & 0 & 0 & 0 & 0 & 0 \\ 0 & 0 & 1 & 0 & 0 & 0 & 0 & 0 \\ 0 & 0 & 0 & -1 & 0 & 0 & 0 & 0 \end{bmatrix}$$

In this design case, we choose

$$Q = \begin{bmatrix} 1 & 0 & 0 & 0 \\ 0 & 1 & 0 & 0 \\ 0 & 0 & 1 & 0 \\ 0 & 0 & 0 & 2 \end{bmatrix}, R = \begin{bmatrix} 0.002 & 0 & 0 & 0 \\ 0 & 0.002 & 0 & 0 \\ 0 & 0 & 0.002 & 0 \\ 0 & 0 & 0 & 0.002 \end{bmatrix}$$

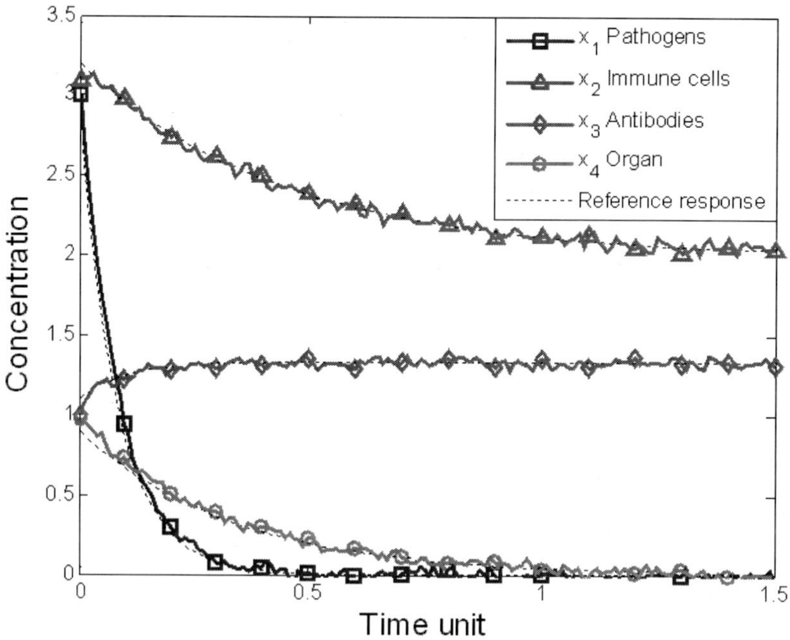

Figure 9.9. The tracking of innate immune system to the desired reference model by the robust minimax matching control under the continuous exogenous pathogens, environmental disturbances and pathogen killer's agent side effects.

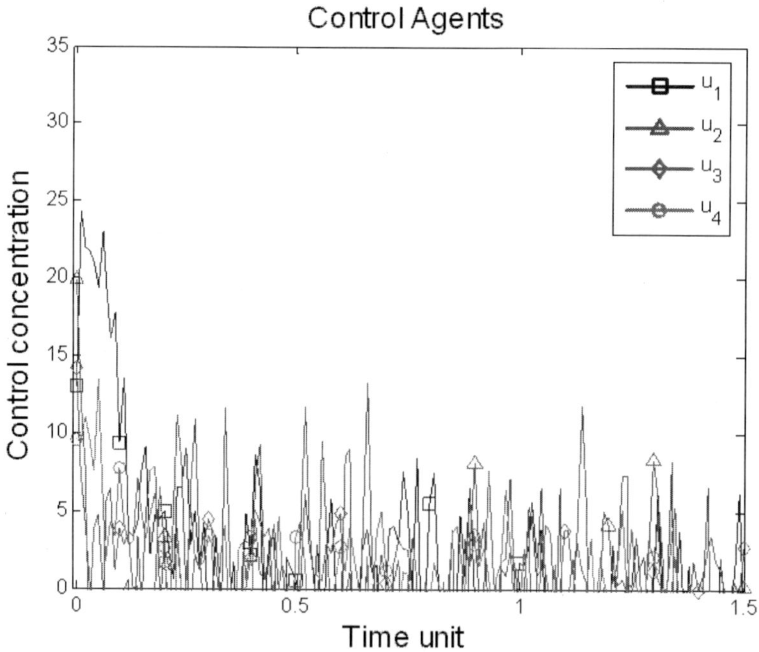

Figure 9.10. The minimax controls with considering the pathogen killer's agent side effects. The drug controls u_1 (blue, solid square line) for pathogens, u_2 for immune cells (green, solid triangle line), u_3 for antibodies (red, solid diamond line) and u_4 for organ (magenta, solid circle line).

Since the organ health state suffers from the impact of the side effects, we should increase the weighting q_{44} to impose a penalty on the organ health tracking error. Further, we relax the weighting on the control signals to improve the tracking efficiency. By the same design procedure, we obtain a minimum attenuation level $\rho_0^2=0.9$ and a common positive definite symmetric matrix P as follows

$$P=\begin{bmatrix} 0.29932 & 0 & 0 & 0 & -0.29932 & 0 & 0 & 0 \\ 0 & 0.39859 & 0 & 0 & 0 & -0.39859 & 0 & 0 \\ 0 & 0 & 0.28408 & 0 & 0 & 0 & -0.28408 & 0 \\ 0 & 0 & 0 & 0.30502 & 0 & 0 & 0 & -0.30502 \\ -0.29932 & 0 & 0 & 0 & 0.35217 & 0 & 0 & 0 \\ 0 & -0.39859 & 0 & 0 & 0 & 0.43261 & 0 & 0 \\ 0 & 0 & -0.28408 & 0 & 0 & 0 & 0.33404 & 0 \\ 0 & 0 & 0 & -0.30502 & 0 & 0 & 0 & 0.33088 \end{bmatrix}$$

The control design with considered drug's side effects is simulated in Figure 9.9 and Figure 9.10. Figure 9.9 shows the responses of the controlled immune system by the proposed minimax model matching control with undergoing the side effects of pathogen killer's agent u_1. From the simulation result, the organ health state x_4 tracks the desired response leisurely at the beginning but it totally matches the desired response at the end. The Figure 9.10 shows the four combined therapeutic control signals with the side effects of pathogen killer's agent u_1. At the beginning, the drug of organ health enhancer u_4 needs to increase to minimize the side effects from the pathogen killer's agent u_1.

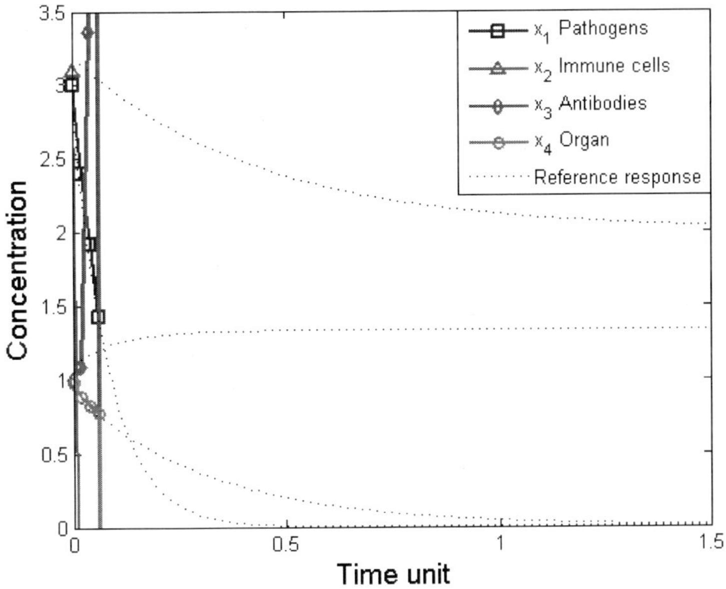

Figure 9.11. In the case of conventional optimal matching control (i.e., $\rho=\infty$ in (9.5)), since the effect of continuous exogenous pathogens intrusion and environmental disturbances is not considered in the design procedure, the states of innate immune system overshoot and diverge and cannot track the desired reference responses.

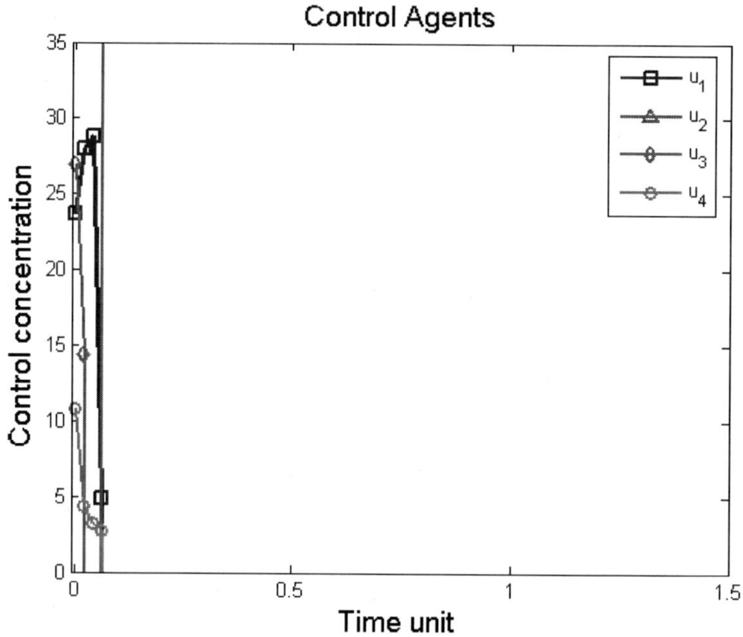

Figure 9.12. The controls of conventional optimal matching control (i.e., ρ=∞ in (9.5)) without considering the effect of continuous exogenous pathogens intrusion and environmental disturbances in the design procedure, the drug controls u_1 (blue, solid square line) for pathogens, u_2 for immune cells (green, solid triangle line), u_3 for antibodies (red, solid diamond line) and u_4 for organ (magenta, solid circle line) are all overshooting and divergent.

9.1.6. Discussion

From the simulation results (Figure 9.7 and Figure 9.8), it is shown that the innate immune system under the continuous intrusion of exogenous pathogens and the corruption of environmental disturbances can be controlled by a robust model matching control design to achieve the desired time response. If we consider the conventional optimal control in (9.3), i.e., the effect of the environmental disturbances is not included in the cost function; the optimal tracking control problem is equivalent to letting $\rho^2=\infty$ in (9.5) and (9.22) [41]. From the simulation results (Figure 9.10, Figure 9.11, and Figure 9.12), the four states of optimal tracking of the immune system are overshooting and diverging without tracking the desired immune time response. Obviously, exogenous pathogens and the environmental disturbances have deteriorated the optimal tracking performance and therefore their effects should be considered in the robust control design procedure. In the situation, the proposed robust matching control design is necessary to achieve a desired time response.

Pathogen killer's agent u_1 has toxic side effect that can damage the organ during the treatment. So, we should consider its detrimental impacts into the innate immune system in the model in (9.31). In the robust model matching control design procedure, the element of weighting matrix q_{44} should be increased to impose a penalty on the tracking error of organ health state to eliminate the side effect of u_1 on x_4. From the simulation results (Figure 9.9 and Figure 9.10), the organ health state x_4 tracks the desired response slowly at the beginning but it almost matches the desired response at the end. It is also observed that the innate immune system can be controlled by a combined therapies design to achieve the desired time response under the continuous intrusion of exogenous pathogens, environmental disturbances and side effects.

The combined therapies design is an important issue for all human diseases [48]. For a long period, the treatment of inflammatory skin diseases such as psoriasis, contact dermatitis and atopic dermatitis has included agents that alleviate symptoms, but these agents have not been aimed at any specific molecular targets involved in the pathogenesis of the disease. Insights into this immune mechanism may facilitate the development of combination therapies that take advantage of the robust model matching design, with the aim of achieving higher efficacy at a lower drug dosage and with a reduced probability of side effect on organ health. The proposed robust model matching design has used four control variables, i.e., pathogen killer's agent u_1, immune cell enhancer u_2, antibody enhancer u_3, and health enhancer u_4, to achieve a minimax matching performance and to efficiently attenuate the effect of exogenous pathogens and environmental disturbances on the immune system.

In this chapter, the model of innate immune dynamic system is taken from the literature, which still needs to compare quantitatively with empirical evidence in practical application. For practical implementation, accurate biodynamic models are required for treatment application. However, model identification is not the topic of this chapter. Furthermore, we have made an assumption that the four states (x_1~x_4) of the concentrations or indices can be measured accurately by the medical equipment. With these detectable signals, we can solve these dynamic game problems for robust tracking control design of innate immune system to obtain the drug administration values in real time through medical instrument readout. If measurement is corrupted by noises in the measurement process, some filter designs should be employed to attenuate these noises to estimate the state variables for control in [49]. Nevertheless, the implementation of filter will increase the complexity of the design problem [50]. Since the proposed robust model matching control design can provide an efficient way to create a real time therapeutic regime to protect suspected patients from the pathogens infection, in the future, we will focus on applications of robust control design to therapy and drug design incorporating with nanotechnology and metabolic engineering scheme.

As a comparison, the similarity and difference between the minimax control method and the optimal control method [34] are given in the following. We all want to minimize the total cost of design, i.e., the weighted sum of the damage caused by pathogens and the cost paid by specific immune cells. On the other hand, the differences are given in the following: (i) A desired model reference is given to be optimally tracked for the enhancement of the immune system in the proposed method and Shudo and Iwasa have designed an optimal control to minimize the cost. (ii) The effect of external disturbances has not been considered in Shudo and Iwasa, but the worst-case effect of external disturbances has been considered and minimized in the proposed design via the minimax Nash game method. (iii) Fuzzy interpolation technique is employed by the proposed method so that linear matrix inequalities (LMIs) technique is used to efficiently solve the nonlinear minimax optimization problem in the proposed design procedure. However, Shudo and Iwasa have used a dynamic programming method to derive an optimal schedule to solve the nonlinear optimization problem of host defense, immune memory and post-infection pathogen levels in mammals.

Robustness is a significant property that allows the innate immune system to maintain its function despite exogenous pathogens, environmental disturbances and system uncertainties.

Based on dynamic game theory, the robust tracking control is formulated as a minimax problem for an innate immune system to achieve a desired time response prescribed prior under environmental disturbances, unknown initial conditions and side effects. In general, the robust model matching control design for innate immune system needs to solve nonlinear Hamilton-Jacobi inequality (HJI), which is generally difficult to solve for this control design. Based on the proposed fuzzy dynamic game scheme, the design of nonlinear dynamic robust matching control problem for innate immune system is transformed to solve a set of equivalent linear dynamic game problem. Such transformation can then allow us an easier approach by solving a LMI-constrained optimization problem for robust minimax control design. With the help of the Robust Control Toolbox in Matlab instead of the HJI, we could solve these linear dynamic game problems for robust matching control of innate immune system efficiently. From the *in silico* simulation examples, the proposed minimax match control of immune system could track the prescribed reference time response robustly, which may lead to potential application in therapeutic drug design for a desired immune response during an infection episode.

9.2. Robust Synthetic Biology Design: Stochastic Game Theory Approach

9.2.1. Introduction of Robust Synthetic Biology Design

Synthetic biology will revolutionize how we conceptualize and approach the engineering of biological systems. The vision and applications of this emerging field will influence many other scientific and engineering disciplines, as well as affect various aspects of daily life and society [51]. Synthetic biology builds living machines from the off-the-shelf chemical ingredients, utilizing many of the same strategies that electrical engineers employ to make computer chips [52]. The main goal of the nascent field of synthetic biology is to design and construct biological systems with the desired behavior [51-60]. By a set of powerful techniques for the automated synthesis of DNA molecules and their assembly into genes and microbial genomes, synthetic biology envisions the redesign of natural biological systems for greater efficiency as well as the construction of functional "genetic circuit" and metabolic pathways for practical purposes [51, 58, 61-66]. Synthetic biology is foreseen to have important applications in biotechnology and medicine [51].

Though the engineering of networks of inter-regulating genes, so-called synthetic gene networks, has demonstrated the feasibility of synthetic biology [63], the design of gene networks is still a difficult problem and most of the newly designed gene networks cannot work properly. These design failures are mainly due to intrinsic perturbations such as gene expression noises, splicing, mutation, uncertain initial states and disturbances such as changing extra-cellular environments, and interactions with cellular context. Therefore, how to design a robust synthetic gene network, which could tolerate uncertain initial conditions, attenuate the effect of all disturbances and function properly on the host cell, will be an important topic for synthetic biology [51-60, 67-73]. Previously, sensitivity analysis has been used for analysis of the dynamic properties of gene networks either in qualitative simulations of coarse-grained models or in extensive numerical simulations of nonlinear differential equation models or stochastic dynamic models [74, 75]. For applications in synthetic biology, these approaches are not satisfying. The local sensitivity analysis can provide only a partial description of all possible behaviors of a nonlinear gene network. In particular, it cannot guarantee that a synthetic gene network behaves as expected for all uncertain initial conditions and disturbances. Moreover, obtaining all convergences of states and parameters by extensive numerical simulations quickly becomes computationally intractable when the size of the synthetic network grows [67].

An approach has recently been developed using semidefinite programming to partition the parameter spaces of polynomial differential equation models into so-called feasible and infeasible regions [76]. Following that, a robustness analysis and tuning approach of synthetic networks was proposed to provide a means to assess the robustness of the expected behavior of a synthetic gene network in spite of parameter variations [67]. This approach has the capability to search for parameter sets for which a given property is satisfied through a publicly available tool called RoVerGeNe. Several gene circuit design networks have been introduced to implement or delete some circuits from an existing gene network so as to modify its structure for improving its robust stability or filtering ability [77-79]. However, robust synthetic gene network design is a different topic. It needs to design a complete man-made gene network to be inserted into a host cell. Therefore, the synthetic gene networks should be designed with enough robustness to tolerate uncertain initial conditions and to resist all possible disturbances on the host cell so that they can function properly in a desired steady state. This is a so-called robust regulation design that can achieve a desired steady state of synthetic gene networks despite uncertain initial conditions and disturbances on the host cell.

In this chapter, a robust regulation design of synthetic gene network is proposed to achieve a desired steady state in spite of uncertain initial conditions, parameter variations and disturbances on the host cell. Because most information of these uncertain factors on the host cell is unavailable, in order to attenuate their detrimental effects, their worst-case effect should be considered by the designer in the regulation design procedure from the worst regulation error perspective. The worst-case effect of all possible initial conditions and disturbances on the regulation error to a desired steady state is minimized for the robust synthetic gene networks, i.e., the proposed robust synthetic gene network is designed from the minimax regulation error perspective. The minimax design scheme is a simple robust synthetic gene network design method because we do not need the precise information of the initial conditions, parameter variations and disturbances on the host cell, which are not easy to measure in the design procedure. This minimax regulation design problem for robust synthetic gene networks could be transformed to an equivalent dynamic game problem [36, 44]. Dynamic game methods have been widely applied to many fields of robust engineering design problems with external disturbances. Recently, the application of dynamic game theory has been used for robust model matching control of immune systems under environmental disturbances [42]. A robust drug administration (control input) is designed to obtain a prescribed immune response under uncertain initial states and environmental disturbances. In this chapter, the stochastic game theory will be used for robust synthetic gene network design so that the engineered gene network can work properly under uncertain initial conditions and environmental disturbances on the host cell. The uncertain initial states and disturbances are considered as a player doing his best to deteriorate the regulation performance from the worst-case point of view, while the system parameters to be designed are considered as another player optimizing the regulation performance under the worst-case deterioration of a former player. Since the synthetic gene networks are highly nonlinear, it is not easy to solve the robust synthetic gene network design problem directly by the nonlinear dynamic game method directly. Recently, fuzzy systems have been employed to efficiently approximate nonlinear dynamic systems to solve the nonlinear control problem [37-40, 45, 80]. A Takagi-Sugeno (T-S) fuzzy model [45] is proposed to interpolate several linearized genetic networks at different operating points to approximate the nonlinear gene network via some smooth fuzzy membership functions. Then with the help of the fuzzy approximation method, a fuzzy dynamic game scheme [44] is developed so that the minimax regulation design of robust synthetic gene networks could be easily solved by the techniques of the linear dynamic game theory, which can be subsequently solved by a constrained optimization scheme via the linear matrix inequality (LMI) technique [41] that can be efficiently solved by the Robust Control Toolbox in Matlab. Because the fuzzy model can approximate any nonlinear system, the proposed robust regulation design method developed from the fuzzy stochastic game theory can be applied to the robust regulation design problem of any synthetic gene network that can be interpolated by a T-S fuzzy model. For comparison, the conventional optimal regulation design method without considering the effect of disturbances is also proposed for the synthetic gene network. Because the effect of disturbances is not attenuated efficiently, the optimal regulation design method of synthetic gene networks is much influenced by the disturbances on the host cell. Finally, an *in silico* example is given to illustrate the design procedure and to confirm the efficiency and efficacy of the proposed minimax regulation design method for robust synthetic gene networks.

9.2.2. Robust Synthetic Gene Network via Dynamic Game

First, for the convenience of problem description, a simple design example of a four-gene network in [67] is provided to give an overview of the design problem of robust synthetic gene networks. A more general design problem of robust synthetic gene networks will be given in the sequel. Let us consider a robust regulation design problem of a cascade loop of transcriptional inhibitions built in *E. coli.* [81]. The synthetic gene network is represented in Figure 9.13. It consists of four genes: *tetR*, *lacI*, *cI* and *eyfp* that code respectively three repressor proteins, TetR, LacI and CI, and the fluorescent potein EYFP (enhanced yellow fluorescent protein) [67]. aTc (anhydrotetracycline) is the input to the system. The fluorescence of the system, due to the protein EYFP, is the measured output. The protein CI inhibits gene *eyfp*. The protein TetR inhibits gene *lacI*. The protein LacI inhibits gene *cI*. The regulatory dynamic equations of the synthetic transcriptional cascade in Figure 9.13 are given as follows [67].

$$\begin{aligned}
\dot{x}_{tetR} &= k_{tetR,0} - \gamma_{tetR} x_{tetR} + w_1 \\
\dot{x}_{lacI} &= k_{lacI,0} + k_{lacI}(r_{lacI}(x_{tetR}) + a_{lacI}(u_{aTc}) - r_{lacI}(x_{tetR}) a_{lacI}(u_{aTc})) - \gamma_{lacI} x_{lacI} + w_2 \\
\dot{x}_{cI} &= k_{cI,0} + k_{cI} r_{cI}(x_{lacI}) - \gamma_{cI} x_{cI} + w_3 \\
\dot{x}_{eyfp} &= k_{eyfp,0} + k_{eyfp} r_{eyfp}(x_{cI}) - \gamma_{eyfp} x_{eyfp} + w_4
\end{aligned} \tag{9.32}$$

with the uncertain initial conditions $x_{tetR}(0)$, $x_{lacI}(0)$, $x_{cI}(0)$ and $x_{eyfp}(0)$ in the host cell. $k_{tetR,0}$, $k_{lacI,0}$, $k_{cI,0}$ and $k_{eyfp,0}$ are basal production rates of the corresponding proteins, which are assumed to be given constants. k_{lacI}, k_{cI} and k_{eyfp} are the production rate parameters while γ_{tetR}, γ_{lacI}, γ_{cI} and γ_{eyfp} are decay rate parameters of the corresponding proteins. The regulatory functions r_{lacI}, r_{cI} and r_{eyfp} are the Hill functions for repressors and a_{lacI} for an activator.

The Hill function can be derived from considering the equilibrium binding of the transcription factor to its site on the promoter region. For a repressor, Hill function is an *S*-shaped curve which can be described in the form $r(x)=\beta_r/[1+(x/K_r)^n]$. β_r is the maximal expression level of promoter. K_r is the repression coefficient. The Hill coefficient n governs the steepness of the input function. For an activator, Hill function can be described in the form $a(x)=\beta_a x^n/(K_a^n+x^n)$. β_a is the maximal expression level of promoter. K_a is the activation coefficient. n determines the steepness of the input function [54]. w_1, w_2, w_3 and w_4 are the disturbances of the synthetic gene network, which denote the total of environmental noises, modeling residuals, intrinsic parameter fluctuations in the host cell. Therefore, w_i, i=1~4 are assumed uncertain but bounded disturbances. The synthetic gene network design is to specify k_{lacI}, k_{cI}, k_{eyfp} and γ_{tetR}, γ_{lacI}, γ_{cI}, γ_{eyfp} such that the system states x_{tetR}, x_{lacI}, x_{cI} and x_{eyfp} can approach the desired states x_{d1}, x_{d2}, x_{d3} and x_{d4}, respectively, in spite of uncertain initial conditions and disturbances.

If a synthetic gene network consists of n genes, then equation (9.32) can be extended to the following n-gene network dynamics.

$$\dot{x} = k_0 + f(x,k,\gamma) + g(u) + w, \;\; x(0) = x_0 \tag{9.33}$$

where the state vector x denotes the concentrations of proteins in the synthetic gene network. k_0 denotes the vector of basal production rates of the corresponding proteins. $f(x,k,\gamma)$ denotes the regulation vector of synthetic gene network, which is the function of production rate parameters k and decay rate parameters γ to be designed. $g(u)$ denotes the input function to the synthetic gene network. w denotes the vector of stochastic disturbances on the host cell, whose statistics may be unavailable. The initial condition x_0 is assumed stochastic with unknown covariance. The robust synthetic gene network design is to select parameters k and γ from feasible ranges so that the state vector x can approach a desired state vector x_d in spite of uncertain initial condition $x(0)$ and disturbances w on the host cell. i.e., $x \rightarrow x_d$ at the steady state despite uncertain $x(0)$ and w. This is a robust regulation problem of synthetic gene networks, i.e., the state vector x of synthetic gene networks is robustly regulated to x_d in the host cell.

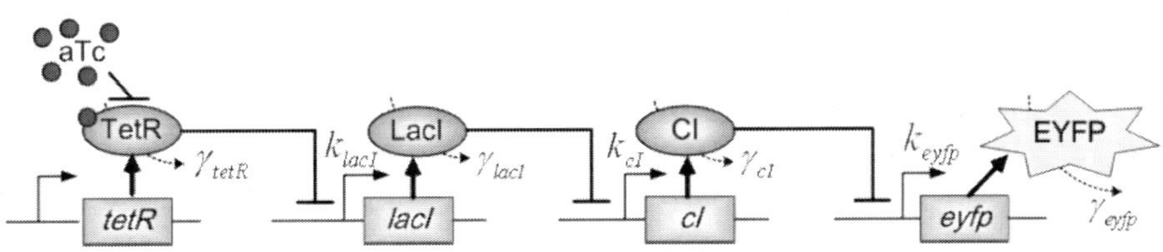

Figure 9.13. Synthetic transcription cascade loop *in silico* design example. aTc represses TetR, TetR represses *lacI*, LacI represses *cI*, CI represses *eyfp*. aTc is the system input and the fluorescent protein EYFP is the output.

Let us denote the regulation error as

$$\tilde{x} = x - x_d \tag{9.34}$$

Then the regulation error dynamic system is given by

$$\dot{\tilde{x}} = f(\tilde{x} + x_d, k, \gamma) + v, \quad \tilde{x}(0) = \tilde{x}_0 \tag{9.35}$$

where $v=k_0+g(u)+w$ denotes the total uncertain disturbance in the regulation error system because these terms always fluctuate in the host cell and are not easily measured correctly. Because of the uncertainty of v and $\tilde{x}(0)$, the minimax regulation design method is an efficient but simple design scheme for robust synthetic gene network. The uncertainty of disturbance v and initial condition $\tilde{x}(0)$ in the following minimax design can be considered as a player maximizing their effects on the regulation error in the following robust design problem of synthetic gene networks [36, 42, 44].

$$\min_{\substack{k\in[k_1,k_2] \\ \gamma\in[\gamma_1,\gamma_2]}} \max_{\tilde{x}(0),v} \frac{E\left[\int_0^{t_f} \tilde{x}^T Q \tilde{x} dt\right]}{E\left[\int_0^{t_f} v^T v dt + \tilde{x}^T(0)\tilde{x}(0)\right]} \tag{9.36}$$

where Q is the weighting matrix. In general, Q is a diagonal weighting matrix with $Q=\text{diag}([q_{11}, q_{22}, \ldots, q_{nn}])$ to denote the punishment on regulation error. If only the last state x_n is required to be regulated to achieve the desired steady state x_{dn}, then we can let $q_{nn}=1$ and $q_{11}=q_{22}=\ldots=q_{n-1n-1}=0$. $[k_1,k_2]$ and $[\gamma_1,\gamma_2]$ denote the allowable ranges of production rate vector k and decay rate vector γ, respectively. The allowable ranges are determined by the engineering biotechnologies of synthetic biology. k and γ to be designed can be considered as another player minimizing the worst-case effect of $\tilde{x}(0)$ and v on the regulation error. If the disturbances v and initial condition $\tilde{x}(0)$ are deterministic, then the expectation operation $E[\ ]$ in (9.36) could be neglected.

The physical meaning of (9.36) is that the worst-case effect of uncertain $\tilde{x}(0)$ and v on the regulation error $\tilde{x}$ must be minimized from the mean energy perspective by k and γ, which are chosen from the allowable ranges. Therefore, for uncertain $\tilde{x}(0)$ and v, the robust synthetic gene network design is to solve the minimax problem in (9.36) subject to the regulation error dynamic system in (9.35). This is the so-called stochastic game problem in the robust synthetic gene network design [36, 82].

In general, it is not easy to solve the nonlinear stochastic game problem in (9.36) subject to (9.35) directly. It is always solved by a sub-minimax method. First, let the upper bound g^2 of (9.36) be [36, 42, 44]

$$\min_{\substack{k\in[k_1,k_2] \\ \gamma\in[\gamma_1,\gamma_2]}} \max_{\tilde{x}(0),v} \frac{E\left[\int_0^{t_f} \tilde{x}^T Q \tilde{x} dt\right]}{E\left[\int_0^{t_f} v^T v dt + \tilde{x}^T(0)\tilde{x}(0)\right]} \le g^2 \tag{9.37}$$

We will first solve the sub-minimax problem in (9.37) and then decrease the upper bound g^2 as much as possible to approach its minimax solution. In general, the minimax problem in (9.37) is equivalent to the following minimax problem [36, 44]

$$\min_{\substack{k\in[k_1,k_2]\\ \gamma\in[\gamma_1,\gamma_2]}} \max_{v} E\left[\int_0^{t_f}\left(\tilde{x}^T Q\tilde{x} - g^2 v^T v\right)dt\right] \le g^2 E\left[\tilde{x}^T(0)\tilde{x}(0)\right], \quad \forall \tilde{x}(0) \tag{9.38}$$

where g^2 is to be minimized because it is the upper bound in (9.37) and should be as small as possible to approach the minimax solution. Let us denote the cost function as

$$J(k,r,v) = E\left[\int_0^{t_f}\left(\tilde{x}^T Q\tilde{x} - g^2 v^T v\right)dt\right] \tag{9.39}$$

9.2.3. Minimax Design Results

Sub-minimax Design for Robust Synthetic Gene Networks

From the above analysis, the dynamic game problem in (9.37) or (9.38) is equivalent to finding the worst-case disturbance v^* which maximizes $J(k,\gamma,v)$ and then the minimax k^* and γ^* which minimize $J(k,\gamma,v^*)$ such that the minimax value $J(k^*,\gamma^*,v^*)$ is less than $g^2E[\tilde{x}^T(0)\tilde{x}(0)]$, i.e.,

$$\begin{aligned} J\left(k^*,\gamma^*,v^*\right) &= \min_{\substack{k\in[k_1,k_2]\\ \gamma\in[\gamma_1,\gamma_2]}} J\left(k,r,v^*\right) \\ &= \min_{\substack{k\in[k_1,k_2]\\ \gamma\in[\gamma_1,\gamma_2]}} \max_{v} J(k,\gamma,v) \le g^2 E\left[\tilde{x}^T(0)\tilde{x}(0)\right], \quad \forall \tilde{x}(0) \end{aligned} \tag{9.40}$$

Hence, if there exist k^*, γ^* and v^* such that the minimax design problem in (9.40) is solved, then they can satisfy the minimax performance of the robust synthetic gene network design in (9.37) as well. Therefore, the first step of robust synthetic gene network design is to solve the following dynamic game problem:

$$\min_{\substack{k\in[k_1,k_2]\\ \gamma\in[\gamma_1,\gamma_2]}} \max_{v} J\left(k,\gamma,v\right) \tag{9.41}$$

subject to the error dynamic equation in (9.35). Since $J(k^*,\gamma^*,v^*) \le g^2E[\tilde{x}^T(0)\tilde{x}(0)]$ according to (9.40) and g^2 is the upper bound of the game in (9.37), the sub-minimax has to make g^2 as small as possible, too.

From the above analysis, we obtain the following sub-minimax result for robust synthetic gene network design.

Proposition 9.3 [82]: The sub-minimax synthetic gene network design is equivalent to solving the following constrained optimization for k^* and γ^*,

$$\min_{\substack{k\in[k_1,k_2]\\ \gamma\in[\gamma_1,\gamma_2]}} g^2 \tag{9.42}$$

subject to the following Hamilton-Jacobi inequality (HJI)

$$\left(\frac{\partial V(\tilde{x})}{\partial \tilde{x}}\right)^T f(\tilde{x}+x_d,k,\gamma)+\tilde{x}^T Q\tilde{x}+\frac{1}{4g^2}\left(\frac{\partial V(\tilde{x})}{\partial \tilde{x}}\right)^T\left(\frac{\partial V(\tilde{x})}{\partial \tilde{x}}\right)<0 \tag{9.43}$$
$$E[V(\tilde{x}(0))]\le g^2 E[\tilde{x}^T(0)\tilde{x}(0)]$$

with $V(\tilde{x})>0$, and the worst-case disturbance is given by

$$v^* = \frac{1}{2g^2}\frac{\partial V(\tilde{x})}{\partial \tilde{x}} \tag{9.44}$$

Proof: see 9.5.4.

Remark 9.4

(1) From (9.37), g^2 is the upper bound of the game. In (9.42), we minimize the upper bound g^2 to achieve the sub-minimax solution for robust synthetic gene networks.
(2) The physical meaning of the constrained minimization in (9.42) and (9.43) is that we want to specify k^* and γ^* from the allowable parameter ranges such that the upper bound g^2 is as small as possible until no positive solution $V(\tilde{x})>0$ of HJI in (9.43) exists.
(3) At present, there exists no efficient analytic or numerical method to solve the HJI in (9.43) for nonlinear stochastic system control or filtering designs [83, 84].

Minimax Robust Synthetic Gene Networks via Fuzzy Interpolation Method

Because it is very difficult to solve the nonlinear HJI in (9.43), no simple approach is available for solving the constrained optimization problem in (9.42) for the minimax robust synthetic gene network design problem. Recently, the Takagi-Sugeno (T-S) fuzzy model has been widely employed [37, 38, 45, 80] to approximate the nonlinear system via interpolating several linearized systems at different operating points so that the nonlinear Nash stochastic problem could be transformed to a fuzzy stochastic game problem [44]. By using such approach, the HJI in (9.43) can be replaced by a set of linear matrix inequalities (LMIs). In this situation, the nonlinear stochastic game problem in (9.41) could be easily solved by the fuzzy dynamic method for the robust design of sub-minimax design problem.

Suppose the nonlinear system in (9.35) could be approximated by a T-S fuzzy system [45]. The T-S fuzzy model is a piecewise interpolation of several linearized models through fuzzy membership functions. The fuzzy model is described by fuzzy *if-then* rules and will be employed to deal with the nonlinear stochastic game problem for robust synthetic gene network design under uncertain initial conditions and disturbances. The ith rule of fuzzy model for nonlinear systems in (9.35) is of the following form [37, 45].

Rule *i*:

$$\begin{aligned}&\text{If } \tilde{x}_1(t) \text{ is } F_{i1},\ldots, \text{ and } \tilde{x}_q(t) \text{ is } F_{iq}\\&\text{then } \dot{\tilde{x}} = \mathbf{A}_i(k,\gamma)\tilde{x} + v, \quad i = 1,\ldots,L\end{aligned} \tag{9.45}$$

where F_{ij} is the fuzzy set. $\mathbf{A}_i(k,\gamma)$ is constant matrix with the elements of k and γ contained in its entries. q is the number of premise variables and $\tilde{x}_1, \ldots, \tilde{x}_q$ are the premise variables. The fuzzy system is inferred as follows [37-40, 45, 80]

$$\begin{aligned}\dot{\tilde{x}}(t) &= \frac{\sum_{i=1}^{L} \mu_i(\tilde{x}(t))[\mathbf{A}_i(k,\gamma)\tilde{x}(t) + v]}{\sum_{i=1}^{L} \mu_i(\tilde{x}(t))}\\&= \sum_{i=1}^{L} h_i(\tilde{x}(t))\left[\mathbf{A}_i(k,\gamma)\tilde{x}(t) + v\right], \ \tilde{x}(0) = \tilde{x}_0\end{aligned} \tag{9.46}$$

where $\mu_i(\tilde{x}(t)) = \prod_{j=1}^{q} F_{ij}(\tilde{x}_j(t))$, $h_i(\tilde{x}(t)) = \mu_i(\tilde{x}(t)) \big/ \sum_{i=1}^{L} \mu_i(\tilde{x}(t))$, and $F_{ij}(\tilde{x}(t))$ is the grade of membership of $\tilde{x}_j(t)$ in F_{ij}.

We assume

$$\mu_i(\tilde{x}(t)) \geq 0 \text{ and } \sum_{i=1}^{L} \mu_i(\tilde{x}(t)) > 0 \tag{9.47}$$

Therefore, we get the following fuzzy basis functions

$$h_i(\tilde{x}(t)) \geq 0 \text{ and } \sum_{i=1}^{L} h_i(\tilde{x}(t)) = 1 \tag{9.48}$$

The T-S fuzzy model in (9.46) is to interpolate L linear systems to approximate the nonlinear system in (9.35) via the fuzzy basis functions $h_i(\tilde{x}(t))$. We could specify system parameter $\mathbf{A}_i(k,\gamma)$ easily so that $\sum_{i=1}^{L} h_i(\tilde{x}(t))\mathbf{A}_i(k,\gamma)\tilde{x}$ can approximate $f(\tilde{x}+x_d,k,\gamma)$ in (9.35) by the fuzzy identification method [45].

After the nonlinear system in (9.35) is approximated by the T-S fuzzy system in (9.46), the nonlinear dynamic game problem in (9.41) is replaced by solving a dynamic game problem in (9.37) subject to the fuzzy system (9.46).

Proposition 9.4 [82]: The sub-minimax robust synthetic gene network design is to solve k^* and γ^* by the following constraint optimization

$$\min_{\substack{k\in[k_1,k_2]\\ \gamma\in[\gamma_1,\gamma_2]}} g^2 \tag{9.49}$$

subject to

$$\begin{aligned} &P\mathbf{A}_i(k,r)+\mathbf{A}_i^T(k,r)P+Q+\frac{1}{g^2}PP\le 0,\ i=1,\cdots,L \\ &P\le g^2I,\ P>0 \end{aligned} \tag{9.50}$$

and the worst-case disturbance v^* is given by

$$v^*=\frac{1}{g^2}\sum_{i=1}^{L}h_i(\tilde{x})P\tilde{x} \tag{9.51}$$

Proof: see 9.5.5.

By the fuzzy approximation, the HJI in (9.43) can be approximated by a set of algebraic inequalities in (9.50). By Schur complement [41], the constrained optimization problem in (9.49)-(9.50) is equivalent to the following LMI-constrained optimization problem

$$g^{*2}=\min_{\substack{k\in[k_1,k_2]\\ \gamma\in[\gamma_1,\gamma_2]}} g^2 \tag{9.52}$$

subject to

$$\begin{aligned} &\begin{bmatrix} P\mathbf{A}_i(k,\gamma)+\mathbf{A}_i^T(k,\gamma)P+Q & P \\ P & -g^2I \end{bmatrix}\le 0,\quad i=1,2,\cdots,L \\ &P\le g^2I,\ P>0 \end{aligned} \tag{9.53}$$

Remark 9.5

(1) The fuzzy basis functions $h_i(\tilde{x}(t))$ in (9.46) and (9.48) can be replaced by other interpolation functions, for example, cubic spline functions.
(2) By the fuzzy approximation, the HJI in (9.43) of nonlinear dynamic game problem can be solved by Robust Control Toolbox in Matlab efficiently. The constrained optimization in (9.49) and (9.50) can be solved by decreasing g^2 to g^{*2}until there is no positive definite solution $P > 0$ in (9.53) with $k^*\in[k_1,k_2]$ and $\gamma^*\in[\gamma_1,\gamma_2]$.
(3) In the LMI-constrained optimization in (9.53) for the robust synthetic gene network design, we do not need the statistics of initial conditions and disturbances on the host cell, which are not easy to be measured. Therefore, the proposed method is simple but robust for synthetic gene networks.

Remark 9.6

For comparison, the conventional optimal regulation design is also proposed for synthetic gene networks. If the effect of external disturbances and uncertain initial conditions on the regulation error is not considered as (9.36) in the design procedure, i.e., only the following optimal regulation design is considered.

$$\min_{\substack{k\in[k_1,k_2]\\ \gamma\in[\gamma_1,\gamma_2]}} E\left[\int_0^{t_f}\tilde{x}^T Q\tilde{x}dt\right] \tag{9.54}$$

subject to (9.35) then we obtain the following sub-optimal regulation design for synthetic gene networks.

Proposition 9.5 [82]: The sub-optimal synthetic gene network design in (9.54) is to solve the following constrained optimization

$$\min_{\substack{k\in[k_1,k_2]\\ \gamma\in[\gamma_1,\gamma_2]}} E\left[V(\tilde{x}(0))\right] \tag{9.55}$$

subject to

$$V(\tilde{x})>0,\quad \frac{\partial V(\tilde{x})}{\partial\tilde{x}}f(\tilde{x}+x_d,k,r)+\tilde{x}^T Q\tilde{x}+\frac{1}{2}\left(\frac{\partial V(\tilde{x})}{\partial\tilde{x}}\right)^T\left(\frac{\partial V(\tilde{x})}{\partial\tilde{x}}\right)<0 \tag{9.56}$$

Proof: see 9.5.6.

Because it is not easy to solve the above HJI-constrained optimization for the sub-optimal regulation design in (9.55) and (9.56), the fuzzy approximation method is needed to simplify the design procedure. If the nonlinear error dynamic equation in (9.35) is represented by the fuzzy interpolation system in (9.46), then the optimal synthetic gene network design in (9.54) is equivalent to the following optimal regulation design problem.

$$\begin{gathered}\min_{\substack{k\in[k_1,k_2]\\ \gamma\in[\gamma_1,\gamma_2]}} E\left[\int_0^{t_f}\tilde{x}^T Q\tilde{x}dt\right]\\ \text{subject to } \dot{\tilde{x}}=\sum_{i=1}^{L}h_i(\tilde{x})\mathbf{A}_i(k,\gamma)\tilde{x}+v\end{gathered} \tag{9.57}$$

Proposition 9.6 [82]: The sub-optimal regulation design problem in (9.57) becomes how to solve the following constrained optimization problem

$$\min_{\substack{k\in[k_1,k_2]\\ \gamma\in[\gamma_1,\gamma_2]}} \mathrm{Tr}\left(PR_0\right) \tag{9.58}$$

subject to

$$P>0,\quad \begin{bmatrix} \mathbf{A}_i^T(k,\gamma)P+P\mathbf{A}_i(k,\gamma)+Q & P \\ P & -2I \end{bmatrix} \le 0,\quad i=1,2,\cdots,L \tag{9.59}$$

where R_0 denotes the covariance matrix $E[\tilde{x}(0)\tilde{x}^T(0)]$.

Proof: similar to the proof of **Proposition 9.4**.

Since the effect of stochastic disturbances on $\tilde{x}$ is not considered as (9.36) in the above sub-optimal synthetic gene network design, the synthesized gene networks will be more sensitive to the external disturbances or other uncertain factors. They will be compared with the sub-minimax robust synthetic gene network in the simulation example.

Remark 9.7

Since the effect of the disturbance v on the regulation error has not been attenuated efficiently on the design procedure of the sub-optimal regulation in **Proposition 9.5** and **Proposition 9.6**, the disturbance will have much effect on the sub-optimal regulation design of synthetic gene network. This property will be discussed and compared with the proposed robust synthetic gene network in the design example in the following section.

According to the analyses above, a design procedure is developed for the proposed robust synthetic gene network.

※ Design Procedure:

(1) Give feasible parameter ranges $[k_1,k_2]$ and $[\gamma_1,\gamma_2]$ for production rate parameters k and decay rate parameters γ, respectively, according to the biotechnology ability.
(2) Give the desired steady state x_d according to the design purpose and develop a regulation error dynamic (9.35) for a synthetic gene network.
(3) Construct a T-S fuzzy model in (9.46) to approximate the regulation error dynamic in (9.35).
(4) Solve the constrained optimization problem from the ranges $k\in[k_1,k_2]$ and $\gamma\in[\gamma_1,\gamma_2]$ in (9.52) and (9.53) for the robust synthetic gene network design k^* and γ^*, respectively according to the sub-minimax scheme or solve the constrained optimization problem in (9.58) and (9.59) for the sub-optimal regulation design.

Example 9.3: minimax design of robust synthetic gene network

Consider the man-made synthetic gene network in the following dynamic equation [67]. The synthetic gene network is shown in Figure 9.13.

$$\begin{aligned}
\dot{x}_{tetR} &= k_{tetR,0}-\gamma_{tetR}x_{tetR}+w_1\\
\dot{x}_{lacI} &= k_{lacI,0}+k_{lacI}(r_{lacI}(x_{tetR})+a_{lacI}(u_{aTc})-r_{lacI}(x_{tetR})a_{lacI}(u_{aTc}))-\gamma_{lacI}x_{lacI}+w_2\\
\dot{x}_{cI} &= k_{cI,0}+k_{cI}r_{cI}(x_{lacI})-\gamma_{cI}x_{cI}+w_3\\
\dot{x}_{eyfp} &= k_{eyfp,0}+k_{eyfp}r_{eyfp}(x_{cI})-\gamma_{eyfp}x_{eyfp}+w_4
\end{aligned} \tag{9.60}$$

where $k_{tetR,0}$, $k_{lacI,0}$, $k_{cI,0}$ and $k_{eyfp,0}$ are basal production rates of the corresponding proteins, which are assumed to be 5000, 587, 210 and 3487, respectively [67, 81]. k_{lacI}, k_{cI} and k_{eyfp} are the production rate parameters while γ_{tetR}, γ_{lacI}, γ_{cI} and γ_{eyfp} are the decay rate parameters of the corresponding proteins in the host cell (i.e., *E. coli.*). In the robust synthetic gene network design, we should select the parameters k and γ from feasible ranges so that the state of synthetic gene network x_i could approach a desired steady state $x_{d,i}$ for some biotechnical purpose. r_{lacI}, r_{cI} and r_{eyfp} are the decreasing Hill functions for regulations of repressors. a_{lacI} is an increasing function since aTc is an activator. The Hill function is a S-shaped curve [54]. u_{atc} is the input to the synthetic gene network system. We assume anhydrotetracycline input concentration to be a constant value 10000 (i.e., u_{aTc} = 10000). For the convenience of simulation, we assume that extrinsic disturbances w_1~w_4 are w_i=$[500n_1\ 10000n_2\ 100n_3\ 100000n_4]^T$, where n_i, i=1,2,3,4 are independent Gaussian white noises with zero mean and unit variance.

From the robust synthetic gene network design procedure, we give the feasible parameter ranges of production rate parameters k and decay rate parameters γ as follows [67]

$$\begin{aligned} &k_{lacI}\in[70,7000], k_{cI}\in[75,8000], k_{eyfp}\in[30,30000], \gamma_{tetR}\in[0.05,5] \\ &\gamma_{lacI}\in[0.01314,0.1517], \gamma_{cI}\in[0.7617,7.2815], \gamma_{eyfp}\in[0.007,0.067] \end{aligned} \tag{9.61}$$

Then we give the desired steady states of the synthetic gene network are $x_{d,i}$=$[1000,50000,300,500000]^T$, i=*tetR*, *lacI*, *cI*, *eyfp*. Then the regulation error dynamic equation in (9.35) is developed for the synthetic gene network. Because it is very difficult to solve the nonlinear HJI in (9.43), no simple approach is available to solve the constrained optimization problem in (9.42) for robust parameters k_i^* and γ_i^*. We construct the T-S fuzzy model in (9.46) to approximate the regulation error dynamic in (9.35) with the regulation error dynamic system's state variables as the premise variables in the following

Rule *i*:

$$\begin{aligned} &\text{If } \tilde{x}_1(t) \text{ is } F_{i1}, \tilde{x}_2(t) \text{ is } F_{i2}, \tilde{x}_3(t) \text{ is } F_{i3}, \text{ and } \tilde{x}_4(t) \text{ is } F_{i4} \\ &\text{then } \dot{\tilde{x}} = \mathbf{A}_i(k,\gamma)\tilde{x} + v, \quad i = 1,\ldots,L \end{aligned}$$

where the parameters $\mathbf{A}_i(k,\gamma)$ and the number of fuzzy rules is L=16. To construct the fuzzy model, we need to find the operating points of the regulation error dynamic system. The operating points for $\tilde{x}_1$ are chosen at $\bar{x}_{11}$=–40 and $\bar{x}_{12}$=4040. Similarly, the operating points of $\tilde{x}_2$, $\tilde{x}_3$, $\tilde{x}_4$ are chosen at $\bar{x}_{21}$=–38510, $\bar{x}_{22}$=381, $\bar{x}_{31}$=–16.7, $\bar{x}_{32}$=1686, $\bar{x}_{41}$=–441590, and $\bar{x}_{42}$=4372, respectively. For the convenience of design, triangle-type membership functions are taken for Rule 1 through Rule 16. We create two triangle-type membership functions for each state (see Figure 9.14).

In order to simplify the nonlinear stochastic game problem of the robust synthetic gene network, we just solve only the sub-minimax problem in (9.37) instead. With the help of fuzzy approximation method and LMI technique, we can easily solve the constrained optimization problem in (9.52) and (9.53) instead of the nonlinear constrained optimization problem in (9.42) and (9.43) for the minimax robust synthetic gene network design. Finally, we obtain the minimum upper bound of the game in (9.52) $g^{*2} = 0.847536$ and a common positive definite symmetric matrix P for (9.53) as follows

$$P = \begin{bmatrix} 0.45842 & -0.0079 & 0.0143 & -0.00068 \\ -0.0079 & 0.07186 & -0.000557 & 0.00268 \\ 0.0143 & -0.000557 & 0.04847 & 0.000718 \\ -0.00068 & 0.00268 & 0.000718 & 0.0578 \end{bmatrix}$$

with the specified robust production rate parameters k_{lacI}^{*}=7000, k_{cI}^{*}=4037.5 and k_{eyfp}^{*}=30000 and robust decay rate parameters γ_{tetR}^{*}=5, γ_{lacI}^{*}=0.1517, γ_{cI}^{*}=4.0216 and γ_{eyfp}^{*}=0.067 of the synthetic gene network. With these design parameters, the parameters $\mathbf{A}_i$ of fuzzy model are described in 9.5.7.

Figure 9.15 presents the simulation result for robust synthetic gene networks by using Monte Carlo method with 50 rounds and with the uncertain initial values. $x_1(0)$~$x_4(0)$ are assumed normal-distributed random numbers with means 5000, 8000, 2000, 10000 and standard deviations 500, 800, 200, 1000, respectively. As can be seen, the synthetic gene network has robust regulation ability to achieve the desired steady state (black dashed line) in spite of uncertain initial states and the disturbances on the host cell. Obviously, the robust synthetic gene network by the proposed sub-minimax regulation design method has robust stability to the uncertain initial conditions and enough filtering ability to attenuate the disturbances on the host cell and can approach the desired steady states.

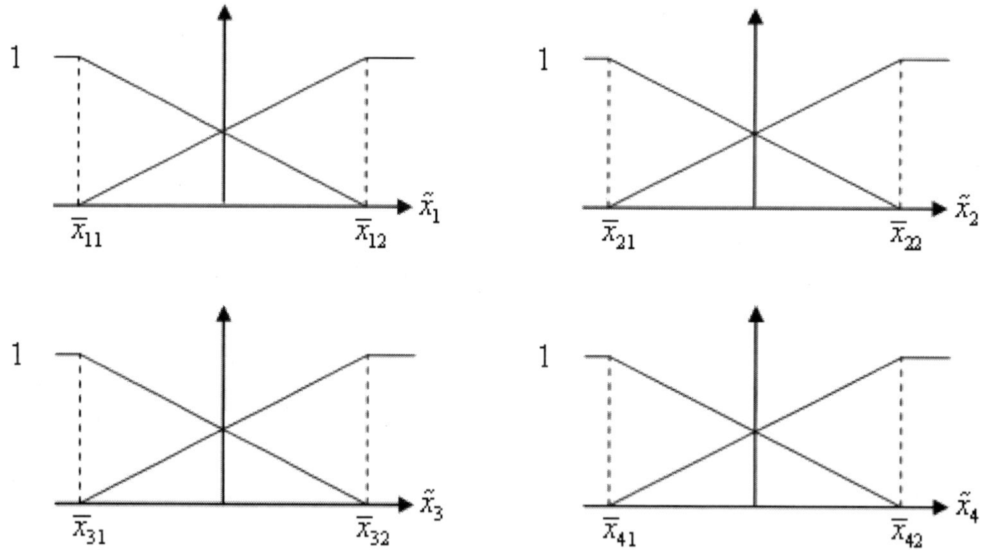

Figure 9.14. Membership functions for four states $\tilde{x}_1$, $\tilde{x}_2$, $\tilde{x}_3$, and $\tilde{x}_4$.

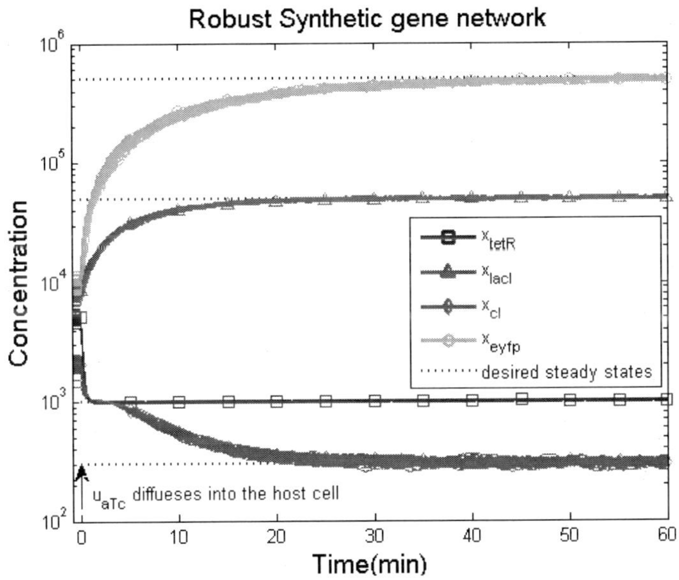

Figure 9.15. The robust synthetic gene network design with uncertain initial values and the desired steady states x_d=[1000,50000,300,500000]^T. And with specified robust production rate parameters k_{lacI}^*=7000, k_{cI}^*=4037.5 and k_{eyfp}^*=30000 while the specified robust decay rate parameters are γ_{tetR}^*=5, γ_{lacI}^*=0.1517, γ_{cI}^*=4.0216 and γ_{eyfp}^*=0.067 of the synthetic gene network. The Monte Carlo simulation method is used with 50 rounds.

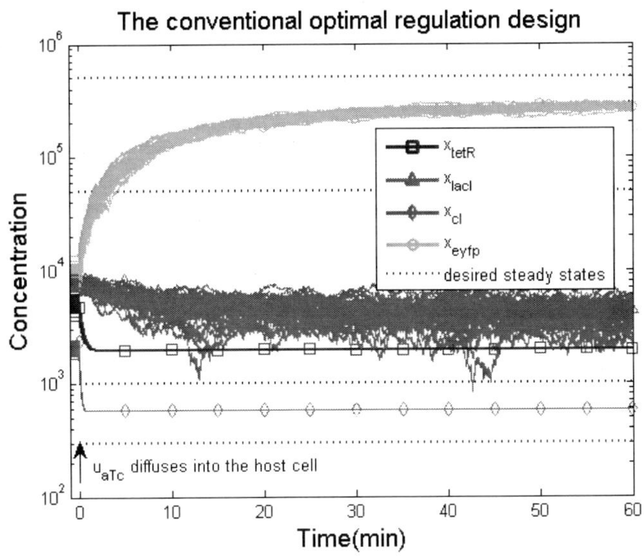

Figure 9.16. The conventional optimal regulation design with uncertain initial values and the desired steady states x_d=[1000,50000,300,500000]^T. And with specified production rate parameters k_{lacI}^*=70, k_{cI}^*=4037.5 and k_{eyfp}^*=15015 while the specified decay rate parameters are γ_{tetR}^*=2.525, γ_{lacI}^*=0.1517, γ_{cI}^*=7.2815 and γ_{eyfp}^*=0.067 of the synthetic gene network. It is seen that the conventional optimal regulation design of the synthetic gene network is sensitive to the initial conditions and disturbances and cannot achieve the desired steady states. The Monte Carlo simulation method is used with 50 rounds.

For comparison, we solve the sub-optimal regulation design problem in (9.58) and (9.59) for the specified production rate parameters k_{lacI}^*=70, k_{cI}^*=4037.5 and k_{eyfp}^*=15015 and decay rate parametersγ_{tetR}^*=2.525, γ_{lacI}^*=0.1517, γ_{cI}^*=7.2815 and γ_{eyfp}^*=0.067 of the synthetic gene network. The simulation result of conventional optimal regulation design is also shown in Figure 9.16. As can be seen, the conventional optimal regulation design of the synthetic gene network is more sensitive to the initial conditions and disturbances and cannot achieve the desired steady state under the uncertain initial conditions and disturbances.

Remark 9.8

The experimental systems in the above example may not be fully observable. If we want to know whether all state variables can approach to the desired states x_d, several fluorescent proteins (red, green and cyan color) should be necessary to observe their protein expressions of all state variables in the experimental design.

9.2.4. Discussion

Because the initial conditions and disturbances on the host cell are uncertain, to simplify the design problem, a robust synthetic biology design is formulated as a stochastic game problem in this chapter. The uncertain initial conditions and disturbances due to intrinsic and extrinsic molecular noises on the host cell are considered as a player maximizing the regulation error and the design parameters are considered as another player minimizing the regulation error. In order to avoid solving HJI in the stochastic game theory-based design problem, a T-S fuzzy interpolation method is introduced to simplify the design procedure of robust synthetic gene networks via only solving a set of LMIs, which can be efficiently solved by Robust Control Toolbox in Matlab.

In this chapter, the weighting matrix is selected as Q=diag(q_{11},q_{22},q_{33},q_{44}) which denotes the punishment on the corresponding tracking error $\tilde{x}$. If we only need to achieve a desired steady state x_{d4} (EYFP), we just assign a value to the fourth diagonal element q_{44} of the weighting matrix Q and set q_{11}=q_{22}= q_{33}=0. The rest of states x_1~x_3 will not approach to the given steady state x_{d1}~x_{d3} because of no any punishment. However, in this case, some infeasible steady states of x_1, x_2, and x_3 may be obtained even an optimal x_4 can be achieved. In this chapter, the desired steady states of x_1, x_2, and x_3 are given because we can avoid obtaining infeasible steady states in x_1, x_2, and x_3 when an optimal x_4 is achieved. Further, the undesired steady states of x_1, x_2, and x_3 may also have metabolic toxicity on host cell and should be avoided. Since the steady states of x_1~x_3 are not that important, the desired steady states x_{d1}~x_{d3} can be adjusted within feasible ranges, so that the desired steady state x_{d4} can still achieve some optimization as possible. This kind of design can avoid hampering the optimization of x_4 when x_1, x_2, and x_3 achieve some feasible steady states.

In our *in silico* design example, we can design the specified robust production rate parameters k_i^* and decay rate parameters γ_i^* within the feasible parameter ranges to achieve the desired steady states of the synthetic gene network. As for the biological implementation, we could refer to standard biological parts in biological device datasheets to construct the genetic circuits with the fine-tuned production rate parameters k_i^* and decay rate parameters γ_i^*. In this way, synthetic biologists can increase efficiency of gene circuit design through registries of biological parts and standard datasheets, which are developed concerned with proper packing and characterizing of 'modular' biological activities so that these biological parts or devices with some desired characteristics may be efficiently assembled into gene circuits [85].

Quantitative descriptions of devices in the form of standardized, comprehensive datasheets are widely used in many engineering disciplines. A datasheet is intended to allow an engineer to quickly determine whether the behavior of a device will meet the requirements of a system in which a device might be used [85]. Such a determination is based on a set of standard characteristics of device behavior, which are the product of engineering theory and experience. In the datasheets of engineering, the characteristics typically reported are common across a wide range of device types, such as sensors, logic elements and actuators. Recently, biological datasheets have been set as standards for characterization, manufacture and sharing of information about modular biological devices for a more efficient, predictable and design-driven genetic engineering science [85, 86]. Because datasheets of biological parts or devices are an embodiment of engineering standard for synthetic biology [85], a good device standard should define sufficient information about biological parts or devices to allow the design of gene circuit systems with the optimal parameters. Datasheets contain a formal set of input-output transfer functions, dynamic behaviors, compatibility, requirements and other details about a particular part or device [85, 86]. Since parameters k_i are combinations of transcription and translation, they could be measured from the input-output transfer functions and dynamic behaviors of biological parts or devices in biological device datasheets. From properly characterized input-output transfer functions and dynamic behaviors of parts or devices in biological device datasheets, an engineer can estimate the corresponding parameters of biological parts or devices. When the biological parts and devices in datasheets become more complete in future, we can rapidly select from a vast list the parts that will meet our design parameters k_i. Therefore we can ensure that devices selected from datasheets can fit the optimal parameters and systems synthesized from them can satisfy the requirements of design specifications for robust synthetic gene networks.

In order to guarantee the biological feasibility of the calculated optimal parameters, the ranges $[k_1,k_2]$ and $[\gamma_1,\gamma_2]$ of parameters should be determined by the whole parameters of biological parts repositories (http://partsregistry.org/) so that the optimal parameters selected within these ranges to minimize g^2 in equations (9.52) and (9.53) have biological meaning, or equivalently from the whole biological parts in biological device datasheets, we can find a set of biological parts whose parameters can minimize the g^2 in equations (9.52) and (9.53) to achieve the robust optimal design of synthetic gene network from the perspective of Nash dynamic game.

In synthetic gene networks, there is much uncertainty about what affects the behavior of biological circuitry and systems. For example, devices will perturb the cellular functions and there are also likely to be parasitic and unpredictable interactions among components as well as with the host. Since k_i is a combination of promoter strength, ribosome binding site and degradation of the transcript, there are some variations or uncertainties on the parameter value k_i. These variations or uncertainties of k_i can be transformed to an equivalent uncertain disturbance w_i in equation (9.32) from the viewpoint of mathematic model. The proposed robust minimax synthetic biology design method can predict the most robust value of k_i from the perspective of stochastic game. In our robust design method, we don't need the statistics of these parameter uncertainties because the proposed synthetic genetic network not only can achieve the desired steady state but also can tolerate the worst-case effect due to these uncertain parameter variations and external noises on the host cell.

For comparison, a sub-optimal regulation design for synthetic gene network is also developed for synthetic gene network. Because the sub-optimal regulation design cannot efficiently attenuate the effect of uncertain initial conditions and disturbances on the regulation, it is not suitable for robust synthetic gene networks with uncertain initial conditions and disturbances on the host cell. As seen in the example *in silico*, the proposed robust synthetic gene network can function properly in spite of uncertain initial conditions and disturbances on the host cell. Design of more robust and complex genetic circuits is foreseen to have important applications in biotechnology, medicine and biofuel production, and to revolutionize how we conceptualize and approach the engineering of biological systems [51]. Therefore, it has much potential for the robust synthetic gene network design in the near future.

9.3. The Stochastic Evolutionary Game on a Population of Biological Networks under Natural Selection

9.3.1. Introduction of Biological Network Evolution

Biological network evolution is the physical, genetic, or behavioral change in populations of biological networks over time. Natural selection is a process that causes biological networks to adapt to their environments. Network robustness is important in ensuring the stability of phenotypic traits of biological networks that are constantly exposed to genetic variations and non-genetic environmental disturbances. A better understanding of network robustness is of paramount importance for understanding biological network evolution. Network robustness can be discussed sensibly only if two cardinal questions have been resolved: what is the phenotypic trait of interest in network evolution, and what is the network perturbation of interest [87-90]. There are three principal kinds of network perturbations to which a biological network needs to be robust: stochastic intrinsic molecular noise, environmental change and genetic variation [91-93]. Stochastic intrinsic molecular noise refers to the stochastic fluctuations that occur in any biological network, for example in the concentration of a biological molecule. Environmental change is the variation in the external environment; for example, a change in temperature, salinity, or nutrient availability. Genetic variation is genetic change, either through de novo mutation or through recombination in the evolutionary process [87]. The ability to buffer genetic variation is called genetic robustness, the noise filtering ability to attenuate the effect of intrinsic molecular noise is called noise robustness, and the ability to resist environmental change is called environmental robustness. The ability to maintain a desired phenotypic trait is called phenotypic robustness. In order to maintain the desired phenotypic trait, a biological network needs an evolutionary strategy with high network robustness to confer genetic robustness to buffer genetic variation, noise robustness to filter stochastic intrinsic molecular noise, and environmental robustness to simultaneously resist environmental change in the evolutionary process.

Network evolvability is defined as the ability to deviate from a phenotypic trait to adapt an environmental change. The network evolvability of a biological network appears to be the reverse of network robustness. If phenotypic traits are robust against genetic variations and external disturbances, a network population may be expected to have difficulty adapting to an environmental change, as several studies have suggested [88, 89, 94]. However, other studies contend that robust networks are more adaptable [90-93]. The relationship between network robustness and network evolvability is complex, because robust network populations harbor a large diversity of neutral genotypes that may be important in adaptation, but the study of the evolvability and robustness of biological networks is still in its infancy. In fact, network robustness allows changes in the structure and components of biological networks, due to genetic variations, while still allowing the desired phenotypic trait to be maintained. Network evolvability allows some environmental changes to influence a biological network, so that a new phenotypic trait of the biological network may evolve by natural selection, to adapt to the new environment [94-97].

In the evolutionary process, a population of evolving biological networks can be represented by a nonlinear stochastic system with heritable random genetic variations and non-heritable random environmental disturbances. The phenotypic trait of a biological network can be represented by an equilibrium point of the nonlinear stochastic system in the fitness landscape (see Figure 9.17). There are many variations of phenotypes (population) around this equilibrium point, due to intrinsic random genetic variations and random environmental disturbances [94-97]. When a phenotypic trait is perturbed so much that it can transit from the domain of its equilibrium point into the domains of other equilibrium points, a new phenotypic trait is generated and presented. A heritable genetic variation that does not directly help a biological network to evolve to a new phenotypic trait is neutral. In evolution, a robust biological network can harbor a large number of neutral genetic variations, before the phenotypic trait transforms from one equilibrium point to another one. The capacity to harbor these neutral genetic variations is a measure of the genetic robustness of the biological network in evolution. If the amount of stochastic neutral genetic variations is larger than the genetic robustness of biological network so that phenotypic trait can transit from one equilibrium point to another, a new phenotype is presented to start another period of network evolution. Although it has been shown that many features of biological network can arise by non-adaptive process [98], the stochastic evolution game scheme will offer a different possibility for biological network adaptive process by natural selection in evolution.

In this chapter, the network evolvability in response to stochastic environmental changes and genetic variations of a nonlinear stochastic network can be formulated for the first time as a stochastic game problem because the environmental changes in the evolutionary process are unpredictable. In this situation, the minimization of the worst-case phenotypic deviation for all possible environmental changes is therefore considered as the phenotypic robustness strategy used by a biological network to resist the effect of environmental changes and to maintain its phenotype in the evolutionary process. Therefore, both the network evolvability and the phenotypic robustness of a nonlinear stochastic biological network in the evolutionary process can be discussed from the nonlinear stochastic game perspective, i.e., the stochastic biological network wants to minimize network evolvability (or maximize network robustness) when the stochastic environmental disturbances try to maximize the network evolvability (or minimize the network robustness). The interplay between genetic robustness, environmental robustness, and network robustness in the phenotypic robustness of an evolutionary biological network [99-101] can also be investigated from the point of view of a stochastic Nash game. In this chapter, by solving the evolutionary game strategy, we find that some genetic variations in the evolutionary process are selected by the evolutionary game strategy to construct new negative feedback loops to efficiently improve network robustness, to maximize the fitness (or to minimize the evolution level) of the biological network in evolution. We also find that with the evolutionary game strategy, a biological network has enough network robustness to cover noise robustness, genetic robustness and environmental robustness to filter stochastic noise, to buffer genetic variation, and to resist environmental disturbance, respectively, and that the stochastic evolutionary biological network becomes more robust near the equilibrium point. This is the so-called `phenotypic robustness criterion' of biological networks in the evolutionary process. However, when random neutral genetic variations harbored by network robustness are accumulated to the extent of violating the phenotypic robustness criterion and providing new material for more evolutionary paths to other phenotypes of gene network by random drift, with the help of a large environmental change, the phenotypic trait of the biological network cannot be maintained again and a phenotype transition may eventually occur by a shift to another equilibrium point. In other words, while the evolutionary game strategy can improve the network robustness for the phenotype at the beginning, the accumulated neutral genetic variations harbored by improved network robustness will finally violate the network phenotypic robustness criterion to break down the network phenotype to make a new phenotypic adaptation in network evolution.

Based on nonlinear stochastic game theory, the study of network robustness and network evolvability needs to solve a very difficult Hamilton-Jacobi-Isaac inequality (HJII) [44]. At present, no good way exists to efficiently solve the HJII analytically or numerically. In this chapter, a global linearization method is employed to interpolate several local linear stochastic systems, at different operation points, to approximate the nonlinear stochastic biological network in evolution. In this situation, linear stochastic game techniques can be employed to solve the nonlinear stochastic evolutionary game problem of the biological network, in which the HJII is replaced by a set of linear matrix inequalities (LMIs). Based on the global linearization technique and linear system theory, the network evolvability and robustness are related to the locations of the eigenvalues of the local linearized stochastic evolutionary biological networks. If the eigenvalues of local linearized evolutionary biological networks are located far in the left-hand side of the s-complex domain, then the nonlinear stochastic biological network is more robust with the phenotypic trait (or the equilibrium point) and less adaptive to other equilibrium points (or other phenotypic traits) in the evolutionary process. On the other hand, if these eigenvalues are near the image axis of the s-complex domain, then they are more easily perturbed to the right-hand side of the s-complex domain. In this situation, the phenotypic trait of the nonlinear stochastic evolutionary biological network is less robustly stable but more adaptable to environmental change in the evolutionary process. In order to maintain the phenotype robustness of the evolutionary biological network, the evolutionary game strategy can select some genetic variations to generate adequate negative feedback loops, so that the eigenvalues of the local linearized biological network can be shifted to the far left-hand side of the s-complex domain. Obviously, the evolutionary game strategy can efficiently enhance the network robustness by developing adequate negative feedback loops to buffer neutral genetic variations and to resist environmental disturbance in the evolutionary process.

In this chapter, we investigate quantitative measures of network evolvability and network robustness from a stochastic game theory point of view. The phenotype robustness criterion for a nonlinear stochastic evolutionary biological network can be measured by solving an evolutionary game strategy based on the LMIs of local linearized stochastic biological networks. That is, if the genetic robustness plus the noise robustness and environmental robustness is less than the network robustness in each local linearized stochastic network, then the phenotypic trait of the nonlinear stochastic biological network is robustly stable in the evolutionary process. If not, the biological network might evolve to another phenotypic trait with the release of accumulated cryptic neutral genetic variations and the aid of strong environmental changes. In this chapter, we first reveal the evolutionary strategy for the adaptive fitness of the biological network in evolution from a stochastic game perspective, and then we investigate the tradeoffs between genetic robustness and environmental robustness and the antagonistic properties between network robustness and network evolvability from a system theory perspective. If there exist M competitive populations and k environmental changes in the network evolutionary process, then the proposed evolutionary game strategy can be easily modified to the n-tuple evolutionary game problem of biological network with M players (competitive populations) and k environmental dynamics. Finally, an in silico example is given to illustrate the stochastic evolutionary game on the stochastic biological networks and to moderate their network evolvability and fitness in evolution.

9.3.2. The Stochastic Evolutionary Game of the Linear Biological Network under Natural Selection

For simplicity of analysis, we first consider a linear biological network as follows

$$\frac{dx}{dt} = Ax(t) \tag{9.62}$$

where the state vector $x(t)=[x_1(t)\ \dots\ x_n(t)]^T$ denotes the concentrations of n genes in a genetic regulatory network, or protein concentrations of n proteins in a protein-protein interaction network, or the population densities of an ecological network with n different species, and $x_i(t)$ represents the number, density, or biomass of individuals of species i. Although we have assumed that there are n different genes, proteins or species, these numbers need not be fixed. In fact, one feature of Darwinian dynamics is that the evolutionary process may determine the number n of genes, proteins, or species as a product of the evolutionary network. In the evolutionary network, the interaction matrix $A \in \mathbb{R}^{n \times n}$ among species is denoted as

$$A = \begin{bmatrix} a_{11} & \cdots & a_{1n} \\ \vdots & \ddots & \vdots \\ a_{n1} & \cdots & a_{nn} \end{bmatrix} \tag{9.63}$$

where a_{ij} denotes the interaction from species j to species i, i.e., A denotes the dynamic interactions between species in a biological network from the present generation to the next generation.

Remark 9.9

(i) The linear biological network in (9.62) can be considered as a linearized system of a nonlinear biological network $dx/dt=f(x(t))$ at an equilibrium point (phenotypic trait) x_e of interest by natural selection in Figure 9.17, i.e., $A=\partial f(x)/\partial x|_{x=xe}$. Without loss of generality, the origin of the nonlinear biological network is always shifted to the equilibrium point (phenotypic trait) x_e, i.e., x_e=0. This detail will be discussed later.

(ii) The interaction matrix A between species may suffer from environmental disturbances due to climate and salinity change, and from random genetic variations due to heritable DNA mutation, genetic transition, and recombination in the evolutionary process.

In the evolutionary process, the biological network (9.62) suffers from both extrinsic disturbances due to environmental changes as well as intrinsic phenotypic variations due to heritable random genetic variations. In order to maintain the favorable phenotypic trait at x_e=0 by natural selection (i.e., phenotypic trait robustness in evolution), the biological network needs to select some heritable genetic variations (the evolutionary strategy of the biological network) to improve network robustness through natural selection, to resist intrinsic variations and environmental disturbances. In this situation, the evolutionary biological network can be represented as follows [102]

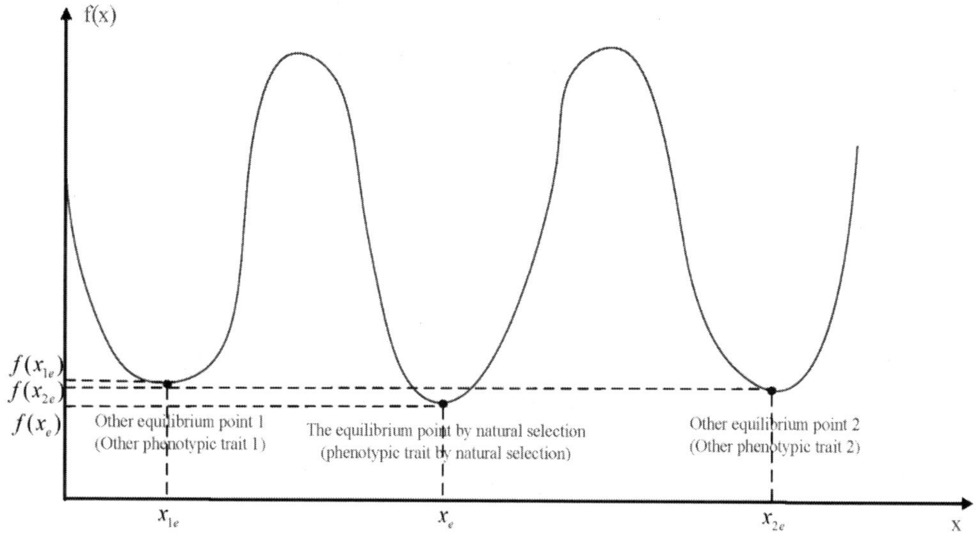

Figure 9.17. The equilibrium points of a nonlinear biological network in a phenotypic landscape. The nonlinear stochastic biological network has many local equilibrium points (phenotypic traits). The phenotypic landscape of three stable equilibrium points with a vertical scale illustrating the relative network robustness of the equilibrium points (phenotypic traits) of the nonlinear stochastic biological network in evolution, i.e., the equilibrium point at x_e with deeper and wider basin is more robust. Here, x_e is the equilibrium point (phenotypic trait) of interest by natural selection.

$$\frac{dx(t)}{dt}=\left(A+\sum_{k=1}^{N(t)} A_k p(t-t_k)\right)x(t)+Bv(t) \tag{9.64}$$

where the weighted Poisson point (counting) process [103]

$$s(t)=\sum_{k=1}^{N(t)} A_k p(t-t_k)=\sum_{k=1}^{N(t)} A_k, t \geq 0 \tag{9.65}$$

denotes heritable network interaction variations due to some genetic variations occurring at t_1, …, t_p; $N(t)$ denotes a Poisson point (counting) process of genetic variation to be selected by natural selection at t_k, where k=0, …, p. Amongst these $s_1(t) \triangleq \Sigma_{k=1}^{N_1(t)} A_k p(t-t_k)$ are selected by natural selection to change the interaction matrix A, in order to improve network robustness to resist an environmental disturbance $v(t)$ due to the surrounding change and interference from other competitive populations. In addition, $s_2(t) \triangleq \Sigma_{k=1}^{N_2(t)} A_k p(t-t_k)$ are due to neutral genetic variations, which have no effect on A to directly change a phenotypic trait, i.e., $s(t)=s_1(t)+s_2(t)$ in the evolutionary process (Figure 9.18). The Poisson counting process $N(t)$ is denoted by

$$N(t)=\sum_{k=1}^{N(t)} p(t-t_k)=\sum_{k=1}^{N(t)} 1 \tag{9.66}$$

with mean λt and variance λt [103]. The matrix $A_k \in \mathbb{R}^{n \times n}$ denotes the random phenotypic variation of the biological network due to the heritable genetic point variation $p(t-t_k)$ occurring at $t=t_k$, $v(t) \in \mathbb{R}^m$ denotes the effect of environmental disturbances, and $B \in \mathbb{R}^{n \times m}$ denotes the coupling matrix between the environmental disturbance and the biological network, i.e.,

$$A_k=\begin{bmatrix} a_{k11} & \cdots & a_{k1n} \\ \vdots & \ddots & \vdots \\ a_{kn1} & \cdots & a_{knn} \end{bmatrix}, B=\begin{bmatrix} b_{11} & \cdots & b_{1m} \\ \vdots & \ddots & \vdots \\ b_{n1} & \cdots & b_{nm} \end{bmatrix}, v(t)=\begin{bmatrix} v_1(t) \\ \vdots \\ v_m(t) \end{bmatrix} \tag{9.67}$$

where a_{kij}=0 if the k-th genetic variation has no phenotypic effect on the interaction a_{ij} between species i and species j in the biological network; b_{ij}=0 if the $v_j(t)$ has no effect on the species i.

Remark 9.10

(i) The weighted Poisson point process $s(t)=\Sigma_{k=1}^{N(t)} A_k p(t-t_k)=\Sigma_{k=1}^{N(t)} A_k$, $t \geq 0$, denotes the heritable network interaction variations accumulated by the biological network, some of which are to be selected by natural selection to resist environmental disturbances in evolution. Since the weighted Poisson point process $\Sigma_{k=1}^{N(t)} A_k p(t-t_k)$ is a random process due to random genetic variations, the stochastic evolutionary network in (9.64) can represent a population of biological networks over all possible Poisson random genetic variations in the evolutionary process, i.e., one possible Poisson genetic variation represents one possible biological network in the network population.

(ii) The accumulated phenotypic variations $s(t)$ may lead to permanent changes in the phenotypic trait.

(iii) $v(t)$ denotes the external disturbances due to environmental changes or stresses, such as temperature or salinity or interferences from other competitive populations. The environmental disturbance can perturb $x(t)$ away from its phenotypic trait at the equilibrium point x_e=0. Since the environmental disturbance $v(t)$ cannot affect the interaction matrix A of the biological network directly, its effect on the biological network is non-heritable. In order to resist the effect of the environmental disturbance, some phenotypic variations $s_1(t)=\sum_{k=1}^{N_1(t)} A_k p(t-t_k)$ are to be selected for the biological network to change the network interaction matrix A by natural selection, to improve the network robustness to resist the environmental disturbance (i.e., the so-called evolutionary strategy of a biological network for the phenotypic trait robustness in the evolutionary process). The remaining genetic variations $s_2(t)= \sum_{k=1}^{N_2(t)} A_k p(t-t_k)$, which have no direct phenotypic effect on the network interaction matrix A are called neutral genetic variations, i.e., $s(t)=s_1(t)+s_2(t)$ and $N(t)=N_1(t)+N_2(t)$ where the Poisson process $N_1(t)$ and $N_2(t)$ have means $\lambda_1 t$ and $\lambda_2 t$, respectively, and $\lambda=\lambda_1+\lambda_2$ (see Figure 9.18).

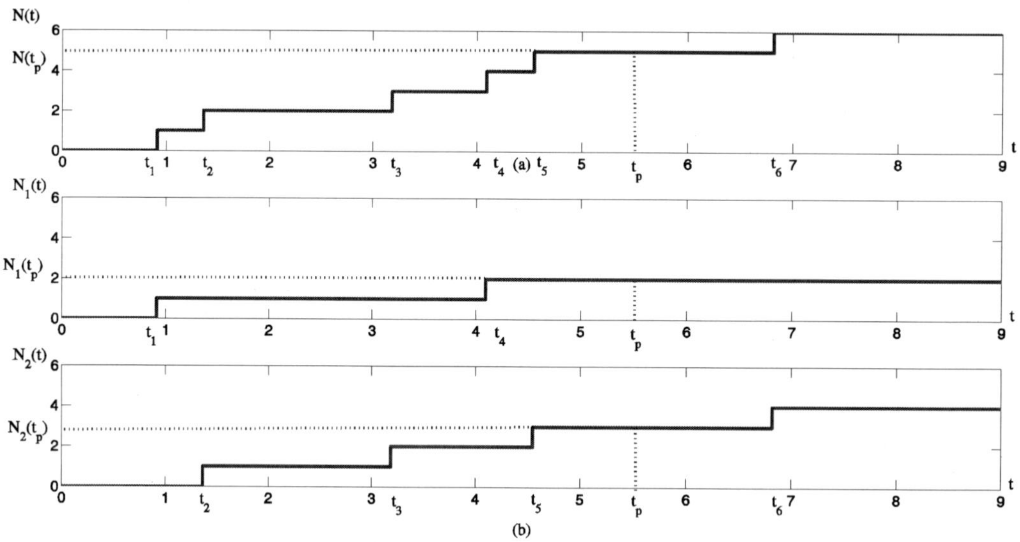

Figure 9.18. (a) The Poisson counting process of genetic variations occurring at t_k in the evolutionary process, where t_p denotes the present time. (b) The Poisson counting process $N(t)$ of genetic variations divided into two parts, (top) genetic variations $N_1(t)$ selected by natural selection, and (bottom) neutral genetic variations $N_2(t)$ in the evolutionary process.

Therefore, the stochastic gene network under environmental disturbances and natural selection in (9.64) can be represented as the following Poisson point process [102]

$$\begin{aligned}\frac{dx(t)}{dt} &= Ax(t)+\sum_{k=1}^{N(t)} A_k p(t-t_k)x(t)+Bv(t) \\ &= Ax(t)+s_1(t)x(t)+s_2(t)x(t)+Bv(t) \\ &= Ax(t)+u(t)+s_2(t)x(t)+Bv(t)\end{aligned} \tag{9.68}$$

where $u(t)=s_1(t)x(t)=\sum_{k=1}^{N_1(t)} A_k p(t-t_k)x(t)$ denotes the evolutionary strategy of the biological network, i.e., how to select adequate phenotypic variations by natural selection to modify the network interaction matrix A, thus improving network robustness to tolerate the neutral genetic variations and resisting the environmental disturbance to maintain the phenotypic trait in the evolutionary process. The main evolutionary problem of the stochastic biological network in (9.68) for natural selection is how to determine the evolutionary strategy $u(t)=s_1(t)x(t)=\sum_{k=1}^{N_1(t)} A_k p(t-t_k)x(t)$, i.e., how to select some adequate genetic variations to enhance system matrix A to improve the network robustness. This enables the biological network to tolerate the accumulated neutral genetic variations $s_2(t)=\sum_{k=1}^{N_2(t)} A_k p(t-t_k)$ and efficiently resist the environmental disturbance $v(t)$ to achieve phenotypic trait robustness in the evolutionary process. Since environmental disturbances are stochastic and unpredictable, the phenotypic robustness strategy $u(t)$ of the biological network to resist the environmental disturbances is based on the worst-case effect of all possible environmental disturbances. Let us denote the set of all possible bounded environmental disturbances as $L_2[0,t_p]$, i.e., $v(t)\in L_2[0,t_p]$ if $\int_0^{t_p} v^T(t)v(t)\,dt<\infty$, where t_p denotes the present time, i.e., $L_2[0,t_p]$ denotes the set of all possible bounded environmental disturbances to the present time (unbounded environmental disturbances are not considered here because they will lead to the extinction of all biological networks).

Remark 9.11

(i) The stochastic system in (9.68) can be considered a Darwinian dynamic to describe a linear dynamic biological network underlying natural selection. Darwinian dynamics are systems of equations that satisfy Darwin's conditions of variability, heritability, and the strategy to survive [104].

(ii) If the stochastic biological network in (9.68) is robustly stable at $x_e=0$ (if $x_e\neq 0$, then the origin of the linear dynamic network in (9.68) must be shifted to x_e for simplicity of analysis, i.e., the phenotypic trait survives at $x_e=0$), then the evolutionary strategy $u(t)=s_1(t)x(t)$ is a robust evolutionary strategy by natural selection. The Poisson genetic variations $\sum_{k=1}^{N_2(t)} A_k p(t-t_k)$ are not selected by natural selection, and are called neutral genetic variations without direct phenotypic effect on the system matrix A of the genetic network, and are further accumulated and harbored in the deeper basin around the phenotypic trait x_e of the robust biological network (see Figure 9.17).

Let us denote the phenotypic deviation around x_e as $\tilde{x}=x(t)-x_e$. Then the evolution level of the stochastic biological network due to the effect of all possible bounded environmental disturbances in the evolutionary process is defined as [102]

$$e = \max_{v(t)\in L_2[0,t_p]} \frac{E\int_0^{t_p}\left(\tilde{x}^T(t)Q\tilde{x}(t)+u^T(t)u(t)\right)dt}{E\tilde{x}^T(0)\tilde{x}(0)+E\int_0^{t_p}v^T(t)v(t)dt} \tag{9.69}$$

where $E(\bullet)$ denotes the expectation of $(\bullet)$, Q is a positive definite weighting matrix, and t_p denotes the present time. The biological meaning of (9.69) is that the worst-case (maximum) effect of all possible bounded but unpredictable environmental disturbances on both phenotypic deviation around x_e and the evolutionary strategy $u(t)$ is considered as the evolution level of the biological network on the basis of the average energy. Since the initial phenotypic deviation $\tilde{x}(0)$ also has an effect on $\tilde{x}(t)$ and $u(t)$, its effect on the evolution level is considered in the denominator of (9.69). If $x=x_e$ or $\tilde{x}(0)=0$, i.e., the initial phenotypic trait $x(0)$ is at the equilibrium point x_e, then the term $E\tilde{x}^T(0)\tilde{x}(0)$ in the denominator should be neglected.

The fitness (phenotypic robustness) of the biological network around the phenotypic trait at the equilibrium point x_e is inversely correlated to the evolution level of the biological network and is defined as

$$f = \frac{1}{e} \tag{9.70}$$

i.e., a biological network with a low evolution level will have a high fitness to the phenotypic trait, and vice versa. Therefore, the fitness maximization of a biological network to the phenotypic trait x_e is equivalent to the minimization of the evolution level as follows

$$\max f = \frac{1}{\min e} \tag{9.71}$$

Combining (9.69) and (9.71), the evolutionary strategy $u(t)$ for the maximization of the fitness of the biological network is formulated for the first time as the following stochastic minimax evolution (evolutionary game) problem [102]

$$e_o \triangleq \min_{u(t)} e = \min_{u(t)} \max_{v(t)\in L_2[0,t_p]} \frac{E\int_0^{t_p}\left(\tilde{x}^T(t)Q\tilde{x}(t)+u^T(t)u(t)\right)dt}{E\tilde{x}^T(0)\tilde{x}(0)+E\int_0^{t_p}v^T(t)v(t)dt} \tag{9.72}$$

The minimum evolution level e_o in (9.72) is called the network evolvability of the biological network in the evolutionary process. The physical meaning of network evolvability is the minimum worst-case effect of environmental disturbances on the biological network with a minimax evolutionary strategy $u(t)$, i.e., the minimax evolutionary strategy is performed for the biological network by natural selection to resist unpredictable environmental disturbances from the point of view of the Nash game in the evolutionary process. The minimax performance in (9.72) has been used for robust control design under intrinsic perturbations and external disturbances [83, 105]. Therefore, the maximum phenotypic robustness (or fitness maximization) problem of the evolutionary biological network in (9.68) can be formulated as a stochastic evolutionary game problem in (9.72), i.e., how to select $u(t)-\Sigma_{k=1}^{N_1(t)} A_k p(t-t_k)x(t)=s_1(t)x(t)$ to achieve the fitness maximization in (9.71) for the biological network in (9.68) with the phenotype at x_e. In other words, the evolutionary strategy of a biological network to maintain its phenotypic trait against intrinsic parameter fluctuations and environmental disturbances is to develop a feasible evolutionary strategy $u(t)=s_1(t)x(t)$ to enhance the biological network with the maximum fitness at x_e, or equivalently, to select some genetic variations $s_1(t)$ to modify the interaction matrix A of the biological network to minimize the worst-case (maximum) effect of environmental disturbance $v(t)$ on both the variation of the phenotypic trait $\tilde{x}(t)$ and the evolutionary strategy $u(t)$ in (9.72) on the basis of mean energy.

Why is $u(t)$ also considered in the numerator of (9.72) or (9.69)? The reason is that the biological network aims to achieve the maximum fitness f_o (or minimum evolution level e_o) with a minimum effort (or cost) (i.e., with a minimum change $\Sigma_{k=1}^{N_1(t)} A_k p(t-t_k)$ on system matrix A of the biological network). There are two players, $u(t)$ and $v(t)$ in the evolutionary game problem in (9.72). The player $v(t)$ will try to maximize the effect of environmental disturbances on the deviation of the phenotype from its equilibrium point x_e and the other player $u(t)$ will try to select some genetic variations to improve network robustness in order to maintain the phenotype around x_e by minimizing the worst-case phenotypic deviation due to all possible environmental disturbances. In general, it is very difficult to solve the stochastic evolutionary game problem in (9.72) for the stochastic gene network in (9.68). Therefore, the following indirect method is employed to solve the stochastic evolutionary game problem [44]

$$e_o = \min_{u(t)} \max_{v(t)\in L_2[0,t_p]} \frac{E\int_0^{t_p}\left(\tilde{x}^T(t)Q\tilde{x}(t)+u^T(t)u(t)\right)dt}{E\tilde{x}^T(0)\tilde{x}(0)+E\int_0^{t_p} v^T(t)v(t)dt} \le e \tag{9.73}$$

where the positive value e denotes the upper bound of the evolvability e_o. The upper bound e of e_o will be given beforehand and then decreased to as small a value as possible to approach e_o and attain the real minimax problem, which is the so-called suboptimal approach to the stochastic evolution game problem in (9.72). The suboptimal stochastic evolution game in (9.73) is equivalent to the following constrained stochastic Nash game [42, 44, 106]

$$\min_{u(t)} \max_{v(t)\in L_2[0,t_p]} E\int_0^{t_p}\left(\tilde{x}^T(t)Q\tilde{x}(t)+u^T(t)u(t)-ev^T(t)v(t)\right)dt \le eE\tilde{x}^T(0)\tilde{x}(0) \tag{9.74}$$

Let us denote

$$J = E\int_0^{t_p} \left(\tilde{x}^T(t)Q\tilde{x}(t) + u^T(t)u(t) - ev^T(t)v(t)\right)dt \tag{9.75}$$

We then need two steps to solve the constrained stochastic game problem in (9.74). The first step is to solve the following stochastic Nash game problem

$$\begin{aligned} J_o &\triangleq \min_{u(t)} \max_{v(t)\in L_2[0,t_p]} J \\ &= \min_{u(t)} \max_{v(t)\in L_2[0,t_p]} E\int_0^{t_p} \left(\tilde{x}^T(t)Q\tilde{x}(t) + u^T(t)u(t) - ev^T(t)v(t)\right)dt \end{aligned} \tag{9.76}$$

and the second step is to solve the following constraint problem

$$J_o \le eE\tilde{x}^T(0)\tilde{x}(0) \tag{9.77}$$

By solving (9.76) and (9.77) for the constrained stochastic Nash game in (9.74), we get the following result [102].

Proposition 9.7 [102]: The stochastic evolutionary game problem of the evolutionary gene network in (9.74) can be solved by the following evolutionary strategy $u^*(t)$ and the worst-case environmental disturbance $v^*(t)$

$$u^*(t) = s^*(t)x = \sum_{k=1}^{N_1(t)} A_k p(t-t_k)x = -Px, v^*(t) = \frac{1}{e}B^T Px \tag{9.78}$$

where the positive definite matrix P is the solution of the following Riccati-like inequalities

$$A^T P + PA - PP + \frac{1}{e}PBB^T P + Q + \lambda_2 \sum_{k=1}^{N_{2p}} A_k^T PA_k \le 0 \tag{9.79}$$

$$\text{and } 0 < P \le eI \tag{9.80}$$

where $N_{2p} \triangleq N_2(t_p)$ represents the total number of neutral genetic variations to the present time t_p.

Proof: see 9.5.8.

The term $\lambda_2 \sum_{k=1}^{N_{2p}} A_k{}^T PA_k$ is the effect of random genetic variations that are not selected by evolutionary strategy, i.e., the effect of the so-called neutral genetic mutations on the gene network in the evolutionary process. The solution of the evolutionary game in (9.79) is that based on the worst-case environmental disturbance $v^*(t)$ on the biological network (due to the uncertainty and unpredictability of environmental disturbances, the evolutionary strategy $u^*(t)$ needs to consider the effect of the worst-case disturbance $v^*(t)$ from the game's perspective), thus the minimax strategy is to select adequate genetic variations as new network interactions amounting to $s^*(t)=\Sigma_{k=1}^{N_1(t)} A_k p(t-t_k)=\Sigma_{k=1}^{N_1(t)} A_k=-P$ to enhance the system matrix A, which can minimize the evolution level or maximize the fitness of the biological network with the phenotype at x_e. Since e is the upper bound of the evolvability e_o in the stochastic evolutionary game in (9.73), the minimum value of e in (9.73) will approach the evolvability $e_o=1/f_o$ where f_o is the maximum fitness of the biological network by the minimax evolutionary strategy. Therefore, we need to solve the following constrained optimization problem for e_o or f_o in the stochastic biological network in the evolution

$$\frac{1}{f_o}=e_o \triangleq \min_{P>0} e \tag{9.81}$$

subject to (9.79) and (9.80)

Remark 9.12

(i) The solution of e_o in (9.81) can be solved by decreasing e until no positive solution $P>0$ exists in (9.80) and within $0<P<eI$ in (9.81). (ii) After solving e_o and f_o from (9.81), then the Riccati-like inequality in (9.79) becomes

$$A^T P + PA - PP + \frac{1}{e_o} PBB^T P + Q + \lambda_2 \sum_{k=1}^{N_{2p}} A_k^T PA_k \le 0 \tag{9.82}$$

The inequality can be considered to be the phenotypic trait robustness criterion for the stochastic biological network (9.68) under the minimax (game) evolutionary strategy in the evolutionary process. For the nominal linear biological network in (9.62), if $A^TP+PA\le 0$ for some $P>0$, then the biological network is stable based on Lyapunov stability, i.e., all the eigenvalues of A have to be in the left-hand side of the complex s-domain (see Figure 9.19). Therefore, $-(A^TP+PA)$ is a measure of network robustness of the nominal network in (9.62). The term $\lambda_2 \Sigma_{k=1}^{N_{2p}} A_k{}^T PA_k$ is the effect of $\Sigma_{k=1}^{N_2(t)} A_k p(t-t_k)$ on the phenotypic trait robustness of the biological network. The term $\frac{1}{e_o} PBB^TP+Q$ is the effect of an environmental disturbance on the phenotypic trait robustness and a measure of the environmental robustness of the biological network. If the evolvability e_o is small, then the environmental robustness of the biological network should be large enough to resist the phenotypic effect of environmental changes and vice versa. The term $-PP$ due to the evolutionary strategy $u^*(t)=-Px$ (i.e., $s^*(t)=\Sigma_{k=1}^{N_1(t)} A_k p(t-t_k)=\Sigma_{k=1}^{N_1(t)} A_k=-P$) could improve the phenotypic trait robustness in (9.82) by enhancing network robustness through negative feedback loops $-Px$ as follows

$$\left(A-\frac{1}{2}P\right)^T P + P\left(A-\frac{1}{2}P\right) + Q + \lambda_2 \sum_{k=1}^{N_{2p}} A_k^T PA_k + f_o PBB^T P \le 0 \tag{9.83}$$

The eigenvalues of $(A-P/2)$ are on the farther left-hand side of the complex s-domain than those of A such that the network robustness of the biological network is improved (see

Figure 9.19). Although the selected phenotypic variations up to the amount $-Px$ caused by the minimax evolutionary strategy can improve the network robustness, the accumulated random neutral genetic variations $\Sigma_{k=1}^{N_2(t)} A_k p(t-t_k)$ harbored in the robust genetic network during the evolutionary process can also lead to a negative effect on the phenotypic trait robustness of the biological network (i.e., the term $\lambda_2 \Sigma_{k=1}^{N_{2p}} A_k{}^T P A_k$ will make the network robustness inequality in (9.83) more difficult to satisfy). In this situation, the phenotypic trait robustness criterion in (9.83) for the stochastic biological network in (9.64) under the minimax (game) evolution strategy in (9.72) can be reformulated as follows

$$\underbrace{\lambda_2 \sum_{k=1}^{N_{2p}} A_k^T P A_k}_{\text{genetic robustness}} + \underbrace{Q + \frac{1}{e_o} PBB^T P}_{\text{environmental robustness}} \le \underbrace{-\left[\left(A - \frac{1}{2}P\right)^T P + P\left(A - \frac{1}{2}P\right)\right]}_{\substack{\text{enhanced network robustness through} \\ \text{the minimaxevolutionary game strategy}}} \tag{9.84}$$

The physical meaning of (9.84) is that if the enhanced network robustness of the biological network through the stochastic minimax (game) evolutionary strategy in (9.72) can confer both genetic robustness to tolerate random Poisson phenotypic variations due to the accumulated neutral genetic variations and environmental robustness to resist the environmental disturbance, then the phenotypic trait of the biological network is maintained in the evolutionary process. Otherwise, the phenotypic trait may not be maintained under environmental disturbance and random genetic variations. In [107] the evolutionary causes of genetic robustness were discussed, using different evolutionary criteria. The correlation with environmental robustness is considered to be the most probable cause of genetic robustness in evolution. From (9.84), based on the congruence scenario, it can be seen that the genetic robustness of a biological network is a by-product of the environmental robustness of the biological network. This is because environmental changes are more frequent than genetic variations in evolution, i.e., when a biological network has developed network robustness to resist environmental disturbances, it would simultaneously provide genetic robustness to tolerate genetic variations. (iii) By the stochastic minimax (game) evolutionary strategy in (9.72), the total accumulated network interaction variations $\Sigma_{k=1}^{N_1(t)} A_k p(t-t_k) = \Sigma_{k=1}^{N_1(t)} A_k$ are chosen by natural selection to the amount of $-P$ to construct negative feedback loops $-Px$ in (9.78). The stochastic evolutionary biological network is thus of the following form

$$\frac{dx(t)}{dt} = (A - P)x(t) + \sum_{k=1}^{N_2(t)} A_k p(t - t_k) x(t) + Bv(t) \tag{9.85}$$

Since the eigenvalues of $A-P$ are in the farther left-hand side of the complex s-domain, the biological network is more robust by the minimax evolutionary strategy to resist all possible environmental disturbances in (9.85). (iv) An evolutionary biological network in (9.85) with greater network robustness in (9.84) will provide a greater buffer to harbor numerous neutral genetic variations (these neutral genetic variations $\Sigma_{k=1}^{N_2(t)} A_k p(t-t_k)$ have no obvious phenotypic effect on the system matrix A of the biological network) simultaneously in the evolutionary process [87, 88, 90]. However, a large amount of neutral genetic variations may provide raw material for new evolutionary possibilities. Therefore, a robust biological network may finally lead to network evolution in the long run when the accumulated neutral genetic variations $\Sigma_{k=1}^{N_2(t)} A_k p(t-t_k)$ are large enough, i.e., the term $\lambda_2 \Sigma_{k=1}^{N_2 p} A_k^T P A_k$ in (9.83) or (9.84) is large enough that the phenotypic trait robustness criterion is violated. (v) In order to provide a buffer against the accumulated neutral phenotypic variations and all possible environmental disturbances, a biological network needs to develop negative feedback loops to improve its network robustness. This is the minimax evolutionary strategy $u^*(t)=-Px$ providing negative feedback loops to enhance network robustness in (9.83) or (9.84). Furthermore, hubs and redundant mechanisms are frequently found in biological networks to provide efficient responses to external stimuli and to attenuate the effect of genetic variations, respectively. They are therefore also favored by natural selection to improve the phenotypic trait robustness of the biological network in the evolutionary process. The evolutionary game strategy through negative feedback interactions $u^*(t)=-Px$ in a genetic network might result in these hubs and redundant mechanisms, in order to improve phenotypic trait robustness in the evolutionary process.

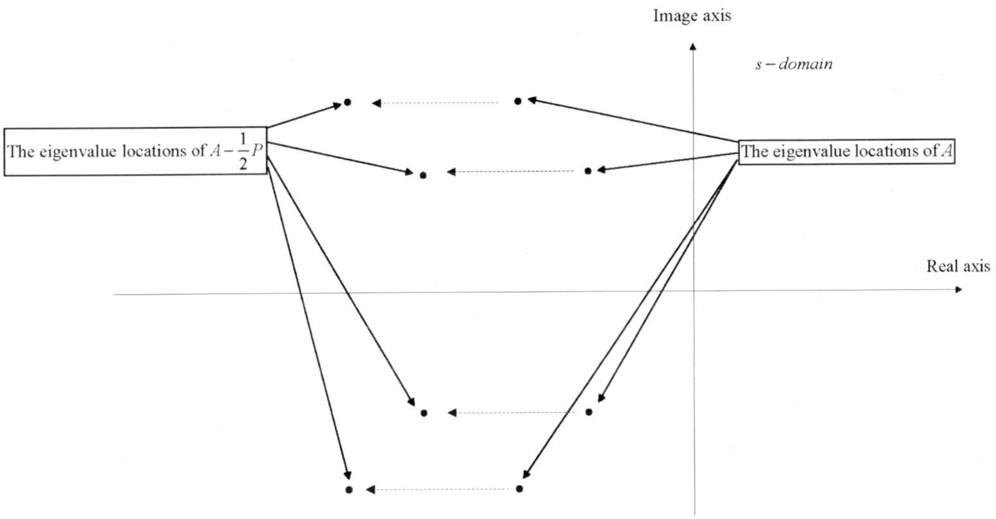

Figure 9.19. The eigenvalues of system matrix A in the evolution network are shifted further to the far left-hand side of the s-domain by the evolutionary strategy $u^*(t)=-Px$, i.e., the eigenvalues of $(A-P/2)$ are farther on the left-hand side of the s-domain.

9.3.3. The Stochastic Game of the Nonlinear Biological Network Evolution

In real biological networks, the system dynamics in (9.62) are always nonlinear and should be modified as follows

$$\frac{dx(t)}{dt} = f(x) \tag{9.86}$$

where $f(x)$ denotes the nonlinear interactive vector among n different species in the biological network. For a nonlinear stochastic biological network with stochastic heritable random parameter fluctuations and non-heritable environmental disturbances in the evolutionary process, the stochastic biological network in (9.68) is modified as the followings Poisson point process

$$\begin{aligned}\frac{dx(t)}{dt} &= f(x)+\sum_{k=1}^{N(t)} f_k(x)\,p(t-t_k)+Bv(t)\\ &= f(x)+\sum_{k=1}^{N_1(t)} f_k(x)\,p(t-t_k)+\sum_{k=1}^{N_2(t)} f_k(x)\,p(t-t_k)+Bv(t)\\ &= f(x)+u(t)+\sum_{k=1}^{N_2(t)} f_k(x)\,p(t-t_k)+Bv(t)\end{aligned} \tag{9.87}$$

with

$$\sum_{k=1}^{N(t)} f_k(x)p(t-t_k)=\sum_{k=1}^{N_1(t)} f_k(x)p(t-t_k)+\sum_{k=1}^{N_2(t)} f_k(x)p(t-t_k)$$

where $f_k(x)$, for k=1, …, $N(t)$ denote the nonlinear stochastic heritable phenotypic variations due to Poisson random genetic variations in the evolutionary process. Sources of these random genetic variations include DNA mutation, deletion, recombination, duplication, inversion, and translocation of chromosomes in the evolutionary process. The nonlinear stochastic system in (9.87) represents a nonlinear biological network under the evolutionary strategy $u(t)=\Sigma_{k=1}^{N_1(t)} f_k(x)p(t-t_k)x(t)$ by natural selection, the neutral genetic variations $\Sigma_{k=1}^{N_2(t)} f_k(x)p(t-t_k)$ and the environmental disturbances $v(t)$ in the evolutionary process.

For the nonlinear stochastic biological network in (9.87), there exist many equilibrium points (phenotypic traits) (see Figure 9.17). Suppose that the phenotypic trait near a stable equilibrium point x_e is of interest in natural selection. For convenience of analysis, the origin of the nonlinear stochastic biological network in (9.87) is shifted to the equilibrium point (phenotypic trait) x_e. In this case, if the shifted nonlinear stochastic network is robustly stable at the origin, then the equilibrium point (phenotypic trait) of interest in natural selection is also robustly stable. Therefore, the analysis of the minimax evolutionary strategy of the nonlinear stochastic biological network in evolution is simplified. Let us denote the phenotypic deviation around the phenotype at x_e as $\tilde{x}(t)=x(t)-x_e$ so then the following shifted nonlinear stochastic biological network is obtained

$$\begin{aligned}\frac{d\tilde{x}(t)}{dt} &= f(\tilde{x})+\sum_{k=1}^{N(t)} f_k(\tilde{x})\,p(t-t_k)+Bv(t)\\ &= f(\tilde{x})+u(t)+\sum_{k=1}^{N_2(t)} f_k(\tilde{x})\,p(t-t_k)+Bv(t)\end{aligned} \tag{9.88}$$

For convenience of notation, we denote $f(\tilde{x}(t)+x_e)$ and $f_k(\tilde{x}(t)+x_e)$ as $f(\tilde{x}(t))$ and $f_k(\tilde{x}(t))$, respectively. In this situation, the origin $\tilde{x}(t)=0$ of the nonlinear stochastic biological network in (9.88) is at the equilibrium point x_e of the original nonlinear stochastic network in (9.87), and $u(t)=s_1(\tilde{x})=\sum_{k=1}^{N_1(t)} f_k(\tilde{x})p(t-t_k)x(t)$ denotes the evolutionary strategy of the biological network to select some phenotypic variations by natural selection to improve the network robustness and thus tolerate the effect of neutral genetic variations $\sum_{k=1}^{N_2(t)} f_k(\tilde{x})p(t-t_k)x(t)$ as well as to resist the environmental disturbances $v(t)$. To maintain the robustness of the phenotypic trait at x_e which is favored by natural selection, the robust evolution of the nonlinear stochastic biological network is to develop an evolutionary strategy $u(t)$ to select some genetic variations to solve the stochastic minimax evolutionary game problem in (9.72) or (9.73). By considering the nonlinear stochastic minimax evolution problem (i.e., the nonlinear evolutionary game problem in (9.72) of the nonlinear stochastic biological network in (9.88), we get the following result [102].

Proposition 9.8 [102]: For the nonlinear stochastic biological network in (9.88), the nonlinear stochastic evolutionary game problem in (9.74) is solved by the following minimax evolutionary strategy $u^*(t)$ and the worst-case environmental disturbance $v^*(t)$

$$\begin{aligned} u^*(t) &= s^*(\tilde{x}) = \sum_{k=1}^{N_1(t)} f_k(\tilde{x})p(t-t_k) = -\frac{1}{2}\left(\frac{\partial V(\tilde{x})}{\partial \tilde{x}}\right), \\ v^*(t) &= \frac{1}{2e}B^T\left(\frac{\partial V(\tilde{x})}{\partial \tilde{x}}\right) \end{aligned} \tag{9.89}$$

where the positive function $V(\tilde{x})>0$ is the positive solution of the following (HJII)

$$\begin{aligned} &\left(\frac{\partial V(\tilde{x})}{\partial \tilde{x}}\right)^T f(\tilde{x}) + \tilde{x}^T Q\tilde{x} - \frac{1}{4}\left(\frac{\partial V(\tilde{x})}{\partial \tilde{x}}\right)^T \left(\frac{\partial V(\tilde{x})}{\partial \tilde{x}}\right) \\ &+\frac{1}{4e}\left(\frac{\partial V(\tilde{x})}{\partial \tilde{x}}\right)^T BB^T\left(\frac{\partial V(\tilde{x})}{\partial \tilde{x}}\right) + \frac{1}{2}\lambda_2\sum_{k=1}^{N_{2P}} f_k^T(\tilde{x})\frac{\partial^2 V(\tilde{x})}{\partial^2 \tilde{x}} f_k(\tilde{x}) \le 0 \end{aligned} \tag{9.90}$$

with

$$V(\tilde{x}(0)) \le e\tilde{x}^T(0)\tilde{x}(0) \tag{9.91}$$

Proof: see 9.5.9.

The last term in (9.90) is the effect of random neutral genetic variations, which are not selected by natural selection but are harbored by network robustness in the evolutionary process [94, 97]. The biological meaning of the evolutionary strategy in (9.89) is that the worst-case environmental disturbance $v^*(t)$ represents the worst-case effect of all possible uncertain environmental disturbances on both the phenotypic deviation and the evolutionary strategy of the biological network in the stochastic game (9.73). In addition, the minimax evolutionary game strategy $u^*(t)$ in (9.89) denotes that, if the number of phenotypic variations selected by natural selection is equal to the number of negative feedback loops $-(\partial V(\tilde{x})/\partial\tilde{x})/2$ then the worst-case effect of all possible environmental disturbances on the phenotypic trait will be minimized; i.e., the phenotypic trait of the biological network will achieve the maximum fitness or the minimum evolution of x_e in (9.73) or (9.81). Since e is the upper bound of the evolvability e_o of the stochastic evolutionary network in (9.73), e_o could be approached for the nonlinear stochastic evolutionary network in (9.87) by minimizing its upper bound e as follows

$$\frac{1}{f_o} = e_o \triangleq \min_{V(\tilde{x})>0} e \tag{9.92}$$

subject to (9.90) and (9.91)

Remark 9.13

(i) In general, the solution of the network evolvability e_o in (9.92) could be obtained by decreasing e until the HJII in (9.90) has no positive solution $V(\tilde{x})>0$ (ii) After solving f_o and e_o in (9.92), then the HJII in (9.90) can be modified as

$$\left(\frac{\partial V(\tilde{x})}{\partial\tilde{x}}\right)^T f(\tilde{x}) + \tilde{x}^T Q\tilde{x} - \frac{1}{4}\left(\frac{\partial V(\tilde{x})}{\partial\tilde{x}}\right)^T\left(\frac{\partial V(\tilde{x})}{\partial\tilde{x}}\right)$$
$$+\frac{1}{4e_o}\left(\frac{\partial V(\tilde{x})}{\partial\tilde{x}}\right)^T BB^T\left(\frac{\partial V(\tilde{x})}{\partial\tilde{x}}\right) + \frac{1}{2}\lambda_2\sum_{k=1}^{N_{2p}} f_k^T(\tilde{x})\frac{\partial^2 V(\tilde{x})}{\partial^2\tilde{x}} f_k(\tilde{x}) \le 0 \tag{9.93}$$

with $V(\tilde{x})\le e_o\tilde{x}^T(0)\tilde{x}(0)$. This HJII can be considered as the phenotypic trait robustness criterion of the nonlinear stochastic biological network in (9.88) with the minimax evolutionary game strategy $u^*(t)$ in the evolutionary process. The term $\frac{1}{4}f_o(\partial V(\tilde{x})/\partial\tilde{x})^T BB^T(\partial V(\tilde{x})/\partial\tilde{x})$ is due to the effect of environmental disturbances on the biological network, and is considered as a measure of environmental robustness. The term $\frac{1}{2}\lambda_2\sum_{k=1}^{N_{2p}} f_k^T(\tilde{x})(\partial V(\tilde{x})/\partial\tilde{x})f_k(\tilde{x})$ is the random effect of the accumulated random neutral genetic variations on the biological network in the evolutionary process. The term $-\frac{1}{4}(\partial V(\tilde{x})/\partial\tilde{x})^T(\partial V(\tilde{x})/\partial\tilde{x})$ is due to the minimax evolutionary game strategy $u^*(t)$, i.e., the robust effect of phenotypic variations $\sum_{k=1}^{N_1(t)} f_k(\tilde{x})p(t-t_k)$ which are selected by natural selection to construct the negative feedback loops $-(\partial V(\tilde{x})/\partial\tilde{x})/2$ to minimize the worst-case effect of all possible environmental disturbances. When the biological network is free of genetic fluctuations, environmental disturbances, and the evolutionary game strategy as in (9.86), i.e., in the nominal nonlinear network case in (9.86), the Lyapunov stability criterion at the equilibrium x_e is $(\partial V(\tilde{x})/\partial\tilde{x})^T f(\tilde{x}(t))\le 0$ and $-(\partial V(\tilde{x})/\partial\tilde{x})^T f(\tilde{x}(t))$ can be considered as the measure of phenotypic stability of x_e for the nominal biological network in (9.86). With the minimax evolutionary game strategy $u^*(t)$ to improve the network robust stability of the biological network in

(9.87) and to maintain the phenotype at x_e in order to resist the effects of genetic variations and environmental disturbances, the phenotypic trait robustness criterion of the evolutionary biological network in (9.93) can be rewritten as follows

$$\underbrace{\frac{1}{2}\lambda_2\sum_{k=1}^{N_{2p}} f_k^T(\tilde{x})\frac{\partial^2 V(\tilde{x})}{\partial^2\tilde{x}} f_k(\tilde{x})}_{\text{genetic robustness}} + \underbrace{\tilde{x}^T Q\tilde{x} + \frac{1}{4e_o}\left(\frac{\partial V(\tilde{x})}{\partial\tilde{x}}\right)^T BB^T\left(\frac{\partial V(\tilde{x})}{\partial\tilde{x}}\right)}_{\text{environmental robustness}}$$
$$\leq \underbrace{-\left(\frac{\partial V(\tilde{x})}{\partial\tilde{x}}\right)^T\left[f(\tilde{x})-\frac{1}{4}\left(\frac{\partial V(\tilde{x})}{\partial\tilde{x}}\right)\right]}_{\text{network robustness through the minimax evolutionary game strategy}} \quad (9.94)$$

From the right-hand side of (9.94), it is apparent that the minimax evolutionary game strategy can improve the network robustness in the evolutionary process by making $(\partial V(\tilde{x})/\partial\tilde{x})^T[f(\tilde{x}(t))-\frac{1}{4}(\partial V(\tilde{x})/\partial\tilde{x})]$ more negative, letting the basin of the phenotypic trait at x_e become much deeper and wider in Figure 9.17. The physical meaning of the phenotypic trait robustness criterion in (9.94) is that if the enhanced network robustness by the minimax evolutionary game strategy can confer enough genetic robustness to buffer the random neutral genetic variations and enough environmental robustness to resist environmental disturbances, then the phenotypic trait of the nonlinear stochastic biological network in (9.88) is maintained in the evolutionary process. By substituting the minimax evolutionary game strategy $u^*(t)$ into the nonlinear stochastic biological network in (9.88), we get [102]

$$\frac{d\tilde{x}(t)}{dt} = \left(f(\tilde{x}) - \frac{1}{2}\frac{\partial V(\tilde{x})}{\partial\tilde{x}}\right) + Bv(t) + \sum_{k=1}^{N_2(t)} f_k(\tilde{x})\,p(t-t_k) \quad (9.95)$$

Since the term $f(\tilde{x}(t))-\frac{1}{2}(\partial V(\tilde{x})/\partial\tilde{x})$ will make the nonlinear stochastic biological network more robustly stable than $f(\tilde{x}(t))$ due to the negative feedback $-\frac{1}{2}(\partial V(\tilde{x})/\partial\tilde{x})$, the biological network with the minimax evolutionary strategy will be more robust to tolerate the random neutral genetic variations and to resist the environmental disturbances $v(t)$ in (9.95) in the evolutionary process. (iii) If the network robustness on the right-hand side of (9.94) is improved by the minimax evolutionary game strategy so that the genetic robustness is increased, then the nonlinear biological network can harbor a large amount of neutral genetic variations. If the stochastic neutral genetic variations are accumulated sufficiently large amount in (9.88) or (9.95) in the evolutionary process, so that the phenotypic trait robustness criterion in (9.94) may finally be violated, then they may provide raw material for a new evolutionary possibility to cause a transition from one phenotypic trait to another through genetic drift [92, 93]; for example, from the basin x_e to another basin x_{1e} in Figure 9.17. At first sight, one might expect an evolutionary strategy to significantly improve network robustness to slow or even stop the evolution of a phenotypic trait in a biological network, but, because the more robust phenotypic trait can harbor a large amount of neutral genetic variations, the robust phenotypic trait of the stochastic biological network might show increased rather than decreased evolutionary potential, in the long term of the biological network evolution. The reasons why increasing network robustness may eventually lead to increasing network evolution include the accumulation of hidden neutral genetic variations that may be useful for later evolution. This is why network robustness is intrinsic to evolution and can eventually improve evolution [91, 92]. This chapter extends these evolution results from the genetic level to the nonlinear stochastic gene network level, using the systematic methods in (9.94) and (9.95) from the point of the view of a stochastic evolutionary game. (iv) The tradeoff between network robustness and the evolvability of the nonlinear stochastic biological network in evolution can be seen from the phenotypic trait robustness criterion in (9.94); i.e., genetic robustness plus environmental robustness is less than or equal to network robustness. In this phenotypic trait robustness criterion, if the biological network has less evolvability (i.e., a small e_o or a large fitness f_o) in evolution, then its environmental robustness becomes larger. In this situation, the network robustness has to be enough large to confer a large environmental robustness to resist environmental disturbances in order to maintain the phenotypic trait robustness, i.e., less evolvability implies a larger network robustness, and vice versa. (v) In general, the cost of the minimax evolutionary strategy in selecting phenotypic variations as negative feedback loops $u^*(t)=-\frac{1}{2}(\partial V(\tilde{x})/\partial\tilde{x})$ in order to improve the network robustness and resist environmental disturbances is much higher [108]. In the case of the stress-avoidance strategy (i.e., $u(t)=0$), the basin of the equilibrium point x_e in Figure 9.17 becomes shallow, i.e., the term $(\partial V(\tilde{x})/\partial\tilde{x})^T[f(\tilde{x}(t))-\frac{1}{4}(\partial V(\tilde{x})/\partial\tilde{x})]$ on the right-hand side of (9.94) is reduced to $(\partial V(\tilde{x})/\partial\tilde{x})^T[f(\tilde{x}(t))-\frac{1}{4}(\partial V(\tilde{x})/\partial\tilde{x})]$ and is less negative. In this case, the network robustness becomes smaller and the environmental robustness must also be smaller in evolution, so that the evolvability e_o will become larger and the biological network will be more flexible (or adaptive) in responding to environmental stresses under rapid environmental changes [108].

9.3.4. Stochastic Evolutionary Game of the Nonlinear Stochastic Biological Network Based on the Global Linearization Method

In general, it is still very difficult to solve the HJII-constrained optimization problem in (9.92) for the stochastic minimax evolutionary strategy in a nonlinear stochastic biological network in evolution or the HJII problem in (9.94) for the phenotypic trait robustness criterion of the nonlinear stochastic biological network under natural selection in evolution. At present, there is no good method to solve the nonlinear partial differential HJII problem either analytically or numerically [83, 84, 109]. In this case, in order to simplify the stochastic evolutionary game problem, the global linearization technique [41, 110] is employed to interpolate a nonlinear stochastic biological network with a set of local stochastic linearized biological networks.

Using the global linearization method, we suppose that all the global linearizations of the nonlinear stochastic biological network in (9.88) are bounded by a polytope, consisting of M vertices as [41, 110]

$$\begin{bmatrix} \frac{\partial f(\tilde{x})}{\partial \tilde{x}} \\ \frac{\partial f_1(\tilde{x})}{\partial \tilde{x}} \\ \vdots \\ \frac{\partial f_{N(t)}(\tilde{x})}{\partial \tilde{x}} \end{bmatrix} \in Co\left[\begin{bmatrix} A_1 \\ A_{11} \\ \vdots \\ A_{N(t)1} \end{bmatrix} \cdots \begin{bmatrix} A_i \\ A_{1i} \\ \vdots \\ A_{N(t)i} \end{bmatrix} \cdots \begin{bmatrix} A_M \\ A_{1M} \\ \vdots \\ A_{N(t)M} \end{bmatrix}\right], \forall \tilde{x}(t) \tag{9.96}$$

where Co denotes the convex hull of the polytope with M vertices. That is, if the local linearized systems at all $\tilde{x}$ are inside the convex hull Co, then the evolutionary trajectory $\tilde{x}$ of the nonlinear stochastic evolutionary network in (9.88) will belong to the convex combination of the stable trajectories of the following M local linearized networks, defined at M vertices of the polytope [41, 110]

$$\frac{d\tilde{x}(t)}{dt} = A_i\tilde{x} + Bv(t) + u(t) + \sum_{k=1}^{N_2(t)} A_{ki}p(t-t_k)\tilde{x}, i = 1,\ldots,M \tag{9.97}$$

Based on the global linearization theory [41], if (9.96) holds, then every evolutionary trajectory of the nonlinear stochastic biological network in (9.88) can be represented by a convex combination of the M local linearized biological networks in (9.97). This combination can be represented by

$$\frac{d\tilde{x}(t)}{dt} = \sum_{i=1}^{M} \alpha_i(\tilde{x})\left(A_i\tilde{x} + Bv(t) + u(t) + \sum_{k=1}^{N_2(t)} A_{ki}p(t-t_k)\tilde{x}\right) \tag{9.98}$$

where the interpolation function $\alpha_i(\tilde{x})$ satisfies $0 \le \alpha_i(\tilde{x}) \le 1$ and $\Sigma_{i=1}^{M} \alpha_i(\tilde{x})=1$, i.e., the evolutionary trajectory of the nonlinear stochastic biological network in (9.88) can be represented by the evolutionary trajectory of the interpolated biological network in (9.98). In (9.98), the evolutionary strategy $u^*(t)=\Sigma_{k=1}^{N_1(t)} f_k(\tilde{x})p(t-t_k)=\Sigma_{i=1}^{M}\Sigma_{k=1}^{N_1(t)} \alpha_i(\tilde{x})A_{ki}p(t-t_k)\tilde{x}$ by the global linearization technique. For the minimax evolutionary strategy problem (9.72) of the interpolated evolutionary biological network in (9.98), we get the following result [102].

Proposition 9.9 [102]: For the interpolated evolutionary biological network in (9.98), the stochastic evolutionary game problem in (9.74) is solved by the following minimax evolutionary strategy $u^*(t)$ and the worst-case environmental disturbance $v^*(t)$

$$u^*(t) = \sum_{i=1}^{M}\sum_{k=1}^{N_1(t)} \alpha_i(\tilde{x})A_{ki}p(t-t_k)\tilde{x} = -P\tilde{x}, v^*(t) = \frac{1}{e}B^T P\tilde{x} \tag{9.99}$$

where the positive definite matrix P is the solution of the following Riccati-like inequalities

$$A_i^T P + PA_i + Q - PP + \frac{1}{e} PBB^T P + \lambda_2 \sum_{k=1}^{N_{2p}} A_{ki}^T PA_{ki} \leq 0, i = 1,\ldots,M \tag{9.100}$$

$$0 < P \leq eI \tag{9.101}$$

Proof: see 9.5.10.

Remark 9.14

(i) In comparison with **Proposition 9.7** of the stochastic linear biological network, the Riccati-like inequality in (9.79) is based only on the local linearized network at $\tilde{x}(t)=0$ when the Riccati-like inequalities in (9.100) are based on M local linearized networks in (9.97) at the M vertices of the polytope in (9.96). Therefore, the result of the minimax evolutionary game strategy in **Proposition 9.9** is more suitable for nonlinear stochastic biological networks in evolution. (ii) The Riccati-like inequalities in (9.100) can be considered as the local linearized HJII in (9.90) at M vertices of the polytope in (9.96). In general, it is very difficult to solve the HJII. However, it is very easy to solve the Riccati-like inequalities in (9.100) by transforming (9.100) into equivalent linear matrix inequalities (LMIs)

$$\begin{bmatrix} XA_i^T + A_i X - I + \frac{1}{e} BB^T & X & XA_{1i}^T & \cdots & XA_{N_{2p}i}^T \\ X & -Q^{-1} & 0 & \cdots & 0 \\ A_{1i} X & 0 & -\frac{1}{\lambda_2} X & \ddots & \vdots \\ \vdots & \vdots & \ddots & \ddots & 0 \\ A_{N_{2p}i} X & 0 & \cdots & 0 & -\frac{1}{\lambda_2} X \end{bmatrix} \leq 0 \tag{9.102}$$

$$\text{and } X \triangleq P^{-1} > 0, \text{ for } i = 1,\ldots,M$$

by using the Schur complement transformation method [41], which can be easily solved by the LMI toolbox in Matlab.

Since e in (9.100) is the upper bound of the evolvability of the evolutionary game in (9.73), the minimum value of e will approach the evolvability $e_o = 1/f_o$. Therefore, we need to solve the following constrained optimization problem for e_o or f_o in the nonlinear stochastic biological network in evolution

$$\frac{1}{f_o} = e_o \triangleq \min_{P>0} e \tag{9.103}$$
subject to (9.101) and (9.102)

Remark 9.15

(i) The e_o can be solved by decreasing e until (9.102) has no positive solution P within $0<P\leq eI$ with the help of the LMI toolbox in Matlab. (ii) After solving e_o and f_o from (9.103), the Riccati-like inequalities in (9.100) and (9.101) then need to be modified as

$$A_i^T P + PA_i + Q - PP + \frac{1}{e_o} PBB^T P + \lambda_2 \sum_{k=1}^{N_{2p}} A_{ki}^T PA_{ki} \leq 0, i = 1, \ldots, M \tag{9.104}$$

$$0 < P \leq e_o I \tag{9.105}$$

After solving P in (9.104), then $u^*(t)=-P\tilde{x}$ and $v^*(t)=f_o B^T P\tilde{x}$ in (9.99) form the minimax evolutionary strategy in (9.72) for the nonlinear stochastic biological network in evolution. (iii) With the minimax evolutionary strategy, the Riccati-like inequalities in (9.104) and (9.105) can be considered as the phenotypic trait robustness criteria of the biological network in evolution. If (9.104) and (9.105) are satisfied, then the phenotypic trait of the nonlinear stochastic biological network in (9.88) or (9.98) is maintained in the basin around the equilibrium point x_e under environmental disturbances and neutral genetic variations in the evolutionary process. If (9.104) and (9.105) have no positive solution P with $0<P\leq e_o I$, then the phenotypic trait at x_e might not exist again or might transit to another phenotypic trait at another equilibrium point under the environmental disturbances and genetic variations in evolution. (iii) The minimax evolutionary strategy $u^*(t)$ in (9.99) can enhance the network robustness by shifting the eigenvalues of the local linear networks farther to the far left-hand side of the s-complex domain as follows

$$\left(A_i - \frac{1}{2}P\right)^T P + P\left(A_i - \frac{1}{2}P\right) + Q + \frac{1}{e_o} PBB^T P + \lambda_2 \sum_{k=1}^{N_{2p}} A_{ki}^T PA_{ki} \leq 0, i = 1, \ldots, M \tag{9.106}$$

That is, every interaction matrix A_i of the local linearized biological networks is shifted by $-\frac{1}{2}P$ so that eigenvalues of $A_i-\frac{1}{2}P$ are on the farther left-hand side of the s-complex domain, and thus the evolutionary network is more robust to simultaneously tolerate the local genetic variations and resist the environmental disturbances. In general, the minimax evolutionary game strategy is to select the total network variations $\sum_{i=1}^{M} \sum_{k=1}^{N_1(t)} \alpha_i(\tilde{x}) A_{ki} p(t-t_k)=-P$, which consist of negative feedback loops for some positive matrix P. This is in order to select some adequate genetic variations to form negative feedback circuits in the biological network, so as to improve network robustness to resist environmental disturbances and to attenuate the effects of neutral genetic variation in the evolutionary process. (iv) The phenotypic trait robustness criterion in (9.106) can be reformulated as follows

$$\underbrace{\lambda_2 \sum_{k=1}^{N_{2p}} A_{ki}^T P A_{ki}}_{\text{local genetic robustness}} + \underbrace{Q + \frac{1}{e_o} P B B^T P}_{\text{localenvironmental robustness}} \leq \underbrace{-\left[\left(A_i - \frac{1}{2}P\right)^T P + P\left(A_i - \frac{1}{2}P\right)\right]}_{\text{local network robustness through the minimaxevolutionary game strategy}} \quad , i = 1, \ldots, M \tag{9.107}$$

The biological meaning in (9.107) is that if the enhanced local network robustness of each local linearized biological network by the minimax evolutionary game strategy can confer both local neutral genetic robustness to buffer local genetic variations as well as local environmental robustness to resist the local environmental disturbance in evolution, then the phenotypic trait of the nonlinear stochastic biological network with the minimax evolutionary game strategy can be maintained in the basin of the equilibrium point at x_e in the evolutionary process. (v) Since the minimax evolutionary game strategy can improve the local network robustness of local linearized networks on the right-hand side of (9.107), the nonlinear stochastic biological network will harbor more neutral genetic variations and will have greater environmental robustness to decrease the evolvability e_o (or increase the fitness f_o) to attenuate the effect of the environmental stimulus. However, when the harbored neutral genetic variations are accumulated by the robust biological network to the extent that the first term on the left-hand side of (9.107) becomes very large and the phenotypic robustness criterion in (9.107) cannot be guaranteed in the evolutionary process, with the help of an environmental disturbance, the phenotype of the biological network might be shifted from one equilibrium point to another equilibrium point in Figure 9.17 and start another period of network evolution at the other equilibrium point (phenotype) favored by natural selection.

Remark 9.16

If we consider network evolution that can be only plausibly studied at a population level, the evolutionary game should naturally invoke an n-tuple game with m-players under k environmental dynamics. In this n-tuple evolutionary game problem, the linear stochastic biological network in (9.68) should be modified as

$$\frac{dx(t)}{dt} = Ax(t) + u(t) + s_2(t)x(t) + \sum_{i=1}^{m} B_i u_i(t) + \sum_{j=1}^{k} D_j v_j(t) \tag{9.108}$$

where $u_i(t)$ denotes the i-th decision variable from other players (i.e., other competitive populations) and $v_j(t)$ denotes the j-th environmental dynamics. Further, the evolutionary game problem in (9.72) should be modified as the following n-tuple evolutionary game [102]

$$\min_{u(t)} \max_{\substack{u_i(t),i=1,\ldots,m \\ v_j(t),j=1,\ldots,k}} \frac{E\int_0^{t_p}\left(\tilde{x}^T(t)Q\tilde{x}(t)+u^T(t)u(t)\right)dt}{E\tilde{x}^T(0)\tilde{x}(0)+E\sum_{i=1}^{m}\int_0^{t_p}u_i^T(t)u_i(t)dt+E\sum_{j=1}^{k}\int_0^{t_p}v_j^T(t)v_j(t)dt} \tag{9.109}$$

i.e., the decision variables $u_i(t)$ from other players and $v_j(t)$ from environmental dynamics are decided to maximize their effect on the phenotypic deviation $\tilde{x}(t)$ and evolutionary strategy $u(t)$ of the evolutionary network while the evolutionary strategy $u(t)$ is decided by the evolutionary network to minimize the worst-case phenotypic deviation and the effort (or cost) of evolutionary strategy. Actually, for the simplicity of solving this n-tuple evolutionary game in the evolutionary network problem, the stochastic network dynamics in (9.108) can be represented by

$$\frac{dx(t)}{dt}=Ax(t)+u(t)+s_2(t)x(t)+BV(t) \tag{9.110}$$

where

$$B\triangleq\begin{bmatrix}B_1 & \cdots & B_m & D_1 & \cdots & D_k\end{bmatrix}$$
$$V(t)\triangleq\begin{bmatrix}u_1(t) & \cdots & u_m(t) & v_1(t) & \cdots & v_k(t)\end{bmatrix}$$

In this situation, the n-tuple game in (9.109) for evolutionary game problem is reduced to the 2-tuple game in (9.72), i.e., the decision variable $v(t)$ of 2-tuple game in (9.72) can represent $V(t)$ of M-players and k environmental dynamics in (9.108) and the n-tuple evolutionary game problem in (9.109) can be replaced by 2-tuple evolutionary game in (9.72). Similarly, the result of nonlinear stochastic evolutionary game problem in (9.87) and (9.72) can be easily extended to an n-tuple game problem of nonlinear stochastic network with m-players (populations) and k environmental dynamics. Therefore, the proposed theoretical approach to evolutionary game strategy naturally invokes an n-tuple game with M competitive populations and k environmental disturbances if B and $V(t)$ in (9.68) can be replaced by $B=[B_1 \ldots B_m\ D_1 \ldots D_k]$ and $V(t)=[u_1(t) \ldots u_m(t)\ v_1(t) \ldots v_k(t)]$ of the evolutionary network in (9.110), respectively.

9.3.5. Simulation Example of Evolutionary Networks with Stochastic Game Strategy

Consider the generic inhibition and activation model in evolution in Figure 9.20, which is a known metabolic pathway and has been widely used for imitating characteristics of a real metabolic pathway [111, 112]. In this metabolic pathway, the metabolite X_3 converted from the metabolite X_2 inhibits an early step in its own production pathway, which is the synthesis of X_1. The metabolite X_2 is converted from the metabolite X_1 which is a divergence branching point. The degradation processes of X_1 into X_2 or X_4, respectively, are independent of each other. Then, the metabolite X_4 modulates downstream for the transformation of X_3 in an activation way. Let's denote $x_1(t)$, $x_2(t)$, $x_3(t)$, and $x_4(t)$ as the concentrations of metabolites X_1, X_2, X_3, and X_4, respectively. Based on the metabolic pathway [111, 112] in evolution in Figure 9.20, we have the following metabolic reaction network

$$\begin{bmatrix} \frac{dx_1(t)}{dt} \\ \frac{dx_2(t)}{dt} \\ \frac{dx_3(t)}{dt} \\ \frac{dx_4(t)}{dt} \end{bmatrix} = \begin{bmatrix} f_1(x) \\ f_2(x) \\ f_3(x) \\ f_4(x) \end{bmatrix} + \sum_{k=1}^{N_1(t)} \begin{bmatrix} \lambda_{11k} & \lambda_{12k} & \cdots & \lambda_{14k} \\ \lambda_{21k} & \lambda_{22k} & \cdots & \lambda_{24k} \\ \vdots & \vdots & \ddots & \vdots \\ \lambda_{41k} & \lambda_{42k} & \cdots & \lambda_{44k} \end{bmatrix} \begin{bmatrix} x_1(t) \\ x_2(t) \\ x_3(t) \\ x_4(t) \end{bmatrix} p(t-t_k)$$

$$+ \sum_{k=1}^{N_2(t)} \begin{bmatrix} \lambda_{1k}\alpha_1 x_3^{g_{13}}(t) \\ \lambda_{2k}\alpha_2 x_1^{g_{21}}(t) \\ \lambda_{3k}\alpha_3 x_2^{g_{32}}(t) \\ \lambda_{4k}\alpha_4 x_1^{g_{41}}(t) \end{bmatrix} p(t-t_k) + \begin{bmatrix} v_1(k) \\ v_2(k) \\ v_3(k) \\ v_4(k) \end{bmatrix} \tag{9.111}$$

with

$$f(x) \triangleq \begin{bmatrix} f_1(x) \\ f_2(x) \\ f_3(x) \\ f_4(x) \end{bmatrix} \triangleq \begin{bmatrix} \alpha_1 x_3^{g_{13}}(t) - \beta_1 x_1^{h_{11}}(t) \\ \alpha_2 x_1^{g_{21}}(t) - \beta_2 x_2^{h_{22}}(t) \\ \alpha_3 x_2^{g_{32}}(t) - \beta_3 x_3^{h_{33}}(t) x_4^{h_{34}}(t) \\ \alpha_4 x_1^{g_{41}}(t) - \beta_4 x_4^{h_{44}}(t) \end{bmatrix}$$

where α_1, …, α_4 and β_1, …, β_4 are the rate constants of net influxes and effluxes with kinetic orders g_{ij} and h_{ij}, $i, j \in S$ and $S \triangleq \{1,2,3,4\}$. The parameters λ_{1k}, …, λ_{4k} are represented as weights of the random phenotypic variations of the biological network which have no direct phenotypic effect on the network. The parameters of this metabolic reaction network are chosen as follows [112]:

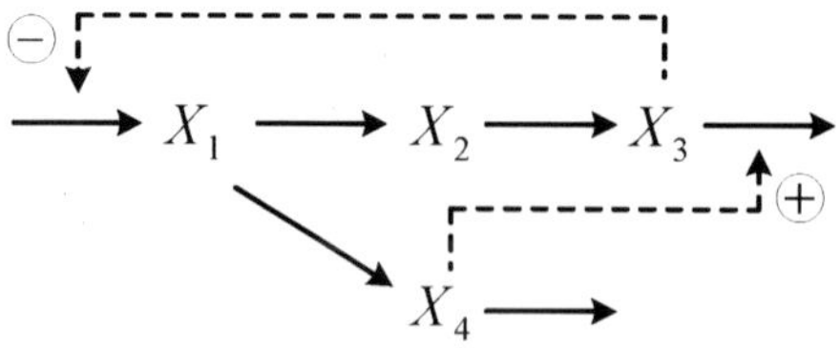

Figure 9.20. A real branched metabolic pathway of the generic inhibition and activation model where X_1, …, X_4 are metabolites.

$$\alpha_1 = 12; \alpha_2 = 8; \alpha_3 = 3; \alpha_4 = 2; \beta_1 = 10; \beta_2 = 3; \beta_3 = 5; \beta_4 = 6; g_{13} = -0.8;$$
$$g_{21} = 0.5; g_{32} = 0.75; g_{41} = 0.5; h_{11} = 0.5; h_{22} = 0.75; h_{33} = 0.5; h_{34} = 0.2;$$
$$h_{44} = 0.8; EV_{ik} = V = 0.1; EV_{ik}^2 = 0.04; E\lambda_{ik} = 0.1; E\lambda_{ik}^2 = 0.04$$

The mean of the Poisson process $p(t)$ is imposed on λ=0.01. The equilibrium point x_e of thc nominal mctabolic rcaction nctwork is $[x_{e1}\ x_{e2}\ x_{e3}\ x_{e4}]^T \simeq [0.400\ 2.006\ 2.228\ 0.143]^T$ simulated from the initial condition $x(0) \triangleq [1.4\ 2.25\ 1.25\ 0.45]^T$. For simplicity of analysis, the state of the network is shifted via the form $\tilde{x}(t)=x(t)-x_e$. Thus, the new equilibrium point of the phenotype concerned is at $\tilde{x}_e=0_{4\times1}$ throughout this simulation example.

Based on the above evolutionary game strategy, the following phenotypic variations in the network in (9.111)

$$\sum_{k=1}^{N_1(t)} \begin{bmatrix} \lambda_{11k} & \lambda_{12k} & \cdots & \lambda_{14k} \\ \lambda_{21k} & \lambda_{22k} & \cdots & \lambda_{24k} \\ \vdots & \vdots & \ddots & \vdots \\ \lambda_{41k} & \lambda_{42k} & \cdots & \lambda_{44k} \end{bmatrix} \begin{bmatrix} \tilde{x}_1(t) \\ \tilde{x}_2(t) \\ \tilde{x}_3(t) \\ \tilde{x}_4(t) \end{bmatrix} p(t-t_k) \quad (9.112)$$

are selected by natural selection to construct negative feedback circuits to modify the metabolic reaction network, in order to resist the environmental disturbance and to tolerate the following effect due to neutral genetic variations

$$\sum_{k=1}^{N_2(t)} \begin{bmatrix} \lambda_{1k}\alpha_1 x_3^{g_{13}}(t) \\ \lambda_{2k}\alpha_2 x_1^{g_{21}}(t) \\ \lambda_{3k}\alpha_3 x_2^{g_{32}}(t) \\ \lambda_{4k}\alpha_4 x_1^{g_{41}}(t) \end{bmatrix} p(t-t_k) \quad (9.113)$$

where the Poisson counting process $N(t)=N_1(t)+N_2(t)$ for genetic variations in the evolutionary process is described in Figure 9.22. The evolutionary game strategy in **Proposition 9.9** is to select genetic variations to construct the following negative feedback loops to improve the network robustness

$$u^*(t) = \sum_{k=1}^{N_1(t)} \begin{bmatrix} \lambda_{11k} & \lambda_{12k} & \cdots & \lambda_{14k} \\ \lambda_{21k} & \lambda_{22k} & \cdots & \lambda_{24k} \\ \vdots & \vdots & \ddots & \vdots \\ \lambda_{41k} & \lambda_{42k} & \cdots & \lambda_{44k} \end{bmatrix} \begin{bmatrix} \tilde{x}_1(t) \\ \tilde{x}_2(t) \\ \tilde{x}_3(t) \\ \tilde{x}_4(t) \end{bmatrix} p(t-t_k) = -P\tilde{x}(t) \quad (9.114)$$

where P is solved from the constrained optimization problem in (9.103). Based on the minimax game, the global linearization technique and the LMI toolbox in Matlab, the following result is obtained

$$P \simeq 10^{-4} \times \begin{bmatrix} 1.352 & 0.039 & -0.290 & 0.116 \\ * & 0.003 & -0.003 & -0.008 \\ * & * & 0.074 & -0.050 \\ * & * & * & 0.064 \end{bmatrix} \quad (9.115)$$

We also find the evolvability $e_o \simeq 0.7843$ and the maximum fitness f_o 1.2750. The simulation result of the metabolic reaction network with the evolutionary game strategy is shown in Figure 9.21. It can be seen that the phenotypic trait of the metabolic reaction network is maintained at the equilibrium point $\tilde{x}_e = 0_{4\times1}$ or $x_e \simeq [0.400\ 2.006\ 2.228\ 0.143]^T$ for a period of time from t=0 to t=383 in Figure 9.21, and the evolvability of the metabolic reaction network by the evolutionary game strategy in this period is estimated with $v_i(t)$, i=1, …, 4 standard white noises, by a Monte Carlo simulation with 10000 runs, as follows

$$\frac{E\int_0^{383}\left(\tilde{x}^T(t)Q\tilde{x}(t)+u^T(t)u(t)\right)dt}{E\tilde{x}^T(0)\tilde{x}(0)+E\int_0^{383}v^T(t)v(t)dt} \simeq 0.7066 \le e_o = 0.7843 \quad (9.116)$$

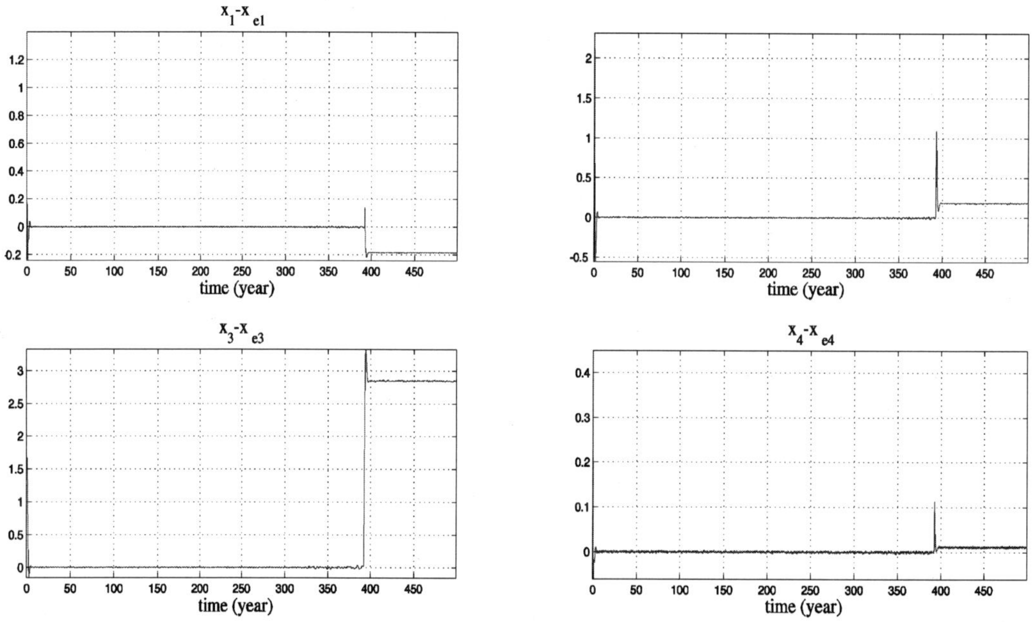

Figure 9.21. The trajectories of the metabolic reaction network with states $\tilde{x}_1, \ldots, \tilde{x}_4$ where $\tilde{x}_i(t)=x_i(t)-x_{ei}$, for i=1, …, 4 suffering from environmental disturbances $v_1(t), \ldots, v_4(t)$ and the Poisson genetic variations with the Poisson counting processes $N(t)=N_1(t)+N_2(t)$ shown in Figure 9.22. It can be seen that the evolutionary game strategy can maintain phenotype stability around x_e for a period of time from t=0 to t=383. However, when the neutral genetic variations are continually accumulated by the robust network, it is seen that the phenotype robustness criterion in (9.107) is violated, and the phenotype of the metabolic reaction network becomes perturbed by neutral genetic variations after t=383, finally shifting to another equilibrium point to start another evolutionary period.

From the calculation of the evolvability of the metabolic reaction network in (9.116), it can be seen that the estimated evolvability e_o by the proposed method is conservative. The main reasons are the conservativeness of the global linearization and the LMIs used to approximate the nonlinear stochastic reaction network and the HJIIs, respectively. Furthermore, the convex optimization method employed by the LMI toolbox in Matlab to solve the LMI constraints for evolvability in (9.103) can also lead to conservative results.

As the neutral genetic variations are continually accumulated, after the negative feedback loops in (9.114) have been constructed by the evolutionary game strategy (see Figure 9.22), from the simulation results in Figure 9.21, it can be seen that by the evolutionary game strategy, the environmental disturbance can be buffered and neutral genetic variations can be tolerated by the metabolic reaction network only for a certain period of evolution time (from t=0 to t=383). However, when the neutral genetic mutations accumulate to the extent that the phenotypic trait robustness criterion of the evolutionary biological network in (9.94) or (9.107) is violated at approximately t=383, the network stability will be destroyed and the phenotype transits to another equilibrium to begin another period of network evolution.

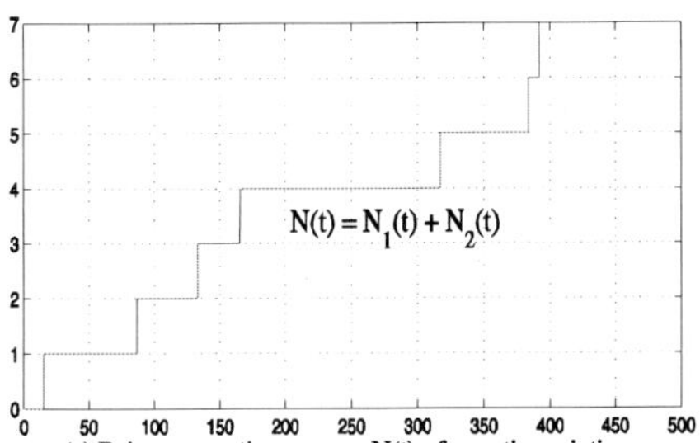

(a) Poisson counting process N(t) of genetic variations.

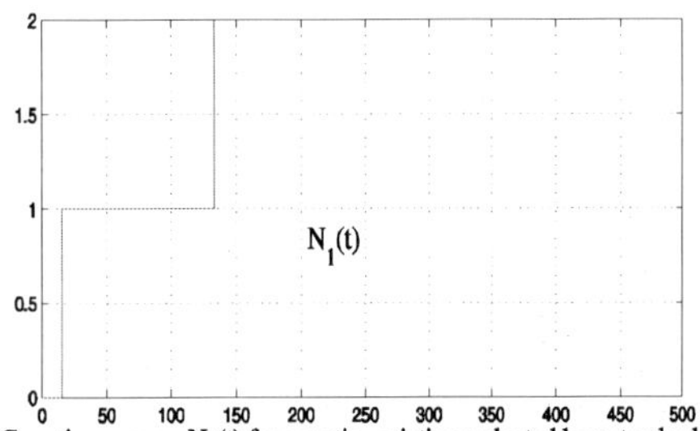

(c) Counting process N_1(t) for genetic variations selected by natural selection and the evolution strategy to construct feedback loops finished at t = 133.

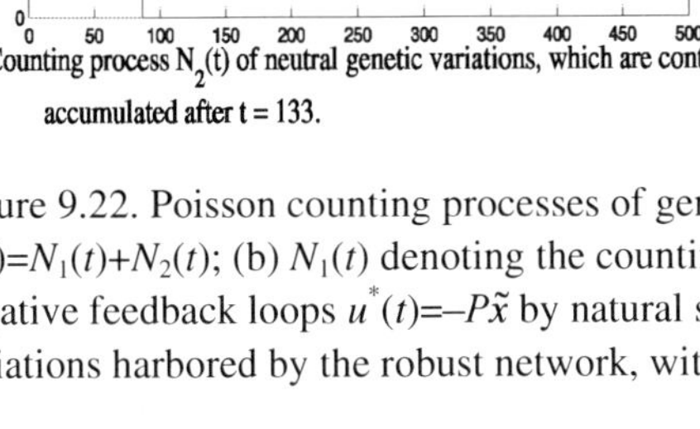

(b) Counting process N_2(t) of neutral genetic variations, which are continually accumulated after t = 133.

Figure 9.22. Poisson counting processes of genetic variations in the evolutionary process, for (a) $N(t)=N_1(t)+N_2(t)$; (b) $N_1(t)$ denoting the counting process of genetic variations selected for constructing negative feedback loops $u^*(t)=-P\tilde{x}$ by natural selection; and (c) $N_2(t)$ denoting the neutral genetic variations harbored by the robust network, without direct effect on the network phenotype.

9.3.6. Discussion

In this chapter, a population of evolutionary biological networks with genetic variations selected by natural selection to resist intrinsic genetic variations and to buffer environmental disturbances to maintain its phenotypic trait is modeled as a robust nonlinear stochastic stabilization system. Since environmental disturbances are uncertain and unpredictable, in order to maintain the robust stability of the phenotypic trait of interest by natural selection, the minimax evolutionary game strategy for phenotypic trait robustness in (9.72) is developed to select some adequate phenotypic variations by natural selection to improve network robustness. This minimizes the worst-case effect of uncertain environmental disturbances on the phenotypic deviation from its phenotypic trait in the evolutionary process. The minimax design scheme has been widely used for robust control in robot systems [44, 106], robust synthetic biology design [42, 82], and robust reference control of economic systems [44] when some system parameters are perturbative and external disturbances are uncertain and unpredictable, so that their worst-case effect on system performance has to be as small a value as possible [42, 44, 82, 104, 106, 107, 113]. With the minimax evolutionary game strategy, the phenotypic trait robustness criterion of the biological network in evolution is found to be: genetic robustness plus environmental robustness is less than or equal to enhanced network robustness. That is, if the enhanced network robustness by the evolutionary game strategy can confer both genetic robustness to buffer the random neutral genetic variations as well as environmental robustness to resist environmental disturbances in the evolutionary process, then the network phenotypic trait is robustly stable in the basin around the equilibrium point at x_e of interest by natural selection. The minimax evolutionary game strategy can improve the network robustness to enhance the phenotypic trait robustness by modifying the network interaction matrix A by $A-\frac{1}{2}P$ in (9.84) (or $f(\tilde{x}(t))$ by $f(\tilde{x}(t))-\frac{1}{4}(\partial V(\tilde{x})/\partial \tilde{x})$ in (9.94) through negative feedback loops, which could decrease the evolvability e_o of the biological network in the evolutionary process. This result extends the idea of Waddington [114] from the gene level to the biological network level. Based on the phenotypic trait robustness criterion in (9.84), (9.94), or (9.107), the correlation between genetic robustness, environmental robustness, and network robustness by recent genomic experiments in yeast [115], i.e., genes with similar robustness to genetic, environmental, and phenotypic perturbations in evolution, can be rationally explained from the network robustness perspective. That is, if the network robustness of a biological network is enhanced by the minimax evolutionary game strategy (i.e., the basin of x_e in Figure 9.17 becomes larger and deeper), then more neutral genetic variations and environmental disturbances can be overcome singly or simultaneously, and the phenotypic trait can be maintained by the minimax evolutionary game strategy through natural selection during the evolutionary process. The fitness f_o of the biological network, which is inversely proportional to the evolvability e_o, is increased by the minimax evolutionary game strategy via natural selection in **Proposition 9.8**, so that the phenotypic trait is more robust in (9.94) due to the larger and deeper basin caused by the adaptive negative feedback loops $u^*(t)=\sum_{k=1}^{N_1(t)} f_k(\tilde{x})p(t-t_k)=-(\partial V(\tilde{x})/\partial \tilde{x})/2$ in the evolutionary process. However, when large amounts of neutral genetic variations are accumulated continuously by the enhanced network robustness through the minimax evolutionary game strategy, they may provide more raw material for new evolutionary possibilities through random genetic drift [114]. Therefore, the minimax evolutionary game strategy may improve the network robustness to maintain the phenotypic trait in a period of evolutionary process, and may eventually improve the evolution of the biological network in the long term. This phenomenon has been illustrated by the *in silico* example of the evolutionary genetic regulatory network in Figure 9.21.

From the phenotypic trait robustness criteria in (9.84) and (9.94), it is observed that there are two ways to improve the phenotypic trait robustness in evolution. One way is to improve the network robustness (i.e., to make the right-hand side of (9.84) or (9.94) as large a value as possible so that the phenotypic trait robustness criterion always holds), and the other way is to reduce the effect of neutral genetic variations and environmental disturbances on the biological network (i.e., to make the left-hand side of (9.84) or (9.94) as small a value as possible). The minimax evolutionary game strategy is to select heritable phenotypic variations such that $u^*(t)=-Px$ in (9.78) for linear biological networks or $u^*(t)=-(\partial V(\tilde{x})/\partial \tilde{x})/2$ in (9.89) for nonlinear biological networks, in order to develop negative feedback loops as the first way to improve the network robustness on the right-hand side of (9.84) or (9.94) to resist environmental disturbances. This is why there are many negative feedback loops in biological networks to improve the network robustness. Furthermore, there are numerous redundant, modular, and scale-free structures in biological networks, which are helpful in attenuating the effect of the intrinsic neutral genetic variations and environmental disturbances on the left-hand side of (9.84) or (9.94), which is the second way to maintain the phenotypic trait in the evolutionary process. Therefore, based on the phenotypic trait robustness criterion of the evolutionary biological network, we can gain a deeper evolutionary insight into the development of negative feedback, redundant, modular, and scale-free structures for biological networks by adaptive or non-adaptive processes [98] in evolution. In different levels of biological networks, two favored strategies can improve phenotypic trait robustness in evolution. One is to improve the network robustness to tolerate more intrinsic neutral genetic variations and to resist environmental disturbances. Negative feedback is the systematic mechanism to improve network robustness, and it is favored by natural selection at different levels of the biological network in evolution, which matches the minimax evolutionary game strategy in this chapter. Another strategy is to reduce the effect of intrinsic neutral genetic variations and environmental disturbances on different biological networks. Redundancies and repairs are the mechanisms of this strategy, and are also favored by natural selection. This is why there are many different redundancies from duplicated genes in gene regulatory networks, redundant pathways in biological chemical networks, and species redundancies in ecological networks [107].

Recently, synthetic biology has focused on the rational construction of biological systems based on engineering principles to fulfill a particular purpose and to allow a better understanding of biological design, using synthetic circuits [116-124]. High-throughput techniques are necessary to determine what is evolved and how engineering genetic networks in vivo mimic the evolutionary processes of genetic networks. An approach involving network evolution has been used to rationally design a synthetic network with a desired behavior in the host cell [121]. Unfortunately, synthetic biologists are faced with the uncertainties of modeling variation, evolutionary genetic variation, and environmental disturbances in vivo (i.e., the phenotypic trait robustness problem of the synthetic biological network) [118-120]. Given this situation, a robust synthetic network design is currently an important topic in the field of synthetic biology [57, 121-124]. As the intrinsic genetic perturbations and environmental disturbances are unpredictable for the synthetic biologists, the proposed minimax evolutionary game strategy should be an efficient robust design method for a synthetic gene network with evolutionary genetic variations, intrinsic uncertainties, and environmental disturbances [57, 121-124].

The contributions of this chapter include the followings: (i) The Poisson point process in (9.68) is used to model genetic variations in stochastic gene network; (ii) the evolution level (9.69) is defined for the stochastic biological network in the evolutionary process, i.e., the worst-case effect of all possible environmental disturbances on the phenotypic deviation around x_e and the effect (cost) of evolutionary strategy $u(t)$. Then, the fitness of biological network around the phenotypic trait is inversely correlated to the evolution level of biological network in (9.70). Therefore, the problem of evolvability or fitness problem can be considered as the minimax evolutionary game problem in (9.72). Based on the evolutionary game strategy, we also find some accumulated genetic variations can provide negative feedbacks via natural selection to maximize the robust stability of the networks; (iii) then the global linearization method is proposed to solve the nonlinear stochastic game problem for the evolutionary game strategy of nonlinear stochastic biological network in the evolutionary process; (iv) based on the evolutionary game strategy in (9.78) or (9.89), we could obtain the phenotypic trait robustness criterion in (9.84) or (9.94) for stochastic biological networks in evolution. Finally, the phenotypic trait robustness criterion is further analysed for evolutionary network to get insight into evolutionary game strategy from the systematic point of view.

In the conventional evolution network [104], the evolvability and fitness of biological network are not well defined. Therefore, the evolutionary game problem is not easily formulated for evolutionary biological networks. Actually, they only solve some evolutionary optimization problem without considering random genetic variations and environmental disturbances. Further, these methods could not provide an efficient method to solve Hamilton-Jacobi equation (HJE) for evolutionary problem. Hence, it is not easy to get insight into the evolutionary mechanism of nonlinear biological network from the systematic point of view.

In this chapter, biological networks with random genetic variations and uncertain environmental disturbances in the evolutionary process are described as nonlinear stochastic systems. In order to maintain the phenotypic trait favored by natural selection, a biological network needs to select some heritable phenotypic variations to improve the network robustness, so as to tolerate neutral genetic variations and to buffer environmental disturbances in the evolutionary process. Since the environmental disturbances are uncertain and unpredictable, the evolutionary strategy of a biological network needs to minimize the worst-case effect of all possible bounded environmental disturbances (i.e., to perform the minimax evolutionary game strategy) in the evolutionary process. We have found that the minimax evolutionary game strategy of a biological network is equivalent to accumulating some phenotypic variations due to some random genetic variations to approach a robust negative feedback scheme, in order to efficiently improve the network robustness, resist the uncertain environmental disturbances, and tolerate neutral genetic variations. However, those harbored random neutral genetic variations also have an effect on network robustness when they accumulate to a large enough amount, serving to provide raw material for new adaptation and evolutionary innovation. If the network robustness can be improved by the minimax evolutionary game strategy, conferring both environmental robustness against the environmental disturbance as well as genetic robustness to tolerate the neutral genetic variations, then the phenotypic trait of the biological network can be maintained in the evolutionary process. Otherwise, the phenotypic trait at an equilibrium point might be destroyed and shifted to another phenotypic trait at the neighboring equilibrium point to commence another period of network evolution.

In general, to gain insight into the minimax evolutionary game strategy of a nonlinear evolutionary biological network, it is necessary to solve a difficult optimization problem constrained by a Hamilton-Jacobi-Isaac inequality (HJII). A global linearization technique is employed to simplify the minimax evolutionary game problem, i.e., the HJII-constrained optimization problem for the minimax evolutionary game strategy can be simplified to an equivalent optimization problem constrained by linear matrix inequalities (LMIs). Based on the global linearization technique, the tradeoff between evolvability, genetic robustness, and network robustness based on the minimax game strategy in evolution can be easily discussed from the local stability robustness and local filtering ability of a set of local evolutionary biological networks on the basis of linear system theory. Finally, the proposed evolutionary game strategy is also extended to an n-tuple evolutionary game problem of a population of biological networks with M players (competitive populations) and k environmental dynamics in the evolutionary process.

9.4. Conclusion

In this chapter, a minimax model matching method based on nonlinear stochastic Nash game theory is proposed for robust control design of immune systems under environmental disturbances. A stochastic minimax design method is developed for robust systematic gene network design under genetic parameter variations and environmental disturbances on the host cell. Finally, a stochastic evolutionary game on a population of biological networks undergoing natural selection is developed to reveal the relationship between network evolution and evolutionary strategies under uncertain environmental disturbances and uncertain genetic mutations.

9.5. Appendix

9.5.1. Proof of Proposition 9.1

Let us denote a Lyapunov energy function $V(\bar{x}(t))>0$. Then (9.9) is equivalent to the following minimax problem

$$\min_{u(t)}\max_{v(t)} J = \min_{u(t)}\max_{v(t)} \left\{ V(\bar{x}(0)) - V(\bar{x}(t_f)) + \int_0^{t_f} \left(\bar{x}^T(t)\mathbf{Q}\bar{x}(t) + u^T(t)Ru(t) - \rho^2 v^T(t)v(t) + \frac{dV(\bar{x}(t))}{dt} \right) dt \right\}, \forall \bar{x}(0) \tag{9.117}$$

By the chain rule, we get

$$\frac{dV(\bar{x}(t))}{dt} = \left(\frac{\partial V(\bar{x}(t))}{\partial \bar{x}(t)}\right)^T \cdot \frac{d\bar{x}(t)}{dt} = \left(\frac{\partial V(\bar{x}(t))}{\partial \bar{x}(t)}\right)^T \cdot (F(\bar{x}(t)) + \mathbf{B}u(t) + Cv(t)) \tag{9.118}$$

Substituting (9.118) into (9.117), we get

$$\min_{u(t)}\max_{v(t)} J = \min_{u(t)}\max_{v(t)} \left\{ V(\bar{x}(0)) - V(\bar{x}(t_f)) + \int_0^{t_f} (\bar{x}^T(t)\mathbf{Q}\bar{x}(t) + u^T(t)Ru(t) - \rho^2 v^T(t)v(t) \right.$$

$$\left. + \left(\frac{\partial V(\bar{x}(t))}{\partial \bar{x}(t)}\right)^T F(\bar{x}(t)) + \left(\frac{\partial V(\bar{x}(t))}{\partial \bar{x}(t)}\right)^T \mathbf{B}u(t) + \left(\frac{\partial V(\bar{x}(t))}{\partial \bar{x}(t)}\right)^T Cv(t))dt \right\}, \forall \bar{x}(0)$$

$$= \min_{u(t)}\max_{v(t)} \left\{ V(\bar{x}(0)) - V(\bar{x}(t_f)) + \int_0^{t_f} \left(\left(\frac{\partial V(\bar{x}(t))}{\partial \bar{x}(t)}\right)^T F(\bar{x}(t)) + \bar{x}(t)^T \mathbf{Q}\bar{x}(t) \right. \right.$$

$$- \frac{1}{4}\left(\frac{\partial V(\bar{x}(t))}{\partial \bar{x}(t)}\right)^T \mathbf{B}R^{-1}\mathbf{B}^T \frac{\partial V(\bar{x}(t))}{\partial \bar{x}(t)} + \frac{1}{4\rho^2}\left(\frac{\partial V(\bar{x}(t))}{\partial \bar{x}(t)}\right)^T CC^T \frac{\partial V(\bar{x}(t))}{\partial \bar{x}(t)}$$

$$+ \left(u^T(t) + \frac{1}{2}R^{-1}\mathbf{B}^T \frac{\partial V(\bar{x}(t))}{\partial \bar{x}(t)}\right)^T R \left(u(t) + \frac{1}{2}R^{-1}\mathbf{B}^T \frac{\partial V(\bar{x}(t))}{\partial \bar{x}(t)}\right)$$

$$\left. \left. - \left(\rho v(t) - \frac{1}{2\rho}C^T \frac{\partial V(\bar{x}(t))}{\partial \bar{x}(t)}\right)^T \left(\rho v(t) - \frac{1}{2\rho}C^T \frac{\partial V(\bar{x}(t))}{\partial \bar{x}(t)}\right)\right) dt \right\}, \forall \bar{x}(0)$$

Therefore, the minimax solution is given as follows

$$J(u^*(t), v^*(t)) = \left\{ V(\bar{x}(0)) - V(\bar{x}(t_f)) + \int_0^{t_f} \left[\left(\frac{\partial V(\bar{x}(t))}{\partial \bar{x}(t)}\right)^T F(\bar{x}(t)) + \bar{x}(t)^T \mathbf{Q}\bar{x}(t) \right. \right.$$

$$\left. \left. - \frac{1}{4}\left(\frac{\partial V(\bar{x}(t))}{\partial \bar{x}(t)}\right)^T \mathbf{B}R^{-1}\mathbf{B}^T \frac{\partial V(\bar{x}(t))}{\partial \bar{x}(t)} + \frac{1}{4\rho^2}\left(\frac{\partial V(\bar{x}(t))}{\partial \bar{x}(t)}\right)^T CC^T \frac{\partial V(\bar{x}(t))}{\partial \bar{x}(t)} \right] dt \right\}, \forall \bar{x}(0)$$

with $u^*(t) = -\frac{1}{2}R^{-1}\mathbf{B}^T \frac{\partial V(\bar{x}(t))}{\partial \bar{x}(t)}$ and $v^*(t) = \frac{1}{2\rho^2}C^T \frac{\partial V(\bar{x}(t))}{\partial \bar{x}(t)}$.

If the equation (9.14) holds, then

$$J(u^*(t), v^*(t)) \le V(\bar{x}(0)) - V(\bar{x}(t_f))$$

From the inequality in (9.11), this minimax solution should be less than $\rho^2\bar{x}^T(0)\mathbf{I}\bar{x}(0)$, and then we get the inequality in (9.15).

$$\min_{u(t)}\max_{v(t)} J \le V(\bar{x}(0)) - V(\bar{x}(t_f)) \le V(\bar{x}(0)) \le \rho^2\bar{x}^T(0)\mathbf{I}\bar{x}(0), \forall \bar{x}(0)$$

9.5.2. Proof of Proposition 9.2

Let us denote a Lyapunov energy function $V(\bar{x}(t))=\bar{x}^T(t)P\bar{x}(t)>0$. Then (9.9) is equivalent to the following

$$
\begin{aligned}
&\min_{u(t)}\max_{v(t)} J \\
&= \min_{u(t)}\max_{v(t)} \Big\{ \bar{x}^T(0)P\bar{x}(0) - \bar{x}^T(t_f)P\bar{x}(t_f) \\
&+\int_0^{t_f}\left(\bar{x}^T(t)\mathbf{Q}\bar{x}(t) + u^T(t)Ru(t) - \rho^2 v^T(t)v(t) + \frac{dV(\bar{x}(t))}{dt} \right) dt \Big\}, \forall \bar{x}(0) \\
&= \min_{u(t)}\max_{v(t)} \Big\{ \bar{x}^T(0)P\bar{x}(0) - \bar{x}^T(t_f)P\bar{x}(t_f) \\
&+\int_0^{t_f}\left(\bar{x}^T(t)\mathbf{Q}\bar{x}(t) + u^T(t)Ru(t) - \rho^2 v^T(t)v(t) + 2\bar{x}^T(t)P\dot{\bar{x}}(t) \right) dt \Big\}, \forall \bar{x}(0) \\
&= \min_{u(t)}\max_{v(t)} \Big\{ \bar{x}^T(0)P\bar{x}(0) - \bar{x}^T(t_f)P\bar{x}(t_f) + \int_0^{t_f}\Big(\bar{x}^T(t)\mathbf{Q}\bar{x}(t) + u^T(t)Ru(t) - \rho^2 v^T(t)v(t) \\
&+2\bar{x}^T(t)P\left(\sum_{i=1}^{L} h_i(x(t))\mathbf{A}_i\bar{x}(t) \right) + 2\bar{x}^T(t)P\mathbf{B}u(t) + 2\bar{x}^T(t)PCv(t) \Big) dt \Big\}, \forall \bar{x}(0) \\
&= \min_{u(t)}\max_{v(t)} \Big\{ \bar{x}^T(0)P\bar{x}(0) - \bar{x}^T(t_f)P\bar{x}(t_f) + \int_0^{t_f}\Big(\bar{x}^T(t)\mathbf{Q}\bar{x}(t) + u^T(t)Ru(t) - \rho^2 v^T(t)v(t) \\
&+\sum_{i=1}^{L} h_i(x(t))\left(2\bar{x}^T(t)P\mathbf{A}_i\bar{x}(t) + 2\bar{x}^T(t)P\mathbf{B}u(t) + 2\bar{x}^T(t)PCv(t) \right) \Big) dt \Big\}, \forall \bar{x}(0) \\
&= \min_{u(t)}\max_{v(t)} \Big\{ \bar{x}^T(0)P\bar{x}(0) - \bar{x}^T(t_f)P\bar{x}(t_f) \\
&+\int_0^{t_f}\Big\{ \bar{x}^T(t)\mathbf{Q}\bar{x}(t) + \sum_{i=1}^{L} h_i(x(t))\bar{x}^T(t)\left[P\mathbf{A}_i + \mathbf{A}_i^T P - P^T\mathbf{B}R^{-1}\mathbf{B}^T P + \frac{1}{\rho^2}P^T CC^T P \right]\bar{x}(t) \\
&+\left(u(t) + R^{-1}\mathbf{B}^T P\bar{x}(t) \right)^T R\left(u(t) + R^{-1}\mathbf{B}^T P\bar{x}(t) \right) \\
&-\left(\rho v(t) - \frac{1}{\rho}C^T P\bar{x}(t) \right)^T \left(\rho v(t) - \frac{1}{\rho}C^T P\bar{x}(t) \right) \Big\} dt \Big\}, \forall \bar{x}(0)
\end{aligned}
$$

The minimax solution is given as follows

$$
\begin{aligned}
&J(u^*(t), v^*(t)) = \Big\{ \bar{x}^T(0)P\bar{x}(0) - \bar{x}^T(t_f)P\bar{x}(t_f) + \int_0^{t_f}\Big\{ \bar{x}^T(t)\mathbf{Q}\bar{x}(t) \\
&+\sum_{i=1}^{L} h_i(x(t))\bar{x}^T(t)\left[P\mathbf{A}_i + \mathbf{A}_i^T P - P^T\mathbf{B}R^{-1}\mathbf{B}^T P + \frac{1}{\rho^2}P^T CC^T P \right]\bar{x}(t) \Big\} dt \Big\}, \forall \bar{x}(0)
\end{aligned}
$$

with $u^*(t) = -\frac{1}{2}R^{-1}\mathbf{B}^T P\overline{x}(t)$ and $v^*(t) = \frac{1}{2\rho^2}C^T P\overline{x}(t)$

In order to simplify the above equation, suppose the inequality in (9.22) holds, then

$$\min_{u(t)} \max_{v(t)} J \le \overline{x}(0)P\overline{x}(0), \forall \overline{x}(0)$$

From the inequality in (9.11), this minimax should be less than $\rho^2 \overline{x}^T(0)\mathbf{I}\overline{x}(0)$, and then

$$\min_{u(t)} \max_{v(t)} J \le \overline{x}(0)P\overline{x}(0) \le \rho^2 \overline{x}^T(0)\mathbf{I}\overline{x}(0), \forall \overline{x}(0), \text{ i.e., } P \le \rho^2\mathbf{I}$$

Since we assume ρ^2 is the upper bound in (9.5), the minimax control becomes how to design $u^*(t)$ in (9.21) by solving the constrained optimization problem in (9.22) and (9.23).

9.5.3. The Parameters $\mathbf{A}_i$, B, *C* of Fuzzy Nonlinear Innate Immune System

The nonlinear innate immune system in (9.28) could be approximated by a Takagi-Sugeno Fuzzy system. By the fuzzy modeling method [45], the matrices of the local linear system $\mathbf{A}_i$, parameters **B** and *C* are calculated as follows:

$$\mathbf{A}_1 = \begin{bmatrix} \begin{matrix} 0.3462 & 0 & 0 & 0 \\ 3.3258 & -1.1155 & 0.3433 & 2.1987 \\ -0.2674 & 1.1816 & -1.7724 & 0.1108 \\ 0.6295 & 0 & 0 & -1.3994 \end{matrix} & 0 \\ 0 & \begin{matrix} -12 & 0 & 0 & 0 \\ 0 & -2.3 & 0 & 0 \\ 0 & 0 & -10 & 0 \\ 0 & 0 & 0 & -3 \end{matrix} \end{bmatrix}$$

$$\mathbf{A}_2 = \begin{bmatrix} \begin{matrix} 0.348 & 0 & 0 & 0 \\ -2.8398 & -0.9076 & -1.4305 & -1.9826 \\ -0.235 & 1.1644 & -1.6635 & 0.1108 \\ 0.295 & -0.0902 & -0.1643 & -1.3994 \end{matrix} & 0 \\ 0 & \begin{matrix} -12 & 0 & 0 & 0 \\ 0 & -2.3 & 0 & 0 \\ 0 & 0 & -10 & 0 \\ 0 & 0 & 0 & -3 \end{matrix} \end{bmatrix}$$

$$\mathbf{A}_3 = \begin{bmatrix} \begin{matrix} -2.3807 & 0 & 0 & 0 \\ 10.616 & -0.9107 & 0.3194 & -1.6078 \\ -2.4789 & 0.5007 & -1.7724 & -0.4134 \\ 0.4117 & 0 & 0 & -0.5537 \end{matrix} & 0 \\ 0 & \begin{matrix} -12 & 0 & 0 & 0 \\ 0 & -2.3 & 0 & 0 \\ 0 & 0 & -10 & 0 \\ 0 & 0 & 0 & -3 \end{matrix} \end{bmatrix}$$

$$\mathbf{A}_4 = \begin{bmatrix} \begin{matrix} -2.2996 & 0 & 0 & 0 \\ -8.4642 & -0.9535 & -0.4587 & 1.1711 \\ -2.3919 & 0.4936 & -1.6635 & -0.4134 \\ 0.1957 & -0.0781 & -0.1643 & -0.5537 \end{matrix} & 0 \\ 0 & \begin{matrix} -12 & 0 & 0 & 0 \\ 0 & -2.3 & 0 & 0 \\ 0 & 0 & -10 & 0 \\ 0 & 0 & 0 & -3 \end{matrix} \end{bmatrix}$$

$$\mathbf{A}_5 = \begin{bmatrix} \begin{matrix} 0.1987 & 0 & 0 & 0 \\ 0.6956 & -0.933 & 0.3733 & -3.5255 \\ 0.3935 & 1.1816 & -0.4106 & 0.1732 \\ 0.2311 & 0 & 0 & -0.3611 \end{matrix} & 0 \\ 0 & \begin{matrix} -12 & 0 & 0 & 0 \\ 0 & -2.3 & 0 & 0 \\ 0 & 0 & -10 & 0 \\ 0 & 0 & 0 & -3 \end{matrix} \end{bmatrix}$$

$$\mathbf{A}_6 = \begin{bmatrix} \begin{matrix} 0.1981 & 0 & 0 & 0 \\ -2.9804 & -1.0591 & -2.0136 & 3.3337 \\ 0.3963 & 1.1644 & -0.3837 & 0.1732 \\ 0.1101 & -0.0902 & -0.0619 & -0.3611 \end{matrix} & 0 \\ 0 & \begin{matrix} -12 & 0 & 0 & 0 \\ 0 & -2.3 & 0 & 0 \\ 0 & 0 & -10 & 0 \\ 0 & 0 & 0 & -3 \end{matrix} \end{bmatrix}$$

$$\mathbf{A}_7 = \begin{bmatrix} -1.0821 & 0 & 0 & 0 & & & & \\ 5.3534 & -0.7742 & -0.6384 & -0.7012 & & 0 & & \\ -0.7249 & 0.5007 & -0.4106 & -0.0417 & & & & \\ 0.1682 & 0 & 0 & -0.1482 & & & & \\ & & & & -12 & 0 & 0 & 0 \\ & & 0 & & 0 & -2.3 & 0 & 0 \\ & & & & 0 & 0 & -10 & 0 \\ & & & & 0 & 0 & 0 & -3 \end{bmatrix}$$

$$\mathbf{A}_8 = \begin{bmatrix} -1.057 & 0 & 0 & 0 & & & & \\ -2.8391 & -1.532 & 0.7299 & 0.2814 & & 0 & & \\ -0.7023 & 0.4936 & -0.3837 & -0.0417 & & & & \\ 0.0812 & -0.0781 & -0.0619 & -0.1482 & & & & \\ & & & & -12 & 0 & 0 & 0 \\ & & 0 & & 0 & -2.3 & 0 & 0 \\ & & & & 0 & 0 & -10 & 0 \\ & & & & 0 & 0 & 0 & -3 \end{bmatrix}$$

$$\mathbf{A}_9 = \begin{bmatrix} 0.3462 & 0.2243 & -2.2977 & 0.087 & & & & \\ 3.0103 & -0.0287 & 7.6732 & -2.1137 & & 0 & & \\ -0.2674 & 0.9206 & -2.1832 & -0.0587 & & & & \\ 0.6295 & 0.1473 & 0.1675 & 0.0737 & & & & \\ & & & & -12 & 0 & 0 & 0 \\ & & 0 & & 0 & -2.3 & 0 & 0 \\ & & & & 0 & 0 & -10 & 0 \\ & & & & 0 & 0 & 0 & -3 \end{bmatrix}$$

$$\mathbf{A}_{10} = \begin{bmatrix} 0.348 & 0.2223 & -2.224 & 0.087 & & & & \\ -2.4956 & -1.8301 & -5.6794 & 0.7798 & & 0 & & \\ -0.235 & 0.9101 & -2.1142 & -0.0587 & & & & \\ 0.295 & 0.0728 & 0.0811 & 0.0737 & & & & \\ & & & & -12 & 0 & 0 & 0 \\ & & 0 & & 0 & -2.3 & 0 & 0 \\ & & & & 0 & 0 & -10 & 0 \\ & & & & 0 & 0 & 0 & -3 \end{bmatrix}$$

$$\mathbf{A}_{11} = \begin{bmatrix} \begin{matrix} -2.3807 & -0.9739 & -2.2977 & -0.5749 \\ 12.767 & 2.4941 & 11.649 & -1.196 \\ -2.4789 & -0.2104 & -2.1832 & -0.598 \\ 0.4117 & 0.1305 & 0.1675 & 0.0489 \end{matrix} & 0 \\ 0 & \begin{matrix} -12 & 0 & 0 & 0 \\ 0 & -2.3 & 0 & 0 \\ 0 & 0 & -10 & 0 \\ 0 & 0 & 0 & -3 \end{matrix} \end{bmatrix}$$

$$\mathbf{A}_{12} = \begin{bmatrix} \begin{matrix} -2.2996 & -0.9641 & -2.224 & -0.5749 \\ -11.901 & -3.8172 & -11.21 & -3.8954 \\ -2.3919 & -0.2087 & -2.1142 & -0.598 \\ 0.1957 & 0.0646 & 0.0811 & 0.0489 \end{matrix} & 0 \\ 0 & \begin{matrix} -12 & 0 & 0 & 0 \\ 0 & -2.3 & 0 & 0 \\ 0 & 0 & -10 & 0 \\ 0 & 0 & 0 & -3 \end{matrix} \end{bmatrix}$$

$$\mathbf{A}_{13} = \begin{bmatrix} \begin{matrix} 0.1987 & 0.2243 & -1.0582 & 0.0495 \\ 0.5515 & -0.6453 & 3.8238 & 2.1633 \\ 0.3935 & 0.9206 & -0.7957 & 0.0991 \\ 0.2311 & 0.1473 & 0.0772 & 0.0275 \end{matrix} & 0 \\ 0 & \begin{matrix} -12 & 0 & 0 & 0 \\ 0 & -2.3 & 0 & 0 \\ 0 & 0 & -10 & 0 \\ 0 & 0 & 0 & -3 \end{matrix} \end{bmatrix}$$

$$\mathbf{A}_{14} = \begin{bmatrix} \begin{matrix} 0.1981 & 0.2223 & -1.0349 & 0.0495 \\ 0.1057 & -1.2516 & -3.148 & -2.882 \\ 0.3963 & 0.9101 & -0.7759 & 0.0991 \\ 0.1101 & 0.0728 & 0.0377 & 0.0275 \end{matrix} & 0 \\ 0 & \begin{matrix} -12 & 0 & 0 & 0 \\ 0 & -2.3 & 0 & 0 \\ 0 & 0 & -10 & 0 \\ 0 & 0 & 0 & -3 \end{matrix} \end{bmatrix}$$

$$\mathbf{A}_{15} = \begin{bmatrix} -1.0821 & -0.9739 & -1.0582 & -0.2642 & & & & \\ 3.6655 & 3.8523 & 4.0278 & 4.4254 & & 0 & & \\ -0.7249 & -0.2104 & -0.7957 & -0.1756 & & & & \\ 0.1682 & 0.1305 & 0.0772 & 0.0203 & & & & \\ & & & & -12 & 0 & 0 & 0 \\ & 0 & & & 0 & -2.3 & 0 & 0 \\ & & & & 0 & 0 & -10 & 0 \\ & & & & 0 & 0 & 0 & -3 \end{bmatrix}$$

$$\mathbf{A}_{16} = \begin{bmatrix} -1.057 & -0.9641 & -1.0349 & -0.2642 & & & & \\ -6.4434 & -4.5651 & -4.4027 & -6.7461 & & 0 & & \\ -0.7023 & -0.2087 & -0.7759 & -0.1756 & & & & \\ 0.0812 & 0.0646 & 0.0377 & 0.0203 & & & & \\ & & & & -12 & 0 & 0 & 0 \\ & 0 & & & 0 & -2.3 & 0 & 0 \\ & & & & 0 & 0 & -10 & 0 \\ & & & & 0 & 0 & 0 & -3 \end{bmatrix}$$

$$\mathbf{B} = \begin{bmatrix} -1 & 0 & 0 & 0 \\ 0 & 1 & 0 & 0 \\ 0 & 0 & 1 & 0 \\ 0 & 0 & 0 & -1 \\ 0 & 0 & 0 & 0 \\ 0 & 0 & 0 & 0 \\ 0 & 0 & 0 & 0 \\ 0 & 0 & 0 & 0 \end{bmatrix}, \quad C = \begin{bmatrix} 3 & 0 & 0 & 0 & & & & \\ 0 & 3 & 0 & 0 & & 0 & & \\ 0 & 0 & 3 & 0 & & & & \\ 0 & 0 & 0 & 3 & & & & \\ & & & & 1 & 0 & 0 & 0 \\ & 0 & & & 0 & 1 & 0 & 0 \\ & & & & 0 & 0 & 1 & 0 \\ & & & & 0 & 0 & 0 & 1 \end{bmatrix}.$$

9.5.4. Proof of Proposition 9.3

Let us consider a Lyapunov energy function $V(\tilde{x})>0$, then the cost function in equation (9.39) is equivalent to

$$J(k,\gamma,v) = E\left[V(\tilde{x}(0)) - V(\tilde{x}(t_f)) + \int_0^{t_f} \left(\tilde{x}^T Q\tilde{x} - g^2 v^T v + \frac{dV(\tilde{x})}{dt} \right) dt \right] \quad (9.119)$$

By the chain rule, we get

$$\frac{dV(\tilde{x}(t))}{dt}=\left(\frac{\partial V(\tilde{x}(t))}{\partial \tilde{x}(t)}\right)^T\cdot\frac{d\tilde{x}(t)}{dt}=\left(\frac{\partial V(\tilde{x}(t))}{\partial \tilde{x}(t)}\right)^T(f(\tilde{x}(t)+x_d(t),k,\gamma)+v(t)) \quad (9.120)$$

Substituting (9.120) into (9.119), we maximize $J(k,\gamma,v)$ by the uncertain disturbance v

$$\begin{aligned}
&\max_v J(k,\gamma,v)\\
&=\max_v E\Big[V(\tilde{x}(0))-V(\tilde{x}(t_f))\\
&+\int_0^{t_f}\left(\tilde{x}^TQ\tilde{x}-g^2v^Tv+\left(\frac{\partial V(\tilde{x})}{\partial \tilde{x}}\right)^T f(\tilde{x}+x_d,k,\gamma)+\left(\frac{\partial V(\tilde{x})}{\partial \tilde{x}}\right)^T v\right)dt\Big]\\
&=\max_v E\Big[V(\tilde{x}(0))-V(\tilde{x}(t_f))+\int_0^{t_f}\Big(\tilde{x}^TQ\tilde{x}+\left(\frac{\partial V(\tilde{x})}{\partial \tilde{x}}\right)^T f(\tilde{x}+x_d,k,\gamma)\\
&-\left(gv-\frac{1}{2g}\frac{\partial V(\tilde{x})}{\partial \tilde{x}}\right)^T\left(gv-\frac{1}{2g}\frac{\partial V(\tilde{x})}{\partial \tilde{x}}\right)+\frac{1}{4g^2}\left(\frac{\partial V(\tilde{x})}{\partial \tilde{x}}\right)^T\left(\frac{\partial V(\tilde{x})}{\partial \tilde{x}}\right)\Big)dt\Big]\\
&=E\Big[V(\tilde{x}(0))-V(\tilde{x}(t_f))+\int_0^{t_f}\Big(\tilde{x}^TQ\tilde{x}+\left(\frac{\partial V(\tilde{x})}{\partial \tilde{x}}\right)^T f(\tilde{x}+x_d,k,\gamma)\\
&+\frac{1}{4g^2}\left(\frac{\partial V(\tilde{x})}{\partial \tilde{x}}\right)^T\left(\frac{\partial V(\tilde{x})}{\partial \tilde{x}}\right)\Big)dt\Big]
\end{aligned} \quad (9.121)$$

with the worst-case disturbance $v^*=\frac{1}{2g^2}\frac{\partial V(\tilde{x})}{\partial \tilde{x}}$.

By the inequality in (9.43), it is seen that $V(\tilde{x}(0))$ is the upper bound of (9.121) i.e., the sub-minimax problem becomes how to solve the following constrained optimization problem

$$\min_{\substack{k\in[k_1,k_2]\\ \gamma\in[\gamma_1,\gamma_2]}}\max_v J(k,\gamma,v)\le\min_{\substack{k\in[k_1,k_2]\\ \gamma\in[\gamma_1,\gamma_2]}}E\left[V(\tilde{x}(0))\right] \quad (9.122)$$

$$\text{subject to (9.43) and } V(\tilde{x})>0$$

By the fact in (9.40), $g^2E[\tilde{x}^T(0)\tilde{x}(0)]$ is the upper bound of $\min_{\substack{k\in[k_1,k_2]\\ \gamma\in[\gamma_1,\gamma_2]}}\max_v J(k,\gamma,v)$.

Therefore $E[V(\tilde{x})]$ in (9.122) should be bounded by $g^2E[\tilde{x}^T(0)\tilde{x}(0)]$, i.e., $E[V(\tilde{x}(0))]\le g^2E[\tilde{x}^T(0)\tilde{x}(0)]$. Therefore the suboptimal solution is to minimize its upper bound. Hence, the sub-minimax problem in (9.122) could be replaced by

$$\min_{\substack{k\in[k_1,k_2]\\ \gamma\in[\gamma_1,\gamma_2]}}\max_v J(k,\gamma,v)\le\min_{\substack{k\in[k_1,k_2]\\ \gamma\in[\gamma_1,\gamma_2]}}E\left[V(\tilde{x}(0))\right]\le\min_{\substack{k\in[k_1,k_2]\\ \gamma\in[\gamma_1,\gamma_2]}}E\left[g^2\tilde{x}^T(0)\tilde{x}(0)\right]=\min_{\substack{k\in[k_1,k_2]\\ \gamma\in[\gamma_1,\gamma_2]}}g^2\mathrm{Tr}(R_0) \quad (9.123)$$

where Tr(R_0) denotes the trace of R_0 and R_0 denotes the covariance of the initial condition $\tilde{x}(0)$ i.e., $R_0=E[\tilde{x}^T(0)\tilde{x}(0)]$, which is independent of the choice of k and γ. Therefore, the sub-minimax design problem is equivalent to solving the following constrained optimization

$$\min_{\substack{k\in[k_1,k_2]\\ \gamma\in[\gamma_1,\gamma_2]}} g^2$$

$$\text{subject to (9.43) and } V(\tilde{x})>0$$

9.5.5. Proof of Proposition 9.4

We replace error dynamic system in (9.35) by its fuzzy interpolation system in (9.46). Then HJI in (9.43) can be represented by

$$\left(\frac{\partial V(\tilde{x})}{\partial \tilde{x}}\right)^T\left(\sum_{i=1}^{L} h_i(\tilde{x})\mathbf{A}_i(k,\gamma)\tilde{x}\right)+\tilde{x}^T Q\tilde{x}+\frac{1}{4g^2}\left(\frac{\partial V(\tilde{x})}{\partial \tilde{x}}\right)^T\left(\frac{\partial V(\tilde{x})}{\partial \tilde{x}}\right)<0 \quad (9.124)$$

Let us choose the Lyapunov function $V(\tilde{x})$ as $V(\tilde{x})=\tilde{x}^T P\tilde{x}$ for some positive definite symmetric matrix P and substitute it into (9.124). Then we get

$$\sum_{i=1}^{L} h_i(\tilde{x})\left\{\tilde{x}^T\left(P\mathbf{A}_i(k,r)+\mathbf{A}_i^T(k,r)P+Q+\frac{1}{g^2}PP\right)\tilde{x}\right\}\leq 0 \quad (9.125)$$

$$P\leq g^2 I$$

where the property in (9.48) is used.

It is seen that the inequalities in (9.50) implies (9.125). Therefore, the sub-minimax design for the fuzzy equivalent system becomes how we solve the constrained optimization in (9.49) and (9.50). By substituting $V(\tilde{x})=\tilde{x}^T P\tilde{x}$ into (9.44), we get the worst-case disturbances v^* in (9.51).

9.5.6. Proof of Proposition 9.5

Again, let us consider a Lyapunov energy function $V(\tilde{x})>0$, then the equation (9.54) is equivalent to

$$\min_{\substack{k\in[k_1,k_2]\\ \gamma\in[\gamma_1,\gamma_2]}} E\left[\int_0^{t_f} \tilde{x}^T Q\tilde{x}dt\right]$$

$$= \min_{\substack{k\in[k_1,k_2]\\ \gamma\in[\gamma_1,\gamma_2]}} E\left[V(\tilde{x}(0)) - V(\tilde{x}(t_f)) + \int_0^{t_f}\left(\tilde{x}^T Q\tilde{x} + \frac{dV(\tilde{x})}{dt}\right)dt\right]$$

$$= \min_{\substack{k\in[k_1,k_2]\\ \gamma\in[\gamma_1,\gamma_2]}} E\left[V(\tilde{x}(0)) - V(\tilde{x}(t_f)) + \int_0^{t_f}\left(\tilde{x}^T Q\tilde{x} + \left(\frac{\partial V(\tilde{x})}{\partial \tilde{x}}\right)^T f(\tilde{x}+x_d,k,\gamma) + \left(\frac{\partial V(\tilde{x})}{\partial \tilde{x}}\right)^T v\right)dt\right]$$

By the fact that $2a^Tb\leq a^Ta+b^Tb$ for any two-vectors a and b, we get

$$\min_{\substack{k\in[k_1,k_2]\\ \gamma\in[\gamma_1,\gamma_2]}} E\left[\int_0^{t_f} \tilde{x}^T Q\tilde{x}dt\right]$$

$$= \min_{\substack{k\in[k_1,k_2]\\ \gamma\in[\gamma_1,\gamma_2]}} E\left[V(\tilde{x}(0)) - V(\tilde{x}(t_f)) + \int_0^{t_f}\left(\tilde{x}^T Q\tilde{x} + \left(\frac{\partial V(\tilde{x})}{\partial \tilde{x}}\right)^T f(\tilde{x}+x_d,k,\gamma) + \frac{1}{2}\left(\frac{\partial V(\tilde{x})}{\partial \tilde{x}}\right)^T\left(\frac{\partial V(\tilde{x})}{\partial \tilde{x}}\right) + \frac{1}{2}v^T v\right)dt\right]$$

By the inequality in (9.56), we get the sub-optimal regulation problem as follows

$$\min_{\substack{k\in[k_1,k_2]\\ \gamma\in[\gamma_1,\gamma_2]}} E\left[\int_0^{t_f} \tilde{x}^T Q\tilde{x}dt\right] \leq \min_{\substack{k\in[k_1,k_2]\\ \gamma\in[\gamma_1,\gamma_2]}} E\left[V(\tilde{x}(0)) + \frac{1}{2}\int_0^{t_f} v^T v dt\right]$$

Since disturbance v is independent of the choice of parameters k and γ, and only the choice of $V(\tilde{x})$ will influence the above minimization, the sub-optimal design becomes how to solve the constrained optimization problem in (9.55) and (9.56).

9.5.7. Parameters $\mathbf{A}_i$ of the T-S fuzzy model with the specified kinetic parameters k^* and decay rates γ^*

$$\mathbf{A}_1 = \begin{bmatrix} -1.6879 & -0.060601 & 0.11879 & -0.0092833 \\ 0.38914 & -0.093297 & 0.010249 & -0.0065119 \\ 0.10826 & -0.02841 & -1.4996 & -0.0060343 \\ 0.00097167 & -0.0025457 & 0.0053402 & -0.066832 \end{bmatrix}$$

$$\mathbf{A}_2 = \begin{bmatrix} -3.5629 & -0.12704 & 0.25074 & -0.0092833 \\ 0.20138 & -0.193 & 0.021476 & -0.0065109 \\ 0.22906 & -0.11458 & -3.1644 & -0.0060447 \\ 0.0014069 & -0.00054073 & 0.0071284 & -0.066833 \end{bmatrix}$$

$$\mathbf{A}_3 = \begin{bmatrix} -1.5351 & -0.060529 & 0.11879 & -0.0092832 \\ 0.40408 & -0.092851 & 0.010249 & -0.0065573 \\ -0.18285 & 0.0322 & -1.4996 & 0.0041303 \\ 0.0012516 & -0.0027325 & -0.0017594 & -0.066801 \end{bmatrix}$$

$$\mathbf{A}_4 = \begin{bmatrix} -3.2403 & -0.12689 & 0.25074 & -0.0092832 \\ 0.23298 & -0.19466 & 0.021741 & -0.0065573 \\ -0.38598 & 0.039126 & -3.1671 & 0.0041304 \\ 0.0019632 & 0.00067731 & -0.00013562 & -0.066801 \end{bmatrix}$$

$$\mathbf{A}_5 = \begin{bmatrix} -3.5287 & -0.060601 & 0.24784 & -0.0093278 \\ 0.19497 & -0.093286 & 0.017006 & 0.0019312 \\ 0.22212 & -0.080273 & -3.1614 & -0.0042428 \\ 0.001744 & -0.0025529 & 0.0072707 & -0.067233 \end{bmatrix}$$

$$\mathbf{A}_6 = \begin{bmatrix} -7.4489 & -0.12704 & 0.52318 & -0.0093278 \\ -0.18351 & -0.1982 & 0.095548 & 0.0014778 \\ 0.21344 & -0.11298 & -7.2861 & 0.00040939 \\ -0.012439 & 0.0026832 & -0.025864 & -0.066952 \end{bmatrix}$$

$$\mathbf{A}_7 = \begin{bmatrix} -3.2061 & -0.060529 & 0.24784 & -0.0093277 \\ 0.22649 & -0.092851 & 0.01727 & 0.0018016 \\ -0.38483 & -0.019544 & -3.1642 & 0.0068314 \\ 0.0023517 & -0.0027325 & 6.7334e\text{-}006 & -0.067149 \end{bmatrix}$$

$$\mathbf{A}_8 = \begin{bmatrix} -6.768 & -0.12689 & 0.52318 & -0.0093277 \\ -0.14191 & -0.19465 & -0.023172 & 0.0018026 \\ -0.81178 & -0.012738 & -6.0679 & 0.0068211 \\ 0.0043172 & 0.00067013 & 0.040657 & -0.06715 \end{bmatrix}$$

$$\mathbf{A}_9 = \begin{bmatrix} -1.6879 & 0.12793 & -0.25078 & 0.019598 \\ -0.727 & -0.07319 & -0.026022 & -0.003619 \\ 0.10826 & -0.031432 & -0.80806 & -0.005567 \\ 0.00097182 & -0.0027504 & 0.0047284 & -0.066801 \end{bmatrix}$$

$$\mathbf{A}_{10} = \begin{bmatrix} -3.5629 & 0.26819 & -0.52934 & 0.019598 \\ -0.91465 & -0.15344 & -0.05495 & -0.003619 \\ 0.22793 & -0.094274 & -1.7058 & -0.005567 \\ 0.0013385 & 0.00063963 & 0.0057541 & -0.066801 \end{bmatrix}$$

$$\mathbf{A}_{11} = \begin{bmatrix} -1.5351 & 0.12778 & -0.25078 & 0.019598 \\ -0.71206 & -0.073303 & -0.026022 & -0.0036189 \\ -0.18285 & 0.034225 & -0.80806 & 0.0041294 \\ 0.0012516 & -0.0026058 & -0.0023716 & -0.066797 \end{bmatrix}$$

$$\mathbf{A}_{12} = \begin{bmatrix} -3.2403 & 0.26787 & -0.52934 & 0.019598 \\ -0.88316 & -0.15367 & -0.054951 & -0.0036189 \\ -0.38597 & 0.043382 & -1.7058 & 0.0041294 \\ 0.0019634 & 0.00094337 & -0.0013455 & -0.066797 \end{bmatrix}$$

$$\mathbf{A}_{13} = \begin{bmatrix} -3.5287 & 0.12793 & -0.52322 & 0.019692 \\ -0.92106 & -0.07319 & -0.058507 & 0.0047537 \\ 0.22099 & -0.083177 & -1.7026 & -0.0029125 \\ 0.0016756 & -0.0027503 & 0.0059171 & -0.067149 \end{bmatrix}$$

$$\mathbf{A}_{14} = \begin{bmatrix} -7.4489 & 0.26819 & -1.1045 & 0.019692 \\ -1.3492 & -0.15343 & -0.12363 & 0.0047547 \\ 0.72194 & -0.14614 & -3.593 & -0.0029229 \\ 0.018292 & 0.00063245 & 0.0083449 & -0.06715 \end{bmatrix}$$

$$\mathbf{A}_{15} = \begin{bmatrix} -3.2061 & 0.12778 & -0.52322 & 0.019692 \\ -0.88965 & -0.073303 & -0.058507 & 0.0047076 \\ -0.38483 & -0.01752 & -1.7026 & 0.0073033 \\ 0.0023519 & -0.0026058 & -0.0011826 & -0.067117 \end{bmatrix}$$

$$\mathbf{A}_{16} = \begin{bmatrix} -6.768 & 0.26787 & -1.1045 & 0.019692 \\ -1.2579 & -0.15367 & -0.12336 & 0.0047076 \\ -0.81291 & -0.0083629 & -3.5957 & 0.0073033 \\ 0.0042487 & 0.00094338 & 0.0010809 & -0.067117 \end{bmatrix}$$

9.5.8. Proof of Proposition 9.7

From

$$J \triangleq E\int_0^{t_p}\left(\tilde{x}^T(t)Q\tilde{x}(t)+u^T(t)u(t)-ev^T(t)v(t)\right)dt$$

and by the fact $\tilde{x}(t)=x(t)$, i.e., $x_e=0$ in the linear biological case, we get

$$\begin{aligned}J &= E\int_0^{t_p}\left(x^T(t)Qx(t)+u^T(t)u(t)-ev^T(t)v(t)\right)dt\\ &= Ex^T(0)Px(0)-Ex^T(t_p)x(t_p)\\ &+E\int_0^{t_p}\left(x^T(t)Qx(t)+u^T(t)u(t)-ev^T(t)v(t)+\frac{d\left(x^T(t)Px(t)\right)}{dt}\right)dt\end{aligned} \quad (9.126)$$

where $V(x(t),t)\triangleq x^T(t)Px(t)$, $P=P^T\geq 0$. By using the Itô formula [44, 77, 78, 82-84, 110, 122], we can get

$$\begin{aligned}J &= Ex^T(0)Px(0)-Ex^T(t_p)x(t_p)+E\int_0^{t_p}\left\{x^TQx+u^Tu-ev^Tv+\left(\frac{\partial V}{\partial x}\right)^T(Ax+Bv+u)\right.\\ &\left.+\lambda_2\left[\sum_{k=1}^{N_{2p}}x^TA_k^T\right]P\left[\sum_{k=1}^{N_{2p}}A_kx\right]\right\}dt\\ &= Ex^T(0)Px(0)-Ex^T(t_p)x(t_p)+E\int_0^{t_p}\left\{x^T\left[Q+A^TP+PA\right]x+u^TPx+x^TPu+u^Tu\right.\\ &\left.-ev^Tv+x^TPBv+v^TB^TPx+\lambda_2\sum_{k=1}^{N_{2p}}x^TA_k^TPA_kx\right\}dt\\ &= Ex^T(0)Px(0)-Ex^T(t_p)x(t_p)+E\int_0^{t_p}\left\{x^T\left[Q+A^TP+PA-PP+\frac{1}{e}PBB^TP\right.\right.\\ &\left.\left.+\lambda_2\sum_{k=1}^{N_{2p}}A_k^TPA_k\right]x+(Px+u)^T(Px+u)-\left(\frac{1}{\sqrt{e}}B^TPx-\sqrt{e}v\right)^T\left(\frac{1}{\sqrt{e}}B^TPx-\sqrt{e}v\right)\right\}dt\end{aligned}$$

(9.127)

$$\begin{aligned}\min_{u(t)}\max_{v(t)}J &= Ex^T(0)Px(0)-Ex^T(t_p)x(t_p)+\min_{u(t)}\max_{v(t)}E\int_0^{t_p}\left\{x^T\left[Q+A^TP\right.\right.\\ &\left.+PA-PP+\frac{1}{e}PBB^TP+\lambda_2\sum_{k=1}^{N_{2p}}A_k^TPA_k\right]x+(Px+u)^T(Px+u)\\ &\left.-\left(\frac{1}{\sqrt{e}}B^TPx-\sqrt{e}v\right)^T\left(\frac{1}{\sqrt{e}}B^TPx-\sqrt{e}v\right)\right\}dt\end{aligned} \quad (9.128)$$

Then we get the solution of minimax game as $u^*(t) = -Px$ (9.129)

$$v^*(t) = \frac{1}{e} B^T Px \tag{9.130}$$

$$\begin{aligned}\min_{u(t)} \max_{v(t)} J = Ex^T(0)Px(0) - Ex^T(t_p)x(t_p) + E\int_0^{t_p} \Big\{ x^T \Big[Q + A^T P \\ + PA - PP + \frac{1}{e} PBB^T P + \lambda_2 \sum_{k=1}^{N_{2p}} A_k^T P A_k \Big] x \Big\} dt\end{aligned} \tag{9.131}$$

By the inequality

$$Q + A^T P + PA - PP + \frac{1}{e} PBB^T P + \lambda_2 \sum_{k=1}^{N_{2p}} A_k^T P A_k \le 0$$

we get $J_o = \min_{u(t)} \max_{v(t)} J \le Ex^T(0)Px(0) - Ex^T(t_p)x(t_p) \le Ex^T(0)Px(0)$. By the inequality

$$J_o \le eE\tilde{x}^T(0)\tilde{x}(0)$$

we get

$$P \le eI \tag{9.132}$$

which is the inequality

$$0 < P \le eI$$

9.5.9. Proof of Proposition 9.8

From

$$J \triangleq E\int_0^{t_p} \left(\tilde{x}^T(t) Q\tilde{x}(t) + u^T(t)u(t) - ev^T(t)v(t) \right) dt$$

for some positive function $V(\tilde{x})$

$$J = EV(\tilde{x}(0)) - EV(\tilde{x}(t_p)) + E\int_0^{t_p} \left(\tilde{x}^T Q\tilde{x} + u^T u - ev^T v \right) dt + EdV(\tilde{x}) \tag{9.133}$$

By the Itô formula [44, 77, 78, 82-84, 110, 122] for nonlinear stochastic network

$$\frac{d\tilde{x}(t)}{dt} = f(\tilde{x}) + Bv(t) + \sum_{k=1}^{N(t)} f_k(\tilde{x})p(t-t_k)$$
$$= f(\tilde{x}) + Bv(t) + u(t) + \sum_{k=1}^{N_2(t)} f_k(\tilde{x})p(t-t_k)$$

and by the fact that $u(t)=$

$\sum_{k=1}^{N_1(t)} f_k(\tilde{x})p(t-t_k)x(t)$ we can get

$$EdV(\tilde{x}) = \left[\left(\frac{\partial V(\tilde{x})}{\partial \tilde{x}}\right)^T (f(\tilde{x}) + u + Bv) + \frac{1}{2}\lambda_2 \sum_{k=1}^{N_{2p}} f_k^T(\tilde{x}) \frac{\partial^2 V(\tilde{x})}{\partial \tilde{x}^2} f_k(\tilde{x})\right] dt \tag{9.134}$$

Substituting (9.134) into (9.133), we get

$$J = EV(\tilde{x}(0)) - EV(\tilde{x}(t_p)) + E\int_0^{t_p} \left\{ \tilde{x}^T Q\tilde{x} + u^T u - ev^T v + \left(\frac{\partial V(\tilde{x})}{\partial \tilde{x}}\right)^T f(\tilde{x}) \right.$$
$$\left. + \left(\frac{\partial V(\tilde{x})}{\partial \tilde{x}}\right)^T u + \left(\frac{\partial V(\tilde{x})}{\partial \tilde{x}}\right)^T Bv + \frac{1}{2}\lambda_2 \sum_{k=1}^{N_{2p}} f_k^T(\tilde{x}) \frac{\partial^2 V(\tilde{x})}{\partial \tilde{x}^2} f_k(\tilde{x}) \right\} dt \tag{9.135}$$

By the fact that

$$u^T u + \left(\frac{\partial V(\tilde{x})}{\partial \tilde{x}}\right)^T u = \left(u + \frac{1}{2}\frac{\partial V(\tilde{x})}{\partial \tilde{x}}\right)^T \left(u + \frac{1}{2}\frac{\partial V(\tilde{x})}{\partial \tilde{x}}\right) - \frac{1}{4}\left(\frac{\partial V(\tilde{x})}{\partial \tilde{x}}\right)^T \left(\frac{\partial V(\tilde{x})}{\partial \tilde{x}}\right) \tag{9.136}$$

$$-ev^T v + \left(\frac{\partial V(\tilde{x})}{\partial \tilde{x}}\right)^T Bv = -\left(\sqrt{e}v - \frac{1}{2\sqrt{e}}B^T\left(\frac{\partial V(\tilde{x})}{\partial \tilde{x}}\right)\right)^T \left(\sqrt{e}v - \frac{1}{2\sqrt{e}}B^T\left(\frac{\partial V(\tilde{x})}{\partial \tilde{x}}\right)\right)$$
$$+ \frac{1}{4e}\left(\frac{\partial V(\tilde{x})}{\partial \tilde{x}}\right)^T B^T B\left(\frac{\partial V(\tilde{x})}{\partial \tilde{x}}\right) \tag{9.137}$$

Substituting (9.136) and (9.137) into (9.135), we get

$$J = EV(\tilde{x}(0)) - EV(\tilde{x}(t_p)) + E\int_0^{t_p} \left\{ \left(\frac{\partial V(\tilde{x})}{\partial \tilde{x}} \right)^T f(\tilde{x}) + \tilde{x}^T Q \tilde{x} \right.$$
$$+ \frac{1}{2}\lambda_2 \sum_{k=1}^{N_{2p}} f_k^T(\tilde{x}) \frac{\partial^2 V(\tilde{x})}{\partial \tilde{x}^2} f_k(\tilde{x}) + \left(u + \frac{1}{2}\frac{\partial V(\tilde{x})}{\partial \tilde{x}} \right)^T \left(u + \frac{1}{2}\frac{\partial V(\tilde{x})}{\partial \tilde{x}} \right)$$
$$- \left(\sqrt{e}v - \frac{1}{2\sqrt{e}} B^T \left(\frac{\partial V(\tilde{x})}{\partial \tilde{x}} \right) \right)^T \left(\sqrt{e}v - \frac{1}{2\sqrt{e}} B^T \left(\frac{\partial V(\tilde{x})}{\partial \tilde{x}} \right) \right) \tag{9.138}$$
$$\left. + \frac{1}{4e} \left(\frac{\partial V(\tilde{x})}{\partial \tilde{x}} \right)^T B^T B \left(\frac{\partial V(\tilde{x})}{\partial \tilde{x}} \right) - \frac{1}{4} \left(\frac{\partial V(\tilde{x})}{\partial \tilde{x}} \right)^T \left(\frac{\partial V(\tilde{x})}{\partial \tilde{x}} \right) \right\} dt$$

Then

$$\min_{u(t)} \max_{v(t)} J = EV(\tilde{x}(0)) - EV(\tilde{x}(t_p)) + \min_{u(t)} \max_{v(t)} E\int_0^{t_p} \left\{ \left(\frac{\partial V(\tilde{x})}{\partial \tilde{x}} \right)^T f(\tilde{x}) + \tilde{x}^T Q \tilde{x} \right.$$
$$+ \frac{1}{2}\lambda_2 \sum_{k=1}^{N_{2p}} f_k^T(\tilde{x}) \frac{\partial^2 V(\tilde{x})}{\partial \tilde{x}^2} f_k(\tilde{x}) + \left(u + \frac{1}{2}\frac{\partial V(\tilde{x})}{\partial \tilde{x}} \right)^T \left(u + \frac{1}{2}\frac{\partial V(\tilde{x})}{\partial \tilde{x}} \right)$$
$$- \left(\sqrt{e}v - \frac{1}{2\sqrt{e}} B^T \left(\frac{\partial V(\tilde{x})}{\partial \tilde{x}} \right) \right)^T \left(\sqrt{e}v - \frac{1}{2\sqrt{e}} B^T \left(\frac{\partial V(\tilde{x})}{\partial \tilde{x}} \right) \right) \tag{9.139}$$
$$\left. + \frac{1}{4e} \left(\frac{\partial V(\tilde{x})}{\partial \tilde{x}} \right)^T B^T B \left(\frac{\partial V(\tilde{x})}{\partial \tilde{x}} \right) - \frac{1}{4} \left(\frac{\partial V(\tilde{x})}{\partial \tilde{x}} \right)^T \left(\frac{\partial V(\tilde{x})}{\partial \tilde{x}} \right) \right\} dt$$

Then we get the minimax solution

$$u^*(t) = s^*(\tilde{x}) = \sum_{k=1}^{N_1(t)} f_k(\tilde{x}) p(t - t_k) = -\frac{1}{2} \left(\frac{\partial V(\tilde{x})}{\partial \tilde{x}} \right), v^*(t) = \frac{1}{2e} B^T \left(\frac{\partial V(\tilde{x})}{\partial \tilde{x}} \right)$$

Further, by the HJII

$$\left(\frac{\partial V(\tilde{x})}{\partial \tilde{x}} \right)^T f(\tilde{x}) + \tilde{x}^T Q \tilde{x} - \frac{1}{4} \left(\frac{\partial V(\tilde{x})}{\partial \tilde{x}} \right)^T \left(\frac{\partial V(\tilde{x})}{\partial \tilde{x}} \right)$$
$$+ \frac{1}{4e} \left(\frac{\partial V(\tilde{x})}{\partial \tilde{x}} \right)^T B^T B \left(\frac{\partial V(\tilde{x})}{\partial \tilde{x}} \right) + \frac{1}{2}\lambda_2 \sum_{k=1}^{N_{2p}} f_k^T(\tilde{x}) \frac{\partial^2 V(\tilde{x})}{\partial \tilde{x}^2} f_k(\tilde{x}) \le 0$$

we get

$$J_o = \min_{u(t)} \max_{v(t)} J \le EV(\tilde{x}(0)) - EV(\tilde{x}(t_p)) \le EV(\tilde{x}(0)) \tag{9.140}$$

By the inequality constraint

$$J_o \le eE\tilde{x}^T(0)\tilde{x}(0)$$

we get

$$EV(\tilde{x}(0)) \le eE\tilde{x}^T(0)\tilde{x}(0) \tag{9.141}$$

which is the inequality

$$V(\tilde{x}(0)) \le e\tilde{x}^T(0)\tilde{x}(0)$$

9.5.10. Proof of Proposition 9.9

Since we replace the nonlinear stochastic biological network

$$\begin{aligned}\frac{d\tilde{x}(t)}{dt} &= f(\tilde{x}) + Bv(t) + \sum_{k=1}^{N(t)} f_k(\tilde{x})p(t-t_k)\\ &= f(\tilde{x}) + Bv(t) + u(t) + \sum_{k=1}^{N_2(t)} f_k(\tilde{x})p(t-t_k)\end{aligned}$$

by the interpolated biological network

$$\frac{d\tilde{x}(t)}{dt} = \sum_{i=1}^{M} \alpha_i(\tilde{x})\left(A_i\tilde{x} + Bv(t) + u(t) + \sum_{k=1}^{N_2(t)} A_{ki}p(t-t_k)\tilde{x}\right)$$

Based on **Proposition 9.8**, the HJII

$$\left(\frac{\partial V(\tilde{x})}{\partial \tilde{x}}\right)^T f(\tilde{x}) + \tilde{x}^T Q\tilde{x} - \frac{1}{4}\left(\frac{\partial V(\tilde{x})}{\partial \tilde{x}}\right)^T \left(\frac{\partial V(\tilde{x})}{\partial \tilde{x}}\right)$$

$$+\frac{1}{4e}\left(\frac{\partial V(\tilde{x})}{\partial \tilde{x}}\right)^T BB^T\left(\frac{\partial V(\tilde{x})}{\partial \tilde{x}}\right) + \frac{1}{2}\lambda_2 \sum_{k=1}^{N_{2p}} f_k^T(\tilde{x})\frac{\partial^2 V(\tilde{x})}{\partial^2 \tilde{x}} f_k(\tilde{x}) \le 0$$

can be replaced by

$$\left(\frac{\partial V(\tilde{x})}{\partial \tilde{x}}\right)^T\left(\sum_{i=1}^{M}\alpha_i(\tilde{x})A_i\tilde{x}\right)+\tilde{x}^TQ\tilde{x}-\frac{1}{4}\left(\frac{\partial V(\tilde{x})}{\partial \tilde{x}}\right)^T\left(\frac{\partial V(\tilde{x})}{\partial \tilde{x}}\right)$$
$$+\frac{1}{4e}\left(\frac{\partial V(\tilde{x})}{\partial \tilde{x}}\right)^T BB^T\left(\frac{\partial V(\tilde{x})}{\partial \tilde{x}}\right)+\frac{1}{2}\lambda_2\sum_{i=1}^{M}\sum_{j=1}^{M}\sum_{k=1}^{N_{2p}}\alpha_i(\tilde{x})\alpha_j(\tilde{x})\tilde{x}^TA_{ki}^TPA_{kj}\tilde{x}$$
$$\leq\left(\frac{\partial V(\tilde{x})}{\partial \tilde{x}}\right)^T\left(\sum_{i=1}^{M}\alpha_i(\tilde{x})A_i\tilde{x}\right)+\tilde{x}^TQ\tilde{x}-\frac{1}{4}\left(\frac{\partial V(\tilde{x})}{\partial \tilde{x}}\right)^T\left(\frac{\partial V(\tilde{x})}{\partial \tilde{x}}\right) \quad (9.142)$$
$$+\frac{1}{4e}\left(\frac{\partial V(\tilde{x})}{\partial \tilde{x}}\right)^T BB^T\left(\frac{\partial V(\tilde{x})}{\partial \tilde{x}}\right)+\frac{1}{2}\lambda_2\sum_{i=1}^{M}\sum_{k=1}^{N_{2p}}\alpha_i(\tilde{x})\tilde{x}^TA_{ki}^TPA_{ki}\tilde{x}\leq 0$$

with $V(\tilde{x}(0))\leq e\tilde{x}^T(0)\tilde{x}(0)$ If we choose $V(\tilde{x})=\tilde{x}^TP\tilde{x}$, by the fact $\sum_{i=1}^{M}\alpha_i(\tilde{x})=1$ and $\partial V(\tilde{x})/\partial\tilde{x}=2P\tilde{x}$ in (9.142), then we get

$$\sum_{i=1}^{M}\alpha_i\tilde{x}^T\left[A_i^TP+PA_i+Q-PP+\frac{1}{e}PBB^TP+\lambda_2\sum_{k=1}^{N_{2p}}A_{ki}^TPA_{ki}\right]\tilde{x}\leq 0 \quad (9.143)$$

The above inequality holds if the Riccati-like inequalities

$$A_i^TP+PA_i+Q-PP+\frac{1}{e}PBB^TP+\lambda_2\sum_{k=1}^{N_{2p}}A_{ki}^TPA_{ki}\leq 0, i=1,\ldots,M$$

hold. Further,

$$u^*(t)=-\frac{1}{2}\left(\frac{\partial V(\tilde{x})}{\partial \tilde{x}}\right)=-P\tilde{x}, v^*(t)=\frac{1}{e}B^TP\tilde{x} \quad (9.144)$$

which are the solution of minimax evolutionary game problem

$$u^*(t)=\sum_{i=1}^{M}\sum_{k=1}^{N_1(t)}\alpha_i(\tilde{x})A_{ki}p(t-t_k)\tilde{x}=-P\tilde{x}, v^*(t)=\frac{1}{e}B^TP\tilde{x}$$

REFERENCES

[1] Stengel, RF; Ghigliazza, RM; Kulkarni, NV. "Optimal enhancement of immune response," *Bioinformatics*, vol. 18, 1227-1235, Sep 2002.

[2] Lydyard, PM; Whelan, A; Fanger, MW. *Instant notes in immunology*. Oxford, UK ; New York: Bios/Springer, 2000.

[3] Janeway, C. *Immunobiology: the immune system in health and disease*, 6th ed. New York: Garland Science, 2005.

[4] Asachenkov, AL. *Disease dynamics*. Boston; Laxenburg, Austria: Birkhäuser; IIASA, 1994.

[5] Rundell, A; Hogenesch, H; Decarlo, R. "Enhanced modeling of the immune system to incorporate natural killer cells and memory," *Proceedings of the 1995 American Control Conference, Vols 1-6*, 255-259, 1995.

[6] Perelson, AS; Weisbuch, G. "Immunology for physicists," *Reviews of Modern Physics*, vol. 69, 1219-1267, Oct 1997.

[7] Nowak, MA; May, RM. *Virus dynamics : mathematical principles of immunology and virology*. Oxford ; New York: Oxford University Press, 2000.

[8] Perelson, AS; Kirschner, DE; Deboer, R. "Dynamics of Hiv-Infection of Cd4+ T-Cells," *Mathematical Biosciences*, vol. 114, 81-125, Mar 1993.

[9] Nowak, MA; May, RM; Phillips, RE; Rowlandjones, S; Lalloo, DG; Mcadam, S; et al. "Antigenic Oscillations and Shifting Immunodominance in Hiv-1 Infections," *Nature*, vol. 375, 606-611, Jun 15 1995.

[10] Perelson, AS; Neumann, AU; Markowitz, M; Leonard, JM; Ho, DD. "HIV-1 dynamics in vivo: Virion clearance rate, infected cell life-span, and viral generation time," *Science*, vol. 271, 1582-1586, Mar 15 1996.

[11] Stafford, MA; Corey, L; Cao, YZ; Daar, ES; Ho, DD; Perelson, AS. "Modeling plasma virus concentration during primary HIV infection," *Journal of Theoretical Biology*, vol. 203, 285-301, Apr 7 2000.

[12] Adler, FR; Karban, R. "Defended Fortresses or Moving Targets - Another Model of Inducible Defenses Inspired by Military Metaphors," *American Naturalist*, vol. 144, 813-832, Nov 1994.

[13] Wiener, N. *Cybernetics; or, Control and communication in the animal and the machine*, 2d ed. New York,: M.I.T. Press, 1961.

[14] Bellman, RE. *Mathematical methods in medicine*. Singapore: World Scientific, 1983.

[15] Swan, GW. "Optimal-Control Applications in Biomedical-Engineering - a Survey," *Optimal Control Applications & Methods*, vol. 2, 311-334, 1981.

[16] Kirschner, D; Lenhart, S; Serbin, S. "Optimal control of the chemotherapy of HIV," *Journal of Mathematical Biology*, vol. 35, 775-792, Aug 1997.

[17] DeBoer RJ; Boucher, CAB. "Anti-CD4 therapy for AIDS suggested by mathematical models," *Proceedings of the Royal Society B-Biological Sciences*, vol. 263, 899-905, Jul 22 1996.

[18] Bonhoeffer, S; May, RM; Shaw, GM; Nowak, MA. "Virus dynamics and drug therapy," *Proceedings of the National Academy of Sciences of the United States of America*, vol. 94, 6971-6976, Jun 24 1997.

[19] Wein, LM; D'Amato, RM; Perelson, AS. "Mathematical analysis of antiretroviral therapy aimed at HIV-1 eradication or maintenance of low viral loads," *Journal of Theoretical Biology*, vol. 192, 81-98, May 7 1998.

[20] Wodarz, D; Nowak, MA. "Specific therapy regimes could lead to long-term immunological control of HIV," *Proceedings of the National Academy of Sciences of the United States of America*, vol. 96, 14464-14469, Dec 7 1999.

[21] Wodarz, D; Nowak, MA. "CD8 memory, immunodominance, and antigenic escape," *European Journal of Immunology*, vol. 30, 2704-2712, Sep 2000.

[22] Maronde, RF. *Topics in clinical pharmacology and therapeutics*. New York: Springer-Verlag, 1986.

[23] Eisenfeld, J; Witten, M. *Modelling of biomedical systems*. Amsterdam ; New York, N.Y., U.S.A.: North-Holland ; Sole distributors for the U.S.A. and Canada, Elsevier Science Pub. Co., 1986.

[24] Bell, DJ; Katusiime, F. "A Time-Optimal Drug Displacement Problem," *Optimal Control Applications & Methods*, vol. 1, 217-225, 1980.

[25] Carson, ER; Cramp, DG. *Computers and control in clinical medicine*. New York: Plenum Press, 1985.

[26] Kwok, GKKKE; Finegan, BA; Shah, SL. "Clinical evaluation of long range adaptive control for mean arterial blood pressure regulation," *Proceedings of the 1995 American Control Conference, Vols 1-6*, 786-790, 1995.

[27] Polycarpou, MM; Conway, JY. "Modeling and control of drug delivery systems using adaptive neural control methods," *Proceedings of the 1995 American Control Conference, Vols 1-6*, 781-785, 1995.

[28] Parker, RS; Doyle, FJ; Harting, JE; Peppas, NA. "Model predictive control for infusion pump insulin delivery," *Proceedings of the 18th Annual International Conference of the Ieee Engineering in Medicine and Biology Society, Vol 18, Pts 1-5*, vol. 18, 1822-1823, 1997.

[29] Gentilini, A; Morari, M; Bieniok, C; Wymann, R; Schnider, TW. "Closed-loop control of analgesia in humans," *Proceedings of the 40th Ieee Conference on Decision and Control, Vols 1-5*, 861-866, 2001.

[30] Stengel, RF; Ghigliazza, R; Kulkarni, N; Laplace, O. "Optimal control of innate immune response," *Optimal Control Applications & Methods*, vol. 23, 91-104, Mar-Apr 2002.

[31] Zhou, K; Doyle, JC; Glover, K. *Robust and optimal control*. Upper Saddle River, N.J.: Prentice Hall, 1996.

[32] Shudo, E; Iwasa, Y. "Inducible defense against pathogens and parasites: Optimal choice among multiple options," *Journal of Theoretical Biology*, vol. 209, 233-247, Mar 21 2001.

[33] Shudo, E; Iwasa, Y. "Optimal defense strategy: Storage vs. new production," *Journal of Theoretical Biology*, vol. 219, 309-323, Dec 7 2002.

[34] Shudo, E; Iwasa, Y. "Dynamic optimization of host defense, immune memory, and post-infection pathogen levels in mammals," *Journal of Theoretical Biology*, vol. 228, 17-29, May 7 2004.

[35] Shudo, E; Haccou, P; Iwasa, Y. "Optimal choice between feedforward and feedback control in gene expression to cope with unpredictable danger," *Journal of Theoretical Biology*, vol. 223, 149-160, Jul 21 2003.

[36] Basar, T; Olsder, GJ. *Dynamic noncooperative game theory*, 2nd ed. Philadelphia: SIAM, 1999.

[37] Chen, BS; Tseng, CS; Uang, HJ. "Robustness design of nonlinear dynamic systems via fuzzy linear control," *IEEE Transactions on Fuzzy Systems*, vol. 7, 571-585, Oct 1999.

[38] Chen, BS; Tseng, CS; Uang, HJ. "Mixed H-2/H-infinity fuzzy output feedback control design for nonlinear dynamic systems: An LMI approach," *IEEE Transactions on Fuzzy Systems*, vol. 8, 249-265, Jun 2000.

[39] Lian, KY; Chiu, CS; Chiang, TS; Liu, P. "LMI-based fuzzy chaotic synchronization and communications," *IEEE Transactions on Fuzzy Systems*, vol. 9, 539-553, Aug 2001.

[40] Li, THS; Chang, SJ; Tong, W. "Fuzzy target tracking control of autonomous mobile robots by using infrared sensors," *IEEE Transactions on Fuzzy Systems*, vol. 12, 491-501, Aug 2004.

[41] Boyd, SP. *Linear matrix inequalities in system and control theory*. Philadelphia: Society for Industrial and Applied Mathematics, 1994.

[42] Chen, BS; Chang, CH; Chuang, YJ. "Robust model matching control of immune systems under environmental disturbances: Dynamic game approach," *Journal of Theoretical Biology*, vol. 253, 824-837, Aug 21 2008.

[43] Astrom, KJ; Wittenmark, Br. *Adaptive control*, 2nd ed. Reading, Mass.: Addison-Wesley, 1995.

[44] Chen, BS; Tseng, CS; Uang, HJ. "Fuzzy differential games for nonlinear stochastic systems: Suboptimal approach," *IEEE Transactions on Fuzzy Systems*, vol. 10, 222-233, Apr 2002.

[45] Takagi, T; Sugeno, M. "Fuzzy Identification of Systems and Its Applications to Modeling and Control," *IEEE Transactions on Systems Man and Cybernetics*, vol. 15, 116-132, 1985.

[46] Milutinovic, D; De Boer, RJ. "Process noise: An explanation for the fluctuations in the immune response during acute viral infection," *Biophysical Journal*, vol. 92, 3358-3367, May 10 2007.

[47] Marchuk, GI; *Mathematical models in immunology*. New York: Optimization Software, Inc. Worldwide distribution rights by Springer-Verlag New York, 1983.

[48] Villadsen, LS; Skov, L; Baadsgaard, O. "Biological response modifiers and their potential use in the treatment of inflammatory skin diseases," *Experimental Dermatology*, vol. 12, 1-10, Mar 2003.

[49] Stengel, RF; Ghigliazza, R. "Stochastic optimal therapy for enhanced immune response," *Mathematical Biosciences*, vol. 191, 123-142, Oct 2004.

[50] Althaus, CL; Ganusov, VV; De Boer, RJ. "Dynamics of CD8(+) T cell responses during acute and chronic lymphocytic choriomeningitis," *Journal of Immunology*, vol. 179, 2944-2951, Sep 1 2007.

[51] Andrianantoandro, E; Basu, S; Karig, DK; Weiss, R. "Synthetic biology: new engineering rules for an emerging discipline," *Molecular Systems Biology*, vol. 2, 2006.

[52] Tucker, JB; Zilinskas, RA. "The promise and perils of synthetic biology," *New Atlantis*, vol. 12, 25-45, Spring 2006.

[53] Alon, U. "Biological networks: The tinkerer as an engineer," *Science*, vol. 301, 1866-1867, Sep 26 2003.

[54] Alon, U. *An introduction to systems biology : design principles of biological circuits*. Boca Raton, FL: Chapman & Hall/CRC, 2007.

[55] Church, GM. "From systems biology to synthetic biology," *Molecular Systems Biology*, vol. 1, 2005.

[56] Endy, D. "Foundations for engineering biology," *Nature*, vol. 438, 449-53, Nov 24 2005.

[57] Hasty, J; McMillen, D; Collins, JJ. "Engineered gene circuits," *Nature*, vol. 420, 224-230, Nov 14 2002.

[58] Heinemann, M; Panke, S. "Synthetic biology - putting engineering into biology," *Bioinformatics*, vol. 22, 2790-2799, Nov 15 2006.

[59] Kobayashi, H; Kaern, M; Araki, M; Chung, K; Gardner, TS; Cantor, CR; et al. "Programmable cells: Interfacing natural and engineered gene networks," *Proceedings of the National Academy of Sciences of the United States of America*, vol. 101, 8414-8419, Jun 1 2004.

[60] Pleiss, J. "The promise of synthetic biology," *Applied Microbiology and Biotechnology*, vol. 73, 735-739, Dec 2006.

[61] Ferber, D. "Synthetic biology: Microbes made to order," *Science*, vol. 303, 158-161, Jan 9 2004.

[62] Forster, AC; Church, GM. "Synthetic biology projects in vitro," *Genome Research*, vol. 17, 1-6, Jan 2007.

[63] Gardner, TS; Cantor, CR; Collins, JJ. "Construction of a genetic toggle switch in Escherichia coli," *Nature*, vol. 403, 339-42, Jan 20 2000.

[64] Isaacs, FJ; Dwyer, DJ; Collins, JJ. "RNA synthetic biology," *Nature Biotechnology*, vol. 24, 545-554, May 2006.

[65] Maeda, YT; Sano, M. "Regulatory dynamics of synthetic gene networks with positive feedback," *Journal of Molecular Biology*, vol. 359, 1107-1124, Jun 16 2006.

[66] Tucker, M; Parker, R. "Mechanisms and control of mRNA decapping in Saccharomyces cerevisiae," *Annual Review of Biochemistry*, vol. 69, 571-595, 2000.

[67] Batt, G; Yordanov, B; Weiss, R; Belta, C. "Robustness analysis and tuning of synthetic gene networks," *Bioinformatics*, vol. 23, 2415-2422, Sep 15 2007.

[68] Goulian, M. "Robust control in bacterial regulatory circuits," *Current Opinion in Microbiology*, vol. 7, 198-202, Apr 2004.

[69] Kaznessis, YN. "Multi-scale models for gene network engineering," *Chemical Engineering Science*, vol. 61, 940-953, Feb 2006.

[70] Kaznessis, YN. "Models for synthetic biology," *BMC Systems Biology*, vol. 1, Nov 6 2007.

[71] Kitano, H. "Systems biology: A brief overview," *Science*, vol. 295, 1662-1664, Mar 1 2002.

[72] Kitano, H. "Biological robustness," *Nature Reviews Genetics*, vol. 5, 826-837, Nov 2004.

[73] Salis, H; Kaznessis, YN. "Computer-aided design of modular protein devices: Boolean AND gene activation," *Physical Biology*, vol. 3, 295-310, Dec 2006.

[74] De Jong, H. "Modeling and simulation of genetic regulatory systems: A literature review," *Journal of Computational Biology*, vol. 9, 67-103, 2002.

[75] Szallasi, Z; Stelling, Jr; Periwal, V. *System modeling in cell biology : from concepts to nuts and bolts*. Cambridge, Mass.: MIT Press, 2006.

[76] Kuepfer, L; Sauer, U; Parrilo, PA. "Efficient classification of complete parameter regions based on semidefinite programming," *BMC Bioinformatics*, vol. 8, Jan 15 2007.

[77] Chen, BS; Chang, YT; Wang, YC. "Robust H infinity-stabilization design in gene networks under stochastic molecular noises: fuzzy-interpolation approach," *IEEE Trans Syst Man Cybern B Cybern*, vol. 38, 25-42, Feb 2008.

[78] Chen, BS; Chen, PW. "Robust Engineered Circuit Design Principles for Stochastic Biochemical Networks With Parameter Uncertainties and Disturbances," *IEEE Transactions on Biomedical Circuits and Systems*, vol. 2, 114-132, Jun 2008.

[79] Chen, BS; Wu, WS. "Robust filtering circuit design for stochastic gene networks under intrinsic and extrinsic molecular noises," *Mathematical Biosciences*, vol. 211, 342-355, Feb 2008.

[80] Hwang, CL. "A novel Takagi-Sugeno-based robust adaptive fuzzy sliding-mode controller," *IEEE Transactions on Fuzzy Systems*, vol. 12, 676-687, Oct 2004.

[81] Hooshangi, S; Thiberge, S; Weiss, R. "Ultrasensitivity and noise propagation in a synthetic transcriptional cascade," *Proceedings of the National Academy of Sciences of the United States of America*, vol. 102, 3581-3586, Mar 8 2005.

[82] Chen, BS; Chang, CH; Lee, HC. "Robust synthetic biology design: stochastic game theory approach," *Bioinformatics*, vol. 25, 1822-1830, Jul 15 2009.

[83] Zhang, WH; Chen, BS. "State feedback H(infinity) control for a class of nonlinear stochastic systems," *SIAM Journal on Control and Optimization*, vol. 44, 1973-1991, 2006.

[84] Zhang, WH; Chen, BS; Tseng, CS. "Robust H-infinity filtering for nonlinear stochastic systems," *IEEE Transactions on Signal Processing*, vol. 53, 589-598, Feb 2005.

[85] Canton, B; Labno, A; Endy, D. "Refinement and standardization of synthetic biological parts and devices," *Nature Biotechnology*, vol. 26, 787-793, Jul 2008.

[86] Arkin, A. "Setting the standard in synthetic biology," *Nature Biotechnology*, vol. 26, 771-774, Jul 2008.

[87] Felix, MA; Wagner, A. "Robustness and evolution: concepts, insights and challenges from a developmental model system," *Heredity*, vol. 100, 132-140, Feb 2008.

[88] Ancel, LW; Fontana, W. "Plasticity, evolvability, and modularity in RNA," *Journal of Experimental Zoology*, vol. 288, 242-283, Oct 15 2000.

[89] Carter, AJR; Hermisson, J; Hansen, TF. "The role of epistatic gene interactions in the response to selection and the evolution of evolvability," *Theoretical Population Biology*, vol. 68, 179-196, Nov 2005.

[90] Bloom, JD; Labthavikul, ST; Otey, CR; Arnold, FH. "Protein stability promotes evolvability," *Proceedings of the National Academy of Sciences of the United States of America*, vol. 103, 5869-5874, Apr 11 2006.

[91] Aldana, M; Balleza, E; Kauffman, S; Resendiz, O. "Robustness and evolvability in genetic regulatory networks," *Journal of Theoretical Biology*, vol. 245, 433-448, Apr 7 2007.

[92] Elena, SF; Sanjuan, R. "The effect of genetic robustness on evolvability in digital organisms," *BMC Evolutionary Biology*, vol. 8, Oct 14 2008.

[93] McBride, RC; Ogbunugafor, CB; Turner, PE. "Robustness promotes evolvability of thermotolerance in an RNA virus," *BMC Evolutionary Biology*, vol. 8, Aug 11 2008.

[94] Draghi, JA; Parsons, TL; Wagner, GP; Plotkin, JB. "Mutational robustness can facilitate adaptation," *Nature*, vol. 463, 353-355, Jan 21 2010.

[95] Chen, BS; Lin, YP. "On the Interplay between the Evolvability and Network Robustness in an Evolutionary Biological Network: A Systems Biology Approach," *Evolutionary Bioinformatics*, vol. 7, 201-233, 2011.

[96] Chen, BS; Wu, WS; Wu, WS; Li, WH. "On the Adaptive Design Rules of Biochemical Networks in Evolution," *Evolutionary Bioinformatics*, vol. 3, 27-39, 2007.

[97] Tawfik, DS. "Messy biology and the origins of evolutionary innovations," *Nature Chemical Biology*, vol. 6, 692-696, Oct 2010.

[98] Lynch, M. "The evolution of genetic networks by non-adaptive processes," *Nature Reviews Genetics*, vol. 8, 803-813, Oct 2007.

[99] Chen, BS; Lin, YP. "A Unifying Mathematical Framework for Genetic Robustness, Environmental Robustness, Network Robustness and their Trade-off on Phenotype Robustness in Biological Networks Part I: Gene Regulatory Networks in Systems and Evolutionary Biology," *Evolutionary Bioinformatics*, vol. 9, 43-68, 2013.

[100] Chen, BS; Lin, YP. "A Unifying Mathematical Framework for Genetic Robustness, Environmental Robustness, Network Robustness and their Tradeoff on Phenotype Robustness in Biological Networks Part II: Ecological Networks," *Evolutionary Bioinformatics*, vol. 9, 69-85, 2013.

[101] Chen, BS; Lin, YP. "A Unifying Mathematical Framework for Genetic Robustness, Environmental Robustness, Network Robustness and their Trade-offs on Phenotype Robustness in Biological Networks. Part III: Synthetic Gene Networks in Synthetic Biology," *Evolutionary Bioinformatics*, vol. 9, 87-109, 2013.

[102] Chen; BS; Ho, SJ. "The stochastic evolutionary game on robust biological network by natural selection," *Evolutionary Bioinformatics*.

[103] Gardner, WA. *Introduction to random processes : with applications to signals and systems*, 2nd ed. New York: McGraw-Hill, 1990.

[104] Vincent, TL; Brown, JS. *Evolutionary game theory, natural selection, and Darwinian dynamics*. Cambridge ; New York: Cambridge University Press, 2005.

[105] Chen, BS; Lee, TS; Feng, JH. "A Nonlinear H-Infinity Control Design in Robotic Systems under Parameter Perturbation and External Disturbance," *International Journal of Control*, vol. 59, 439-461, Feb 1994.

[106] Chen, BS; Uang, HJ; Tseng, CS. "Robust tracking enhancement of robot systems including motor dynamics: A fuzzy-based dynamic game approach," *IEEE Transactions on Fuzzy Systems*, vol. 6, 538-552, Nov 1998.

[107] de Visser, JAGM; Hermisson, J; Wagner, GP; Meyers, LA; Bagheri, HC; Blanchard, JL; et al. "Perspective: Evolution and detection of genetic robustness," *Evolution*, vol. 57, 1959-1972, Sep 2003.

[108] Badyaev, AV. "Stress-induced variation in evolution: from behavioural plasticity to genetic assimilation," *Proceedings of the Royal Society B-Biological Sciences*, vol. 272, 877-886, May 7 2005.

[109] B. S. Chen and Y. C. Wang, "On the attenuation and amplification of molecular noise in genetic regulatory networks," *BMC Bioinformatics*, vol. 7, Feb 2 2006.

[110] Chen, BS; Chen, WH; Wu, HL. "Robust H-2/H-∞ Global Linearization Filter Design for Nonlinear Stochastic Systems," *IEEE Transactions on Circuits and Systems I-Regular Papers*, vol. 56, 1441-1454, Jul 2009.

[111] Sriyudthsak, K; Shiraishi, F; Hirai, MY. "Identification of a Metabolic Reaction Network from Time-Series Data of Metabolite Concentrations," *Plos One*, vol. 8, Jan 10 2013.

[112] Voit, EO. *Computational analysis of biochemical systems : a practical guide for biochemists and molecular biologists*. New York: Cambridge University Press, 2000.

[113] Kitano, H. "Biological robustness," *Nat Rev Genet*, vol. 5, 826-37, Nov 2004.

[114] Waddington, CH. "Canalization of development and the inheritance of acquired characters," *Nature*, vol. 150, 563-565, Jul-Dec 1942.

[115] Lehner, B. "Genes Confer Similar Robustness to Environmental, Stochastic, and Genetic Perturbations in Yeast," *Plos One*, vol. 5, Feb 3 2010.

[116] Basu, S; Gerchman, Y; Collins, CH; Arnold, FH; Weiss, R. "A synthetic multicellular system for programmed pattern formation," *Nature*, vol. 434, 1130-4, Apr 28 2005.

[117] Hasty, J; Dolnik, M; Rottschafer, V; Collins, JJ. "Synthetic gene network for entraining and amplifying cellular oscillations," *Physical Review Letters*, vol. 88, Apr 8 2002.

[118] Kaern, M; Elston, TC; Blake, WJ; Collins, JJ. "Stochasticity in gene expression: From theories to phenotypes," *Nature Reviews Genetics*, vol. 6, 451-464, Jun 2005.

[119] Isaacs, FJ; Hasty, J; Cantor, CR; Collins, JJ. "Prediction and measurement of an autoregulatory genetic module," *Proceedings of the National Academy of Sciences of the United States of America*, vol. 100, 7714-7719, Jun 24 2003.

[120] Kaern, M; Blake, WJ; Collins, JJ. "The engineering of gene regulatory networks," *Annual Review of Biomedical Engineering*, vol. 5, 179-206, 2003.

[121] Chen, BS; Hsu, CY; Liou, JJ. "Robust Design of Biological Circuits: Evolutionary Systems Biology Approach," *Journal of Biomedicine and Biotechnology*, 2011.

[122] Chen, BS; Wu, CH. "Robust Optimal Reference-Tracking Design Method for Stochastic Synthetic Biology Systems: T-S Fuzzy Approach," *IEEE Transactions on Fuzzy Systems*, vol. 18, 1144-1159, Dec 2010.

[123] Wu, CH; Lee, HC; Chen, BS. "Robust synthetic gene network design via library-based search method," *Bioinformatics*, vol. 27, 2700-2706, Oct 1 2011.

[124] Chen, BS; Chen, PW. "GA-based Design Algorithms for the Robust Synthetic Genetic Oscillators with Prescribed Amplitude, Period and Phase," *Gene Regul Syst Bio*, vol. 4, 35-52, 2010.

Chapter 10

CONCLUDING REMARKS AND FUTURE CHALLENGES IN SYSTEMS BIOLOGY

ABSTRACT

In this systems biology book, we have first used reverse engineering methods to construct GRNs and PPI networks from microarray data and database mining. Based on system theory and the H_∞ robust noise attenuation method, we then discussed the system characteristics of network robustness, network filtering, and transduction ability, with applications for robust design in synthetic biology and metabolic engineering. Finally, a stochastic dynamic game approach was employed for robust control of immune systems, robust synthetic circuit design, and evolution of biological networks. These multi-facetted applications illustrate that systems biology sits at the heart of new integrative paradigms. This chapter discusses challenges and future prospects for systems biology.

INTRODUCTION

In Chapter 2 of this book, some related mathematical methods were introduced based on bioinformatics data sets to construct dynamic models and analyze system characteristics at different levels of biological networks. Several reverse-engineering methods to identify parameters and system order of biological network dynamic models from protein or gene microarray data were introduced in Chapter 3, so that in Chapter 4 we could construct a GRN from corresponding gene microarray data and the PPI network from protein microarray data using bioinformatics databases. In Chapter 5, we measured system characteristics like robustness and sensitivity based on constructed dynamic models of biological networks. In Chapter 6 we progressed to calculating network robustness, noise filtering ability, and signal transduction ability for more general nonlinear stochastic biological networks, characteristics that are very useful for the systematic design of biological networks with desired design specifications. These systematic design methods for biological networks were applied to engineer desired synthetic gene networks for systems synthetic biology in Chapter 7. Given the synthetic circuit topology and desired design specifications, we were also able to select adequate devices or components based on the dynamic model and develop systematic methods from corresponding device libraries to meet these design specifications. Based on the S-system model and proposed systematic methods we designed robust metabolic circuits for metabolic systems engineering in Chapter 8. In order to achieve desired system behavior and to overcome uncertain intrinsic fluctuations and unpredictable (or hostile) environmental disturbances for the stochastic biological network, several robust systematic network designs based on dynamic Nash game theory were proposed in Chapter 9 to efficiently attenuate the effect of intrinsic noise and environmental disturbance.

10.1. Future Challenges in Systems Biology

Systems biology approaches to synthetic biology and metabolic engineering build on molecular and genetic findings as well as study results in -omics fields such as genomics, proteomics, and metabolomics. The main concerns in systems biology surround the complexity and dynamic character of biological systems, the vast quantities of biological data, and the fragmented nature of biological knowledge. This fragmented information needs to be integrated with nonlinear stochastic dynamic models at different system levels with suitable computational tools before they can be applied to systems synthetic biology and systems metabolic engineering.

The first challenge for systems biology is how to enrich -omics data. This includes updating and improving protein microarrays, ChIP-chip databases, and pathway databases for systematic studies (discussed in Chapters 4–9), and requires powerful computational tools for sophisticated data handling, advanced modeling, integrated analysis, and knowledge integration. A systems biology approach based on the integration of enriched -omics data and advanced computational models will allow us to predict the behavior of biological systems more precisely. This will improve our understanding of underlying molecular mechanisms and our predictions concerning the effects of designed genetic circuits in metabolic engineering and the impact of perturbations on biological systems during drug treatment.

The second challenge for systems biology is how to enable innovation by allowing researchers to navigate and relate various data and knowledge resources using an integrated platform including bioinformatics, systems synthetic biology, and system metabolic engineering. Connecting genomics, molecular networks, and physiology will provide us with a deeper understanding of how individual differences in the genome affect physiological processes through alterations in molecular networks. Predictive and preventive medicine based on network-based biomarkers is likely to lead to personalized medicine, revolutionizing healthcare in the future [1].

The third challenge for systems biology is how to integrate bioinformatics, gene circuit design, and metabolic engineering technologies for systems drug design in predictive and preventative medicine. In pharmacology, systems biology may face its most practical opportunity and challenge. Although drug action is both a quantitative and multicomponent problem, drug targets at present are not pathways but individual proteins. But if the ultimate drug targets are products of individual genes, are the qualitative pictures that we currently derive from biochemistry and genetics sufficiently informative to allow effective identification of those targets? Could a quantitative understanding of a dynamic model of the metabolic pathways at a high level of integration facilitate this process? If pathway understandings derived from systems biology merge with traditional areas in pharmacology, a high level of integration may dramatically increase search efficiency for new and better medicines.

Current global concerns include drastically increasing oil consumption, exhaustion of natural resources, and global warming. These issues have spurred research in areas of bioenergy and biorefining, as well as in microbial production of bulk chemicals and materials. Systems biology will play an important role in these undertakings, collaborating with systems synthetic biology and systems metabolic engineering to generate bio-based products to replace some, if not most, currently used chemicals and materials. The fourth challenge of systems biology involves making these bio-based processes competitive and valid [2].

At present, systems biology has had great success in describing small biological systems in detail. However, it is in a period of awkward adolescence. Biologists still differ sharply over the desired emphasis of systems biology and its place within the wider world of biology. In fact, some researchers still question the validity of a systems biology approach. The fifth challenge is how to phrase questions, and design and interpret experiments, that will illustrate how complexity is achieved in larger biological systems. More is different – a concept that insinuates that completely new concepts will emerge from the study of increasingly large biological systems. Different biological networks at different scales allow for formulation of dynamic models for systematic conceptual frameworks of each biological system, whether it is an organelle, unicellular organism, metazoan, or plant. These will help us understand how the complexity of a biological system is achieved not only by outlining basic design principles but also by understanding how simple building modules function together dynamically in space and time, especially concerning systems synthetic biology and systems metabolic engineering [3]. This will require continuous development and application of unbiased and high-throughput technologies to define and perturb large biological systems to obtain bioinformatic data sets for the construction of dynamic models of biological networks for systems biology. Then, a deep and comprehensive systematic knowledge of homeostasis maintenance and response to outside stimuli can be developed, which will be essential to ultimately understand how biological systems change during aging and disease.

However, systematic approaches in systems biology depend much more on data quality than data quantity. Data sets useful for systems biology need to quantify each component of the biological system in each perturbed state with minimal error. This means that data sets need to be quantitatively accurate, reproducible, and complete. Except for transcription profiling (i.e., microarray data), data sets with these attributes are rare. Therefore, the final challenge for systems biology is how to generate accurate, reproducible, and complete data sets so that systems biologists can construct an accurate and reliable dynamic model for different levels of biological systems. Fortunately, new biotechnologies and tools are advancing rapidly, generating large data sets. In particular, new mass spectrometry approaches are accurately quantifying predetermined sets of molecules (proteins, phosphoproteins, metabolites, lipids) with very high sensitivity, reproducibility, and dynamic range, and seem poised to alleviate systems biology's desire for high quality data sets.

10.2. The Future Prospect of Systems Biology

What can we expect from the interplay between systems biology and bioinformatics, synthetic biology, and metabolic engineering in the near and distant future? Already companies are promising to deliver new fuels and carbon-based products (such as plastics). Within five years, this could be a reality, thereby beginning to remove petroleum from the production loop. In ten years, we believe that many high value commodities like drugs will be produced biologically due to advances in systems synthetic biology and metabolic engineering. In twenty to fifty years, we hope that systems biology-based synthetic biology and metabolic engineering might bring new insights into pharmacology, leading to new cell-based therapies, expansion of immunotherapy, synthetic organs and tissues, and reconstruction of damaged environments and ecosystems. In the more distant future, it could involve reprogramming entire organisms, engineering ecosystems to support sustainable agriculture, environmental remediation, and pathogen removal to treat human disease. Finally, we feel that the cycle of data analysis, modeling, understanding, and design as displayed on the cover page serves well to highlight the relationship between systems biology and bioinformatics, synthetic biology, and metabolic engineering. Systems biology adds layers of information and new technologies that will empower future efforts to design synthetic biological systems and test our understanding of natural biological systems. By integrating their contributions into this rapidly evolving field, scientists and engineers together will be well positioned to transform health, well-being, and the environment in the years to come [2].

10.3. Conclusion

The increasing use of high-throughput technologies and computational modeling reveals new levels of biological function and organization. These features of systems biology influence our view of the organelle, cell, tissue, organ, and organism. This book has provided system modeling methods, reverse engineering methods, network construction methods, robust gene circuit design methods, stochastic game technologies, and database mining methods to introduce systems biology as an integrated platform combining bioinformatics, synthetic biology, and metabolic engineering in a new era of biology.

References

[1] Walhout, AJM; Aebersold, R; Meyer, T; Bork, P; Kirschner, MW. "Systems Biology: What's the Next Challenge?" *Cell*, vol. 144, 837-838, 3/18/ 2011.
[2] Nurse, P; Hayles, J. "The Cell in an Era of Systems Biology," *Cell*, vol. 144, 850-854, 2011.
[3] Chen, BS; Wang, YC. *Synthetic gene network: modeling, analysis, and robust design methods*. United States: CRC Press, 2014.

AUTHORS' CONTACT INFORMATION

Dr. Bor-Sen Chen
Distinguished Chair Professor
Lab of Control and Systems Biology
Department of Electrical Engineering
National Tsing Hua University
Hsinchu, Taiwan
Tel: 886-3-573-1155
E-mail: bschen@ee.nthu.edu.tw

Dr. Chia-Chou Wu
National Tsing Hua University
Hsinchu, Taiwan
E-mail: d9761820@oz.nthu.edu.tw

INDEX

A

accurate models, 45
acid, 62, 186, 202, 205
activity level, 51, 54
actuators, 263
adaptation(s), 98, 265, 266, 294, 318
adenocarcinoma, 109, 110, 118
ADH, 202, 209, 210, 211, 212
adhesion, 57, 102, 120
adjustment, 205
ADP, 186
age, 96, 99, 102, 116
aging process, 6, 99, 101, 102, 104
agriculture, 324
AIDS, 314
alcohols, 222, 223
aldehydes, 222
algorithm, 17, 33, 34, 35, 36, 40, 42, 43, 44, 45, 64, 82, 138, 164, 165, 169
alternative hypothesis, 28
amino, 221, 222
amino acid(s), 221, 222
ammonium, 220
amplitude, 28, 123, 129, 133, 140, 164
anchorage, 119
angiogenesis, 103, 110
annotation, 107, 108
ANOVA, 113
antibody, 226, 228, 239, 247
anti-cancer, 110
antigen, 104, 227, 228
apoptosis, 9, 47, 51, 52, 53, 61, 86, 87, 101, 102, 103, 110, 116, 120, 130, 226
applied mathematics, 225
Arabidopsis thaliana, 11
arrest, 101, 103, 110
arthritis, 98
assimilation, 319
atherosclerosis, 99
atopic dermatitis, 247
ATP, 186, 275
Austria, 314
automated synthesis, 4, 248

B

Bacillus subtilis, 205, 218
bacteria, 50, 121, 204, 206, 214, 221, 226
bacterial cells, 213
bacterial strains, 206, 214
base, 4, 176, 222, 262
behavioral change, 226, 264
behaviors, vii, 2, 4, 5, 6, 64, 154, 156, 165, 167, 169, 249, 263
benefits, 61, 62
biochemistry, 181, 323
bioenergy, 323
biofuel, 175, 201, 203, 215, 221, 223, 264
bioinformatics, viii, 1, 2, 3, 5, 7, 13, 321, 322, 324
biological activities, 263
biological circuits, viii, 90, 172, 316
biological functions, viii, 4, 6, 163
biological networks, vii, 3, 5, 6, 7, 47, 74, 93, 113, 121, 122, 141, 220, 225, 226, 264, 265, 266, 267, 269, 271, 277, 282, 283, 284, 285, 291, 293, 294, 295, 321, 323
biological processes, 93, 99
biological systems, viii, 4, 13, 21, 31, 37, 45, 75, 121, 129, 153, 248, 264, 293, 322, 323, 324
biological understanding, viii
biomarkers, 52, 93, 94, 104, 111, 113, 322
biomass, 267
biopolymers, 4
biosensors, 7
biosynthesis, 175, 204, 205, 221, 222

biotechnological applications, 175
biotechnology, 4, 153, 210, 221, 222, 223, 248, 258, 264
blastoderm, 63, 64, 82
blood, 129, 130, 181, 226, 315
blood clot, 129, 181
blood plasma, 226
blood pressure, 315
bone, 120
bounds, 184
brain, 102
branching, 287
breast cancer, 7, 104, 110, 120
Brownian motion, 123, 146
budding, 59, 87

C

Ca^{2+}, 111, 129, 150
caloric restriction, 116
calorie, 96, 117
CAM, 102
cancer, 3, 9, 47, 49, 52, 86, 89, 93, 94, 103, 104, 105, 106, 108, 109, 110, 112, 113, 117, 118, 119, 120, 151, 153
cancer cells, 52, 94, 111, 120
cancerous cells, 3
candidates, 66, 81
carbon, 223, 324
carcinogen, 120
carcinogenesis, 104, 105, 106, 107, 108, 109, 110, 111, 119
carcinoma, 118, 119
cardiac muscle, 117
cascades, 59, 61, 130, 150
catalysis, 128, 214
causal relationship, 100
CBP, 119
CDK inhibitor, 117, 151
cell biology, 150, 317
cell culture, 207, 215
cell cycle, 3, 8, 47, 49, 86, 89, 101, 103, 110, 111, 119
cell death, 62, 101, 110, 226
cell division, 81
cell fate, 121
cell invasion, 120
cell line(s), 61
cell surface, 57, 58, 87
cellular homeostasis, 6
cellular regulation, 175
cellular signaling pathway, 129
challenges, viii, 1, 6, 318, 321
chaos, 151
chemical(s), 121, 122, 129, 130, 153, 191, 196, 199, 200, 204, 210, 214, 218, 223, 227, 248, 293, 323
chemical reactions, 129, 130
chemotherapy, 314
choriomeningitis, 316
chromatograms, 207
chromatography, 219
chromatography analysis, 219
chromosome, 33, 34, 35, 36, 81, 164
chronic diseases, 98
cigarette smoking, 113
classes, 64, 84, 109
classification, 7, 11, 66, 113, 172, 205, 317
cleavage, 66, 74
climate, 268
clone, 218
cloning, 206, 220
clustering, 113
CO2, 186
coding, 33, 164, 203
coenzyme, 186
collisions, 52
color, 78, 81, 262
communication, 7, 129, 151, 314
compatibility, 263
complement, 125, 136, 226, 229, 256, 284
complexity, 4, 9, 39, 43, 74, 82, 120, 240, 247, 322, 323
composition, 129, 141
compounds, 227
computation, 37, 66, 82, 138, 201
computational modeling, 324
computer, vii, 4, 238, 248
condensation, 186
conflict of interest, 225
congruence, 276
connective tissue, 64
connectivity, 58
construction, vii, 4, 37, 47, 48, 51, 52, 53, 55, 56, 60, 82, 86, 93, 104, 108, 132, 153, 173, 191, 200, 204, 218, 248, 293, 323, 324
consumption, 323
contact dermatitis, 247
control condition, 26
convergence, 24, 32
correlation(s), 22, 23, 52, 57, 74, 84, 107, 113, 212, 276, 292
correlation coefficient, 22, 113
corruption, 246
cost, 203, 210, 227, 230, 231, 240, 246, 247, 253, 273, 282, 287, 294, 302
cross-validation, 113

cues, 90, 199
cultivation, 218
culture, 206, 207, 210, 214, 215, 218, 219
culture media, 207
culture medium, 207, 214, 215, 218, 219
cycles, 49, 150
cyclins, 49
cyclooxygenase, 103
cytokines, 101, 102, 103, 110, 226
cytoplasm, 56, 102
cytoskeleton, 121

D

danger, 315
data analysis, 13, 89, 324
data mining, 3, 43, 51, 55, 56, 176
data set, 2, 5, 6, 7, 48, 105, 112, 113, 321, 323
database, 7, 8, 53, 64, 83, 91, 108, 188, 321, 324
decay, 67, 82, 155, 157, 158, 159, 160, 161, 162, 163, 251, 252, 258, 259, 260, 261, 262, 305
decoding, 33
decomposition, 17, 19, 179
deficiencies, 84
degradation, 48, 51, 67, 68, 111, 129, 139, 155, 166, 177, 180, 182, 263, 287
degradation process, 287
degradation rate, 68, 166
dependent variable, 177, 178
derivatives, 32
detectable, 210, 214, 247
detection, 13, 37, 44, 49, 55, 60, 65, 74, 78, 90, 319
detergents, 87
detoxification, 101, 102
deviation, 54, 265, 272, 273, 278, 279, 287, 292, 294
diet, 96
differential equations, 7, 13, 36, 65, 176, 182
diffusion, 36, 63, 64, 65, 67, 68, 74, 76, 78, 82, 85, 122, 191, 192
diffusion mechanisms, 63, 64, 65, 74, 82
diffusion process, 36, 68
digestion, 206
dilation, 117
diploid, 87
discontinuity, 31
disease gene, 7
diseases, 63, 99, 103, 104, 111, 113, 153, 247
distribution, 21, 22, 25, 26, 27, 35, 42, 149, 316
distribution function, 21, 25
divergence, 121, 287
diversity, 93, 265
DNA, 1, 2, 3, 4, 5, 81, 90, 101, 103, 104, 110, 111, 117, 121, 141, 166, 218, 248, 268, 278
DNA damage, 103, 110
DNA repair, 101, 103, 104, 111
dosage, 247
down-regulation, 117
drawing, 58
Drosophila, vi, 63, 64, 65, 66, 67, 68, 69, 70, 74, 75, 78, 79, 81, 82, 83, 88, 89, 90, 91, 92, 99, 104, 116, 137
drug action, 322
drug delivery, 227, 315
drug design, 113, 141, 175, 176, 199, 215, 247, 248, 322
drug discovery, 227
drug targets, 9, 86, 94, 322
drug therapy, 79, 314
drug treatment, 322
drugs, 57, 153, 227, 239, 241, 243, 324
dynamic equations, viii, 37, 55, 68, 74, 123, 159, 209, 250
dynamic systems, 93, 134, 225

E

economic systems, 292
ecosystems, viii, 324
ectoderm, 64
electromyography, 89
electron, 186
electrophoresis, 206
elongation, 223
embryogenesis, 63, 64, 65, 66, 67, 68, 74, 81, 82
EMG, 89
emission, 219
encephalitis, 62, 88
encoding, 33, 34, 87, 140, 185
endocrine, 129
endothelial dysfunction, 117
endothelium, 117
energy, 20, 29, 95, 113, 129, 133, 141, 142, 153, 157, 197, 204, 222, 231, 252, 272, 273, 295, 297, 302, 304
energy density, 204
energy expenditure, 129
environment, 4, 52, 62, 63, 128, 153, 265, 324
environmental change, 3, 8, 52, 140, 163, 164, 183, 202, 226, 264, 265, 266, 267, 268, 270, 275, 276, 282
environmental conditions, 52
environmental stress(es), 3, 103, 109, 202, 282
environments, 154, 156, 248, 264, 324
enzymatic activity, 214

enzyme(s), 102, 103, 111, 128, 137, 175, 176, 177, 180, 182, 186, 191, 197, 198, 201, 202, 203, 204, 205, 206, 209, 212, 213, 214, 215, 222
epidermis, 64
epinephrine, 130
epitopes, 226
equality, 142, 145, 147
equilibrium, 6, 30, 225, 229, 251, 265, 266, 268, 270, 272, 273, 278, 279, 280, 282, 285, 286, 289, 290, 291, 292, 294
estimation process, 146, 148, 149
ethanol, 11, 204
eukaryotic, 8, 85, 186
evidence, 62, 102, 104, 110, 111, 117, 247
evolution, 116, 121, 138, 163, 177, 191, 225, 226, 264, 265, 266, 267, 268, 269, 272, 273, 275, 276, 277, 278, 279, 280, 281, 282, 284, 285, 286, 287, 291, 292, 293, 294, 295, 318, 319, 321
excitation, 181, 219
experimental condition, 21
experimental design, 82, 262
exposure, 98
external environment, 264
extinction, 271
extracts, 207

F

false positive, 43, 47, 49, 51, 52, 53, 55, 57, 60
Fata, 120
fermentation, 176, 203, 204
fertilization, 63
fibroblast growth factor, 109
fibroblasts, 51
film thickness, 219
filters, 122
filtration, 122
fitness, 33, 34, 35, 44, 45, 164, 226, 229, 265, 266, 267, 272, 273, 275, 280, 282, 286, 290, 292, 294
flame, 219
fluctuations, 4, 6, 68, 98, 121, 122, 124, 128, 154, 155, 156, 157, 159, 160, 161, 162, 175, 192, 196, 197, 198, 199, 251, 264, 273, 278, 280, 316, 322
fluid, 226
fluorescence, 82, 159, 167, 203, 204, 205, 212, 214, 218, 219, 250
force, 89
formation, 1, 3, 62, 64, 65, 68, 76, 78, 79, 81, 172, 209, 210, 222, 320
formula, 30, 129, 142, 147, 169, 171, 308, 309
fragility, 5, 179
free radicals, 99
fungi, 87
fuzzy membership, 228, 250, 254
fuzzy sets, 135, 138

G

game theory, 10, 154, 163, 172, 225, 228, 231, 248, 249, 262, 266, 295, 315, 318, 319, 322
GC-FID, 206, 207, 209, 215
gel, 206
gene expression, 4, 8, 48, 52, 53, 56, 61, 64, 81, 85, 86, 88, 89, 96, 98, 101, 102, 104, 106, 113, 117, 122, 126, 151, 154, 155, 212, 221, 222, 248, 315, 320
gene regulation, vii, 56, 116, 181
gene therapy, 99, 199
genetic components, 169
genetic drift, 281, 292
genetic engineering, 1, 263
genetic mutations, 95, 98, 169, 274, 291, 295
genetics, 2, 3, 5, 33, 323
genome, 1, 2, 8, 11, 64, 85, 90, 175, 176, 203, 322
genomic stability, 110
genomics, 322
genotype, 33, 34
global warming, 323
glucose, 130, 197, 205, 219
glycolysis, 202
growth, 52, 58, 59, 60, 87, 99, 103, 107, 108, 109, 110, 111, 112, 118, 119, 120, 129, 130, 150, 206, 210, 212, 213, 214, 221
growth factor, 99, 103, 109, 118, 150
growth hormone, 150
growth rate, 214
growth temperature, 221

H

healing, 227, 228, 229
health, 104, 228, 229, 239, 241, 243, 245, 246, 247, 314, 324
heat shock protein, 212, 223
Helicobacter pylori, 61, 87
heritability, 271
heterogeneity, 121
histamine, 227
histone, 81
HIV, 227, 314
HIV-1, 314
homeostasis, 98, 129, 323
hormonal control, 181
hormones, 96, 129

host, 2, 47, 52, 57, 58, 60, 62, 63, 83, 86, 87, 154, 156, 160, 162, 163, 169, 175, 201, 202, 204, 226, 227, 247, 248, 249, 251, 252, 256, 259, 260, 262, 263, 264, 293, 295, 315
human, 7, 87, 88, 89, 104, 109, 110, 111, 116, 118, 119, 120, 223, 227, 247, 324
human neutrophils, 87, 88
hybrid, 89
hydrogen, 102
hydrogen peroxide, 102
hypertrophy, 99
hypothesis test, 13, 24, 26

I

ICAM, 102
identification, vii, 2, 3, 5, 8, 25, 36, 37, 43, 45, 55, 60, 64, 66, 68, 69, 73, 74, 75, 76, 82, 84, 85, 86, 89, 94, 104, 110, 111, 117, 135, 138, 235, 247, 255, 323
identification problem, 68
identity, 15, 37, 96, 125, 233
IL-8, 103
image, 266
immune defense, 62, 63, 227
immune memory, 227, 247, 315
immune response, 57, 61, 63, 138, 225, 226, 227, 229, 230, 231, 233, 234, 238, 239, 240, 241, 248, 249, 313, 315, 316
immune system, 9, 49, 50, 61, 225, 226, 227, 229, 230, 231, 232, 234, 237, 238, 239, 240, 242, 244, 245, 246, 247, 249, 295, 298, 314, 316, 321
immunity, 151, 227
immunodeficiency, 227
immunoprecipitation, 2
immunotherapy, 324
imprinting, 81
in vitro, 120, 317
in vivo, 120, 155, 156, 158, 160, 162, 163, 293, 314
inbreeding, 116
independent variable, 22, 177, 178, 186
inducer, 167, 169, 214
induction, 101, 103, 206, 207
industry, 203
inequality, 29, 30, 31, 95, 96, 114, 115, 122, 124, 125, 134, 136, 142, 143, 144, 145, 146, 148, 170, 172, 193, 194, 216, 217, 218, 228, 233, 235, 248, 250, 254, 266, 275, 276, 284, 295, 296, 298, 303, 305, 309, 312, 313
infancy, 265
infection, 3, 47, 57, 58, 59, 60, 62, 63, 87, 99, 138, 227, 239, 241, 247, 248, 314, 315
infectious agents, 199
inflammation, 3, 10, 49, 52, 61, 86, 98, 99, 109, 110, 118, 138, 151
inflammatory responses, 61
information processing, 129
ingredients, 153, 248
inheritance, 91, 320
inhibition, 61, 110, 118, 120, 128, 129, 154, 155, 287, 288
inhibitor, 101, 102, 103, 139, 191
initial state, 163, 226, 227, 231, 248, 249, 260
input signal, 131, 133, 137, 138
insertion, 204
insulin, 103, 104, 109, 118, 223, 315
integration, viii, 3, 4, 104, 113, 175, 176, 227, 230, 322, 323
interface, 54, 55, 56
interference, 87, 269
interferons, 227
interleukin-8, 103
intervention, 118, 227, 229
intervention strategies, 118
intima, 186
inversion, 17, 42, 278
ionization, 219
ionizing radiation, 138
isomerization, 223
iteration, 33, 36

K

K^+, 129, 150
kidney, 102
kill, 61, 226
kinase activity, 117
kinetic model, 11
kinetic parameters, 140, 157, 158, 159, 160, 161, 162, 163, 177, 180, 181, 182, 192, 198, 199, 305
kinetics, 130

L

lactose, 197
landscape(s), 6, 265, 268
larvae, 82
lead, 4, 43, 98, 101, 137, 146, 148, 149, 227, 230, 248, 270, 271, 276, 277, 282, 291, 314, 322
leucine, 210
leukemia, 116
life cycle, 89, 108
life sciences, 221
ligand, 101, 104, 116
light, 4, 11, 101, 129, 202, 223

linear systems, 136, 163, 225, 235, 255
lipids, 323
liquid chromatography, 220, 223
Listeria monocytogenes, 222
liver, 102, 130
living organisms, vii
localization, 139
locus, 119
longevity, 102, 104, 118
low temperatures, 203
lung cancer, 11, 86, 93, 104, 105, 108, 110, 111, 112, 113, 118, 119, 120
Luo, 223
Lyapunov function, 29, 30, 31, 114, 141, 142, 169, 170, 194, 215, 304
lymphocytes, 226

M

machinery, 3
macromolecules, 63
macrophages, 87, 229
magnitude, 108
majority, 214
malignancy, 118
mammalian cells, 137
mammals, 227, 247, 315
manipulation, 204, 210, 218
mapping, 33
mass spectrometry, 223, 323
materials, 323
mathematical methods, 13, 321
matrixes, 135
matter, 130
measurement(s), 6, 9, 26, 37, 43, 45, 57, 68, 93, 94, 99, 100, 103, 115, 134, 135, 136, 138, 139, 140, 141, 146, 218, 238, 247, 320
media, 215, 219
mediation, 229
medical, 57, 229, 247
medicine, vii, 4, 63, 153, 210, 248, 264, 314, 315, 322
meiosis, 81
MEK, 137
melting, 222
membership, 135, 234, 235, 237, 240, 255, 259
membranes, 63
memory, 91, 226, 314
mesoderm, 64
messengers, 129
metabolic, v, vii, 4, 129, 175, 176, 183, 191, 221, 222, 223, 319
metabolic disorder(s), 98
metabolic engineering, vii, viii, 1, 2, 3, 4, 5, 7, 9, 11, 13, 56, 175, 176, 199, 201, 202, 215, 220, 223, 247, 321, 322, 323, 324
metabolic pathways, vii, 2, 3, 4, 13, 52, 53, 55, 56, 121, 141, 153, 175, 176, 194, 201, 202, 248, 323
metabolism, 2, 3, 4, 5, 11, 220, 221, 222
metabolites, 3, 13, 176, 177, 180, 186, 223, 288, 323
metabolizing, 197
metabolome, 2, 3
metastasis, 7, 109, 110, 118, 119
methanol, 220
methodology, 222
methylation, 81
microorganisms, 52
migration, 107, 108, 110, 112
mitochondria, 186
mitogen, 108, 118, 130, 150
mobile robots, 316
model system, 146, 204, 318
models, 2, 3, 4, 5, 6, 13, 37, 45, 63, 65, 67, 74, 75, 81, 89, 132, 150, 154, 167, 227, 234, 247, 248, 249, 254, 314, 316, 317, 321, 322, 323
modules, 8, 79, 86, 89, 90, 121, 323
molecular biology, 221
molecular dynamics, 66, 67
molecules, viii, 4, 7, 61, 96, 98, 102, 128, 129, 153, 186, 191, 192, 202, 209, 248, 323
Monte Carlo method, 260
morphogenesis, 59, 87
mortality, 99, 116, 239
mortality rate, 116
Moses, 118
motif, 91
MR, 172, 222
mRNA, 48, 52, 54, 63, 64, 65, 66, 67, 68, 69, 74, 75, 76, 81, 82, 83, 84, 94, 104, 165, 166, 197, 317
mRNAs, 166, 191, 192, 202
multidimensional, 176
muscarinic receptor, 150
mutant, 48, 78, 110, 118
mutation(s), 33, 34, 35, 66, 82, 98, 104, 110, 116, 118, 121, 153, 154, 155, 164, 178, 248, 264, 268, 278

N

Na^+, 150
NaCl, 218
NAD, 186, 205
NADH, 187
nanotechnology, 200, 247
Nash equilibrium, 225

National Academy of Sciences, 8, 36, 91, 92, 116, 118, 119, 120, 150, 151, 221, 223, 314, 317, 318, 320
natural killer cell, 314
natural resources, 323
natural selection, 33, 99, 127, 163, 164, 165, 226, 265, 268, 269, 270, 271, 273, 276, 277, 278, 279, 280, 282, 286, 289, 291, 293, 294, 295, 319
negative compensation, 185
nematode, 104
nervous system, 64
network levels, vii, 6
neural network(s), 65, 89
neutral, 265, 266, 267, 269, 270, 271, 274, 276, 278, 279, 280, 281, 285, 286, 289, 290, 291, 292, 293, 294
neutrophils, 229
New England, 118, 218
next generation, 35, 267
nitric oxide, 103
nitric oxide synthase, 103
nodes, 51, 53, 58
nonlinear dynamic systems, 134, 150, 151, 220, 221, 228, 250, 315
nonlinear systems, 255
normal distribution, 25, 28
nucleotides, 203, 205
nucleus, 56, 57, 63, 102, 118, 130, 137, 139
null hypothesis, 26
nutrient, 214, 264

O

oil, 323
old age, 103, 104
olfaction, 129
oncogenes, 110
operations, 33, 34, 35
operon, 197
optimization, 13, 32, 33, 34, 35, 36, 41, 95, 97, 122, 126, 134, 136, 138, 148, 168, 169, 177, 194, 195, 196, 220, 227, 228, 234, 236, 237, 238, 242, 247, 248, 250, 253, 254, 255, 256, 257, 258, 259, 262, 275, 282, 284, 289, 291, 294, 295, 298, 303, 304, 305, 315
optimization method, 41, 177, 227, 291
organ, 227, 228, 229, 230, 238, 239, 240, 241, 242, 243, 244, 245, 246, 247, 324
organelle, 323, 324
organism, 3, 7, 323, 324
organs, 227, 324
oscillation, 6, 129, 130
oscillators, 11, 164
osmotic pressure, 129, 202
osmotic stress, 129
ovarian cancer, 110
oxidation, 60, 62
oxidative stress, 96, 98, 100, 102, 103, 110, 116, 117
oxygen, 61, 102

P

P13K, 97
p53, 99, 100, 101, 103, 104, 110, 117, 118
Pacific, 92, 120
palliative, 227
pancreatic cancer, 119
parallel, 99, 100, 127
parameter estimates, 146
parameter estimation, 13, 17, 37, 38, 39, 40, 41, 42, 43, 44, 45, 49, 52, 65, 70, 73, 74, 75, 83, 84, 85, 94, 137, 138, 146, 148, 149, 177
parasites, 315
parents, 35
partial differential equations (PDEs), 65, 67, 91, 92
partition, 249
pathogenesis, 60, 62, 63, 247
pathogens, 61, 62, 226, 227, 229, 230, 231, 234, 238, 239, 240, 241, 242, 243, 244, 245, 246, 247, 315
pathophysiological, 93, 96, 98
pathways, viii, 3, 4, 6, 7, 8, 45, 52, 56, 58, 59, 60, 63, 65, 86, 87, 118, 122, 128, 129, 130, 132, 133, 135, 138, 140, 141, 150, 151, 153, 176, 200, 201, 222, 223, 293, 322
PCR, 218
percentile, 22, 26, 28
peroxide, 102
petroleum, 324
pH, 60, 202
phagocytic cells, 229
pharmacokinetics, 227
pharmacology, 315, 322, 323, 324
phenotypic variations, 268, 270, 271, 276, 277, 278, 279, 280, 282, 288, 289, 292, 293, 294
Philadelphia, 116, 149, 173, 220, 315, 316
phosphorylation, 117, 138, 139, 186
photosynthesis, 3
Physiological, 89, 118
physiology, 2, 11, 91, 122, 175, 176, 181, 202, 221, 322
plants, 62
plasma cells, 226
plasma membrane, 56, 226
plasmid, 204, 207
plasticity, 319
plastics, 324

platform, viii, 2, 5, 6, 7, 223, 322, 324
point mutation, 206
polyacrylamide, 206
polypeptides, 4
population, 11, 33, 34, 35, 36, 104, 121, 215, 226, 227, 265, 267, 269, 286, 291, 295
positive feedback, 317
precipitation, 79
primary data, 108, 112, 113
primary tumor, 111
principal component analysis, 19
principles, vii, 90, 121, 130, 151, 153, 172, 199, 204, 293, 314, 316, 323
prior knowledge, 66, 130
probability, 21, 22, 23, 24, 25, 26, 27, 28, 34, 35, 40, 41, 52, 142, 158, 170, 247
probability density function, 21, 22, 25, 26, 40, 41
probability distribution, 21, 23, 26
probability theory, 24
prognosis, 104
programming, 11, 154, 172, 247, 249, 317
prokaryotes, 186, 210
proliferation, 52, 101, 103, 110, 117, 118
promoter, 4, 10, 65, 67, 103, 151, 163, 165, 166, 167, 168, 169, 171, 172, 191, 198, 202, 204, 205, 206, 212, 214, 222, 251, 263
propagation, 64, 84, 214, 318
prostate cancer, 116
protective mechanisms, 99
protein family, 109
protein folding, 212
protein interactions, vii, 7, 11, 44, 51, 53, 56, 62, 68, 81, 87, 113, 118
protein kinases, 110
protein synthesis, 67, 79, 82
protein-protein interactions, 7, 11, 51, 53, 81, 87, 113, 118
proteome, 2
proteomics, 322
prototypes, 10
pruning, 48, 49, 51, 53, 74, 82
psoriasis, 247
PTEN, 97, 116
punishment, 231, 252, 262
purity, 219
P-value, 49, 107, 108

Q

quantification, 223

R

radiation, 118, 139
radicals, 61
random numbers, 260
rate of change, 177
reaction rate, 121, 209
reaction rate constants, 209
reactions, 98, 122, 176, 186, 209
reactive oxygen, 61, 87
reactivity, 239
real numbers, 14
real time, 247
receptors, 129, 137
recognition, 91
recombination, 163, 264, 268, 278
reconstruction, 8, 43, 84, 86, 324
recovery, 84, 239
reductionism, 1
redundancy, 6
reference system, 238
registries, 263
regression, 15, 37, 38, 39, 40, 42, 48, 52, 70, 91, 105
regression equation, 48, 52
regression model, 37, 39, 105
regulations, 11, 43, 49, 54, 56, 63, 64, 65, 66, 67, 68, 74, 78, 79, 81, 82, 87, 118, 259
regulatory changes, 101, 183
regulatory systems, 53, 317
remediation, 324
repair, 101, 102, 104
replication, 104, 121, 153
repression, 48, 64, 251
repressor, 67, 154, 159, 166, 167, 212, 250, 251
reproduction, 33, 34, 35, 99, 239
requirements, 185, 263
residuals, 251
residues, 229
resistance, 101, 102, 109, 110
resolution, 222
resources, 212, 322
response, 4, 5, 6, 8, 9, 61, 79, 81, 92, 93, 94, 96, 97, 98, 99, 100, 101, 102, 103, 104, 109, 110, 115, 116, 117, 122, 125, 126, 129, 130, 138, 140, 150, 162, 197, 200, 204, 221, 225, 226, 227, 229, 230, 231, 239, 245, 246, 248, 265, 316, 318, 323
restriction enzyme, 218
retinoblastoma, 117
ribosome, 4, 202, 203, 221, 263
risk, 22, 28, 104, 212
RNA, 91, 202, 203, 204, 221, 222, 317, 318
RNAi, 81

robust design, 153, 181, 182, 183, 225, 227, 252, 254, 263, 293, 321, 324
root(s), 19, 133, 138

S

salinity, 264, 268, 270
salmonella, 203, 222
sample mean, 23, 27
sample variance, 23, 27
saturation, 175, 176
SDS-PAGE, 206, 207
secretion, 3, 62, 129
self-regulation, 185
senescence, 103, 117
sensing, 4, 11, 52, 129, 222
sensitivity, 5, 6, 28, 93, 94, 96, 112, 113, 129, 154, 175, 177, 183, 184, 185, 186, 191, 248, 321, 323
sensors, 4, 153, 202, 221, 263, 316
serum, 104
shock, 150
short supply, 197
side effects, 227, 243, 244, 245, 247, 248
signal transduction, 6, 7, 52, 59, 103, 104, 118, 121, 128, 129, 132, 133, 134, 135, 136, 137, 138, 139, 140, 141, 146, 147, 148, 149, 150, 151, 321
signaling pathway, 59, 60, 61, 102, 128, 129, 130, 132, 133, 134, 135, 136, 137, 138, 139, 140, 141, 146, 147, 148, 149
signalling, 118
signals, 5, 17, 31, 90, 96, 100, 103, 110, 113, 114, 121, 122, 125, 128, 129, 130, 132, 133, 137, 138, 139, 140, 153, 164, 183, 238, 242, 245, 247, 319
significance level, 24
signs, 65
simulation, 76, 137, 138, 139, 140, 141, 154, 162, 175, 183, 199, 201, 203, 210, 228, 229, 238, 242, 243, 245, 246, 248, 258, 259, 260, 261, 262, 289, 290, 291, 317
simulations, vii, 154, 209, 214, 238, 248
Singapore, 314
skin diseases, 247, 316
sodium, 206
solubility, 204, 223
solution, 19, 21, 30, 32, 33, 34, 35, 36, 40, 73, 74, 95, 96, 123, 124, 125, 137, 146, 158, 159, 164, 196, 199, 203, 227, 233, 234, 235, 237, 253, 254, 256, 274, 275, 279, 280, 284, 285, 296, 297, 303, 309, 311, 313
space-time, 82
species, 3, 13, 14, 57, 61, 87, 121, 267, 268, 269, 278, 293
specifications, 10, 153, 154, 157, 158, 159, 160, 161, 166, 169, 172, 222, 263, 321, 322
spinal cord, 96, 100, 102, 103
squamous cell carcinoma, 110
stability, vii, 5, 6, 13, 29, 30, 31, 69, 73, 75, 82, 85, 93, 95, 114, 125, 154, 162, 175, 180, 181, 192, 194, 199, 203, 249, 260, 264, 275, 280, 290, 291, 292, 294, 295, 318
stabilization, 6, 9, 121, 141, 158, 159, 172, 175, 193, 292, 317
stable states, 166, 167
standard deviation, 156, 157, 161, 238, 260
standardization, 318
states, vii, 4, 106, 162, 198, 227, 240, 241, 242, 245, 246, 247, 249, 251, 259, 260, 261, 262, 290
statistical inference, 13, 21
statistics, 13, 23, 24, 36, 74, 192, 251, 256, 263
stem cells, 82
stimulation, 114, 118
stimulus, 79, 129, 286
stochastic model, 122, 159
stochastic processes, 13, 36, 68, 82
strain improvement, 176
strategy use, 265
stress, 47, 49, 52, 55, 56, 87, 101, 102, 103, 110, 118, 129, 150, 213, 282
stress response, 49
structure, vii, 1, 2, 105, 106, 129, 130, 137, 140, 154, 163, 202, 249, 265
subnetworks, 54, 57, 58, 108, 112
substrate(s), 180, 176, 177, 202, 209, 210, 214
sulfate, 206
Sun, 119
suppression, 61, 91, 101, 102, 103, 110, 116, 117, 239
survival, 35, 52, 63, 87, 101, 104, 107, 108, 110, 111, 112, 213, 241
susceptibility, 87
symptoms, 247
synchronization, 11, 129, 316
synthesis, 67, 129, 202, 209, 211, 222, 287
synthetic biology, vii, viii, 1, 2, 3, 4, 5, 6, 7, 10, 11, 13, 56, 128, 141, 154, 165, 169, 172, 175, 176, 201, 220, 223, 226, 248, 252, 262, 263, 292, 293, 316, 317, 318, 321, 322, 323, 324
system analysis, 36
system control methods, vii
system design, vii, 13, 141, 154, 180
system dynamics, vii, 4, 5, 6, 277
system states, vii, 251
systematic design methods, vii, 321
systematic mechanisms, vii, 3, 93, 113
system-level understanding, vii

T

T cell, 316
target, 48, 49, 51, 52, 55, 60, 63, 65, 67, 68, 74, 82, 99, 101, 105, 106, 109, 111, 118, 128, 140, 163, 164, 203, 212, 227, 316
tau, 119
techniques, 5, 32, 65, 127, 129, 153, 163, 201, 248, 250, 266, 293
technologies, viii, 2, 13, 322, 323, 324
technology, 5, 163, 175
temperature, 87, 175, 201, 202, 203, 204, 205, 206, 208, 210, 212, 213, 214, 215, 219, 220, 222, 223, 264, 270
test statistic, 27
testing, 13, 26, 27, 28, 33, 99, 133, 140
TGF, 118
therapeutic agents, 227, 232
therapeutic targets, vii
therapeutics, 119, 315
therapy, 7, 109, 113, 118, 128, 247, 314, 316
threonine, 9
thymus, 96, 100, 102, 103
time series, 57
tissue, 98, 104, 109, 226, 324
tissue homeostasis, 104
TNF, 102, 103, 139
TNF-α, 102, 103, 139
topology, 166, 167, 321
toxic side effect, 246
toxicity, 213, 262
TP53, 107, 110
tracks, 245, 246
training, 64, 66, 74, 84
traits, 163, 226, 264, 265, 266, 268, 278
trajectory, 89, 159, 283
transcription, 8, 10, 11, 47, 48, 49, 53, 54, 55, 56, 61, 63, 64, 65, 67, 68, 74, 78, 79, 81, 82, 86, 87, 89, 104, 110, 116, 117, 118, 120, 128, 137, 138, 139, 140, 167, 201, 251, 263, 323
transcription factors, 8, 10, 47, 48, 49, 54, 56, 63, 65, 86, 116, 117, 128, 137, 138, 140
transcriptomics, 63
transducer, 110, 150
transduction, vii, 6, 7, 91, 93, 100, 101, 102, 103, 121, 122, 128, 129, 132, 133, 134, 135, 136, 138, 139, 140, 141, 146, 148, 151, 321
transfection, 198
transformation, 69, 110, 120, 130, 209, 248, 284, 288
transformation processes, 209
translation, 54, 55, 56, 65, 67, 68, 79, 81, 82, 122, 166, 201, 263
translocation, 139, 278
transmission, 31
transport, 191, 198
treatment, vii, 26, 227, 229, 238, 239, 241, 246, 247, 316
tricarboxylic acid, 175, 186
tricarboxylic acid cycle, 175
triggers, 163
tumo(u)r(s), 101, 102, 103, 104, 109, 110, 116, 118
tumor development, 101
tumo(u)r growth, 109, 110, 117, 118
tumor necrosis factor, 102, 116
tumorigenesis, 93, 101, 102, 103, 104
two-sided test, 27
Type I error, 28
type II error, 28
tyrosine, 109, 110, 118

U

ubiquitin, 111, 139
UK, 313
unification, 8, 86
United States (USA), 8, 36, 91, 92, 116, 118, 119, 120, 150, 151, 221, 222, 223, 314, 317, 318, 320, 324

V

validation, 75, 82, 113
valine, 210
vapor, 204
variables, 21, 22, 23, 25, 26, 27, 32, 35, 38, 65, 67, 135, 138, 139, 140, 155, 177, 178, 229, 234, 240, 247, 255, 259, 262, 287
variations, 3, 4, 5, 6, 75, 99, 104, 122, 127, 154, 175, 178, 179, 183, 185, 249, 263, 264, 265, 266, 267, 268, 269, 270, 271, 273, 274, 276, 278, 279, 280, 281, 285, 286, 289, 290, 291, 293, 294, 295
vascular endothelial growth factor (VEGF), 103
vector, 14, 15, 17, 20, 22, 32, 35, 36, 38, 39, 41, 48, 94, 122, 131, 132, 144, 156, 164, 191, 207, 229, 231, 251, 252, 267, 278
viral gene, 314
viral infection, 316
virology, 314
viruses, 226
vision, 248

W

water, 102, 186, 204
wavelet, 88

X

xanthophyll, 223

Y

yeast, 8, 47, 49, 52, 86, 87, 89, 129, 137, 150, 180, 185, 214, 292
yield, 34, 175, 201, 202, 204, 205, 206, 208, 209, 210, 214, 215, 220, 226